WITHDRAWN
UTSA LIBRARIES

THE GEORGE FISHER BAKER NON-RESIDENT LECTURESHIP
IN CHEMISTRY AT **CORNELL UNIVERSITY**

Integrated Chemical Systems

Integrated Chemical Systems

A Chemical Approach to Nanotechnology

Allen J. Bard

Department of Chemistry and Biochemistry
The University of Texas at Austin
Austin, Texas

A Wiley-Interscience Publication
John Wiley & Sons, Inc.

New York • Chichester • Brisbane • Toronto • Singapore

> A NOTE TO THE READER
> This book has been electronically reproduced from digital information stored at John Wiley & Sons, Inc. We are pleased that the use of this new technology will enable us to keep works of enduring scholarly value in print as long as there is a reasonable demand for them. The content of this book is identical to previous printings.

This text is printed on acid-free paper.

Copyright © 1994 by John Wiley & Sons, Inc.

All rights reserved. Published simultaneously in Canada.

Reproduction or translation of any part of this work beyond that permitted by Section 107 or 108 of the 1976 United States Copyright Act without the permission of the copyright owner is unlawful. Requests for permission or further information should be addressed to the Permissions Department, John Wiley & Sons, Inc., 605 Third Avenue, New York, NY 10158-0012.

Library of Congress Cataloging in Publication Data:
Bard, Allen J.
   Integrated chemical systems: a chemical approach to nanotechnology / Allen J. Bard.
     p.   cm. — (Baker lecture series)
   "A Wiley-Interscience publication."
   Includes index.
   ISBN 0-471-00733-1 (cloth)
   1. Electrochemistry, Industrial. 2. Nanotechnology. I. Title. II. Series: George Fisher Baker non-resident lectureship in chemistry at Cornell University.
TP256.B37   1994
660'.297—dc20   93-39694

CONTENTS

Preface	vii
Abbreviations	ix
CHAPTER 1. An Introduction to Integrated Chemical Systems	1
CHAPTER 2. Construction of Integrated Chemical Systems	35
CHAPTER 3. Characterization of Integrated Chemical Systems	96
CHAPTER 4. Chemically Modified Electrodes	127
CHAPTER 5. Electrochemical Characterization of Modified Electrodes	184
CHAPTER 6. Photoelectrochemistry and Semiconductor Materials	227
CHAPTER 7. Future Integrated Chemical Systems	289
Index	313

PREFACE

This book is based on the Baker Lectures that I gave at Cornell University in the spring of 1987. Most of the outline of the book and the main parts of several of the chapters were completed during my stay in Ithaca, but unfortunately, the pressures of teaching, research, and the myriad other duties and distractions that characterize the modern academic prevented my completion of this until now. An additional factor that delayed completion was the continued growth and the publication of numerous interesting new papers in this area. Many of these, not available at the time of the lectures, have been incorporated into this book, which is current through 1992.

The book begins with a discussion and some examples of integrated chemical systems and analogies between man-made systems and biological ones (a theme continued through most of the book). The next chapters represent an elementary general treatment of the methods available for the construction and characterization of such systems. Two chapters are devoted to modified electrodes and electrochemical methods of characterizing these. The next discusses semiconductor materials (which may be key components in many systems) and their use in photoelectrochemical systems. Finally, the book closes with speculations and suggestions about possible future applications.

I became interested in the concept of integrated chemical systems mostly through my research on modified electrode surfaces and the fabrication of semiconductor-based systems for photoelectrochemistry. These topics were thus heavily stressed in my lectures and make up a large part of this volume. I hope that the chapters on these topics will serve as a useful introduction to these fields, although the work here should not be considered as an advanced treatment or a detailed review. Moreover, electrochemical methods are probably overstressed compared to other powerful methods of surface characterization (e.g., spectroscopic, NMR), mainly because these other methods are dealt with more extensively elsewhere.

Many of the examples used to illustrate a system in this book are those from our laboratory and publications, not because these are necessarily the best examples, but mainly because they were the easiest ones for me to use. There has been a vast amount of work published in areas that are discussed here, including sensors, modified electrodes, and photoelectrochemical systems, and I have not attempted to reference all of the work, or even all of the best work, in these fields. I hope my colleagues whose contributions have not been included will understand.

I appreciate the helpful comments and suggestions of Tom Mallouk and Larry Faulkner. I especially want to thank Rose Buettner for her careful and valuable assistance in the preparation of the final version of the manuscript.

ALLEN J. BARD

Austin, Texas
June 1994

ABBREVIATIONS

ADP	Adenosine diphosphate
AES	Auger electron spectroscopy
AFM	Atomic force microscopy
AOT	Bis(2-ethylhexyl)sodium sulfosuccinate (Aerosol-OT) (p. 281)
ATP	Adenosine triphosphate
ATRIR	Attenuated total reflectance infrared spectroscopy
BEDT-TTF	Bis(ethylenedithiolo)tetrathiafulvalene*
BET	Brunauer–Emmett–Teller method
BLM	Bilayer lipid membrane
BPQ^{2+}	Polymer derived from N,N'-bis[(p-trimethoxysilyl)benzyl]-4,4'-bipyridinium*
bpy	2,2'-Bipyridine*
BQ	Benzoquinone*
CB	Conduction band
CD	Cyclodextrin
CEC	Process in which a chemical reaction precedes and follows the electron-transfer process
CHEMFET	Chemical field effect transistor
CME	Chemically modified electrodes
Cp_2FeTMA^+	[(Trimethylammonio)methyl]ferrocene*
CV	Cyclic voltammetry
CVD	Chemical vapor deposition
DCC	Dicyclohexyl carbodiimide (p. 47)
DHP	Dihexadecyl phosphate

*Structures of selected compounds follow; structures of some compounds are given on page numbers indicated in ().

DMTCNQ	Dimethyltetracyanoquinodimethane*
DNA	Deoxyribonucleic acid
DODAC	Dioctadecyldimethylammonium chloride (p. 229)
DSA	Dimensionally stable anode
EBIC	Electron beam induced current
EDTA	Ethylenediaminetetraacetic acid*
EEL	Electron energy loss
EL	Electroluminescence
EMP	Electron microprobe
ESCA	Electron spectroscopy for chemical analysis
ESR	Electron spin resonance
et	Electron transfer
EXAFS	Extended X-ray absorption fine structure
FAD	Flavin adenine dinucleotide
$FADH_2$	Reduced form of flavin adenine dinucleotide
Fc	Ferrocene (dicyclopentadienyliron)*
FET	Field effect transistor
FTIR	Fourier transform infrared spectroscopy
GC	Glassy carbon (electrode)
GO	Glucose oxidase
HOPG	Highly oriented pyrolytic graphite
HQ	Hydroquinone*
ICS	Integrated chemical system
IPE	Inverse photoemission
ITO	Indium tin oxide
LB	Langmuir–Blodgett (films)
LC	Liquid crystal
LED	Light emitting diode
LEED	Low energy electron diffraction
LPCVD	Low pressure CVD
MBE	Molecular beam epitaxy
MOCVD	Metal-organic chemical vapor deposition
MOSFET	Metal oxide semiconductor field effect transistor
MPTH	N-Methylphenothiazine*
MV^{2+}	Methyl viologen (N,N'-dimethyl-4,4'-bipyridinium)*
$NADP^+$	Nicotinamide adenine dinucleotide phosphate

ABBREVIATIONS

NADPH	Reduced NADP
NAF	Nafion (pp. 18, 142)
NFSOM	Near-field scanning optical microscopy
NMR	Nuclear magnetic resonance
OTS	n-Octadecyltrichlorosilane (p. 160)
PANI	Polyaniline (pp. 42, 143)
PAQ	9,10-Phenanthraquinone*
PAS	Photoacoustic spectroscopy
PB	Prussian Blue (p. 154)
PBV	Poly(benzyl viologen) (p. 18)
PEC	Photoelectrochemical
PL	Photoluminescence
PP	Polypyrrole (pp. 42, 143)
PPV	Poly(p-phenylenevinylene)
PQ^{2+}	Polymerized viologen organosilane (pp. 15, 142)
PT	Polythiophene (pp. 42, 143)
PTC	Perovskite type ceramic
PTS	Photothermal spectroscopy
PVA	Poly(vinylalcohol)*
PVF	Poly(vinylferrocene) (p. 142)
PVP or PVPy	Poly(vinylpyridine) (p. 143)
$PVPyH^+$	Poly(vinylpyridinium) (pp. 38, 143)
PVS	Poly(styrenesulfonate) (pp. 38, 142)
pzc	Point of zero charge
RBS	Rutherford backscattering
RDE	Rotating disk electrode
RNA	Ribonucleic acid
SAM	Scanning auger microscopy
SAM	Self-assembled monolayer
SCE	Saturated calomel electrode
SECM	Scanning electrochemical microscopy
SEM	Scanning electron microscopy
SERS	Surface enhanced Raman spectroscopy
SHG	Second harmonic generation

SIMS	Secondary ion mass spectroscopy
SPE	Solid polymer electrolyte
SPM	Scanning probe microscopy
SSCE	Saturated sodium chloride calomel electrode
STEM	Scanning transmission electron microscopy
STM	Scanning tunneling microscopy
TBAP	Tetra-n-butylammonium perchlorate*
TBEA	2,2'-Thiobis(ethylacetoacetate) (p. 160)
TCNQ	Tetracyanoquinodimethane*
TEAP	Tetraethylammonium perchlorate*
TEM	Transmission electron microscopy
THF	Tetrahydrofuran
TISES	Texas Instruments Solar Energy System
TMPD	N,N,N',N'-Tetramethyl-p-phenylenediamine*
TMS	Trimethylsilyl
TMTSF	Tetramethyltetraselenafulvalene*
TMTTF	Tetramethyltetrathiafulvalene*
TMV	Tobacco mosaic virus
TPAI	Tetrapropylammonium iodide*
$TPPS_4$	Tetra(p-sulfophenyl)porphyrin
TTF	Tetrathiafulvalene*
TTT	Tetrathiatetracene*
UHV	Ultrahigh vacuum
UPD	Underpotential deposition
UPS	Ultraviolet photoelectron spectroscopy
UV	Ultraviolet
VB	Valence band
vpy	Vinylpyridine*
XPS	X-ray photoelectron spectroscopy
XRD	X-ray diffraction
XRF	X-ray fluorescence
ZnODEP	Zinc octakis(β-decoxyethyl)porphyrin (p. 305)
$ZnTMPyP^{4+}$	Zinc tetra(N-methyl-4-pyridyl)porphyrin (p. 273)

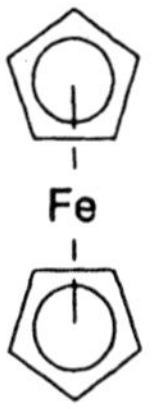

Ferrocene (Fc)

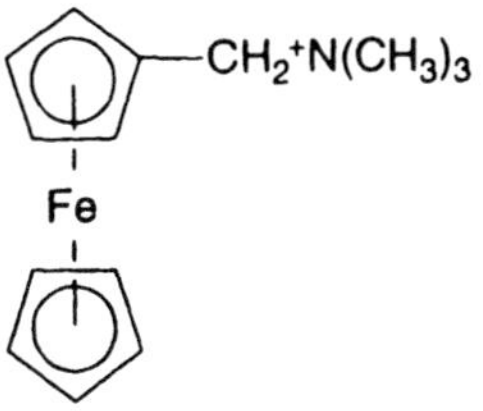

[Trimethylammonio)methyl]ferrocene (Cp$_2$FeTMA$^+$)

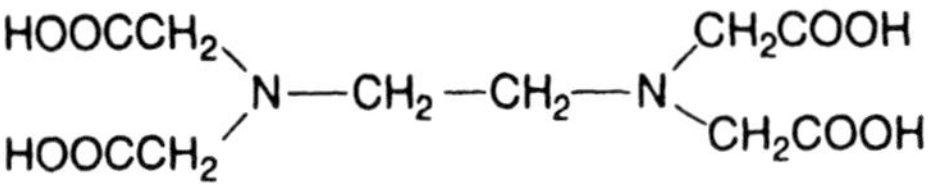

Ethylenediaminetetraacetic acid (EDTA)

$(CH_3CH_2CH_2CH_2)_4 N^+ \; Cl^-$

Tetra-n-butylammonium perchlorate (TBAP)

$(CH_3CH_2)_4 \; N^+ \; Cl^-$

Tetraethylammonium perchlorate (TEAP)

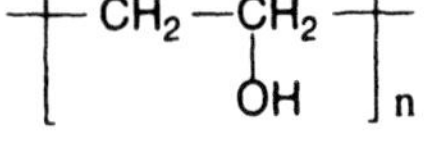

Poly(vinylalcohol) (PVA)

$(CH_3CH_2CH_2)_4 N^+ \; I^-$

Tetrapropylammonium iodide (TPAI)

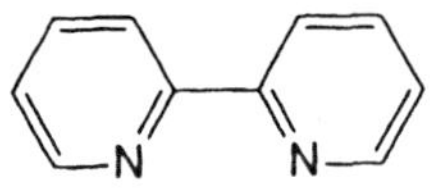

2,2'-Bipyridine (bpy)

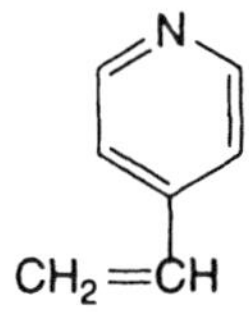

Vinylpyridine (vpy)

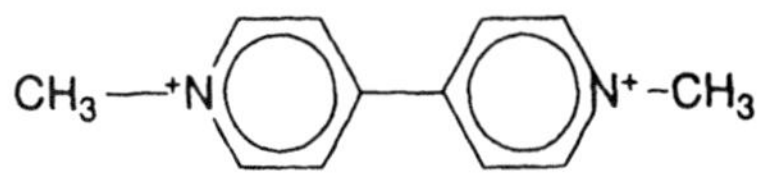

N,N'-dimethyl-4,4'-bipyridinium (MV^{2+})

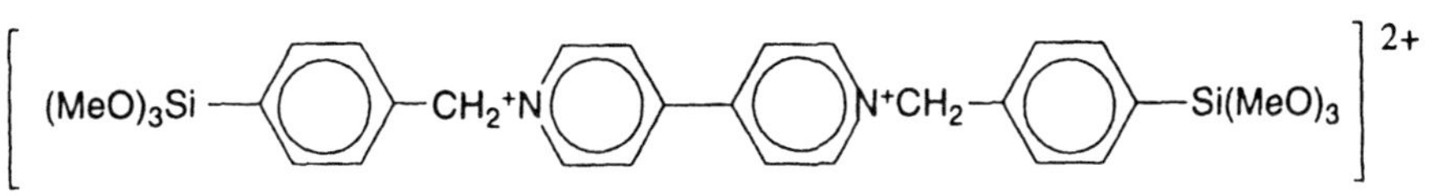

N,N'-Bis[(p-trimethoxysilyl)benzyl]-4,4'-bipyridinium (BPQ^{2+})

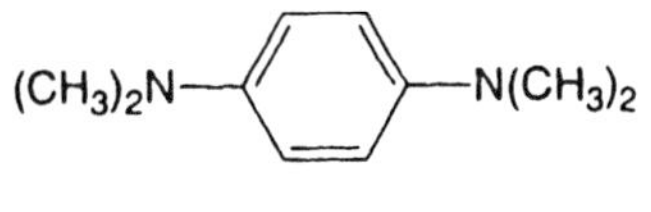

N,N,N',N'-Tetramethyl-p-phenylenediamine (TMPD)

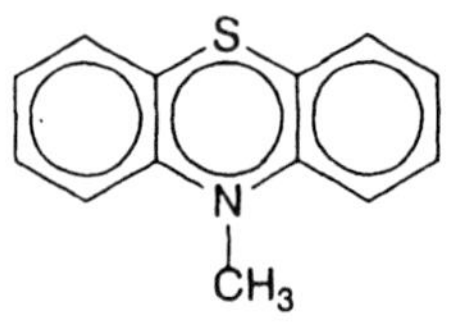

N-methylphenothiazine (MPTH)

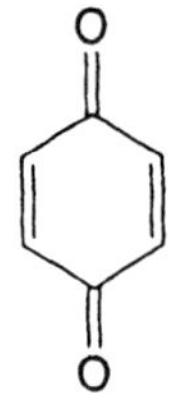

Benzoquinone (BQ)

Hydroquinone (HQ)

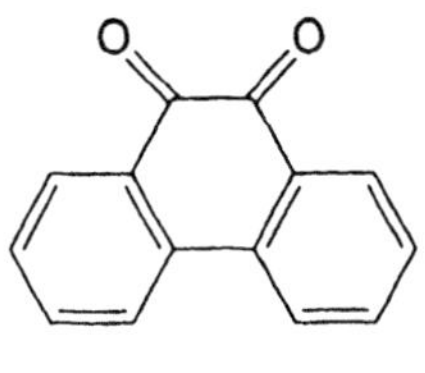

9,10-Phenanthraquinone (PAQ)

Tetrathiafulvalene (TTF): R = H, X = S
Tetraselenafulvalene (TSF): R = H, X = Se
Tetramethyltetrathiafulvalene (TMTTF): R = CH$_3$, X = S
Tetramethyltetraselenafulvalene (TMTSF): R = CH$_3$, X = Se

Tetracyanoquinodimethane (TCNQ): R = H
Dimethyltetracyanoquinodimethane (DMTCNQ): R = CH$_3$

Tetrathiatetracene (TTT)

Bis(ethylenedithiolo)tetrathiafulvalene (BEDT-TTF)

Integrated Chemical Systems

Chapter 1
AN INTRODUCTION TO INTEGRATED CHEMICAL SYSTEMS

The chemist interested in catalysis, stereochemical specificity, polymer structure, metallo-organic reactions, surface effects and a hundred other facets of chemistry will be challenged and instructed by the myriad forms in biology.

—Arthur Kornberg
1988 Aharon Katzir Memorial Lecture
Weizmann Institute
Rehovot, Israel

1.1. INTRODUCTION

Large complex systems (macrosystems) generally have hierarchical structures; that is, they are assembled from smaller units that are, in turn, built from smaller and simpler ones, finally down to units with atomic and molecular dimensions. This general structure is illustrated in Figure 1.1.1, along with the approximate sizes of the different units. Consider, for example, a biological system (Fig. 1.1.2). At the molecular level are the simple chemical species (e.g., Cl^-, water, phosphates) and molecular building blocks, the amino acids, sugars, nucleotides, and so on. These are components of the macromolecules at the next level, for example, polypeptides, proteins, polysaccharides, and nucleic acids. The structures that actually carry out the reactions are more complex, however, and involve the next level (the *integrated chemical system* level), where the different macromolecules and other components are assembled in a unique way to form a particular structure or carry out a certain process. For example, it might consist of several different

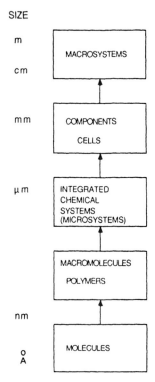

FIGURE 1.1.1. Schematic representation of the buildup of a general macrosystem from smaller and simpler components.

enzymes contained in a membrane in contact with a liquid phase with dissolved reactants. These subsystems form the cells that are the building blocks of the tissues and finally the organism.

A complicated electronic system, such as a digital computer, is built up in an analogous manner (Fig. 1.1.3). At the molecular level are the atoms and molecules, including those that form the semiconductors Si and GaAs. The next level involves larger aggregates, up to bulk semiconductor crystals. However, the actual device depends on the construction of complex integrated chemical systems, such as transistors and integrated circuits, that require dopants, junctions, metal leads, contacts, and encapsulants. These are assembled into still larger units of different types, such as circuit boards, that contain the power supply, microprocessor, memory, and interconnections that make up the macrosystem—

FIGURE 1.1.2. Buildup of a general biological system following the scheme in Figure 1.1.1.

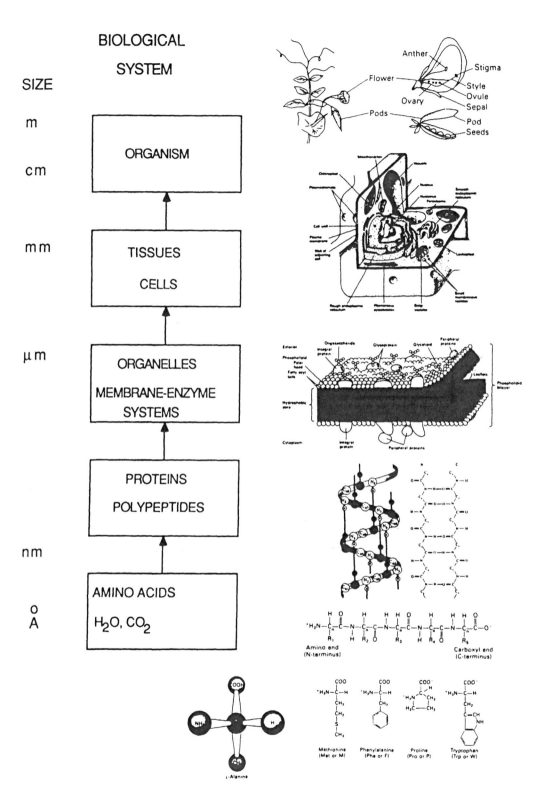

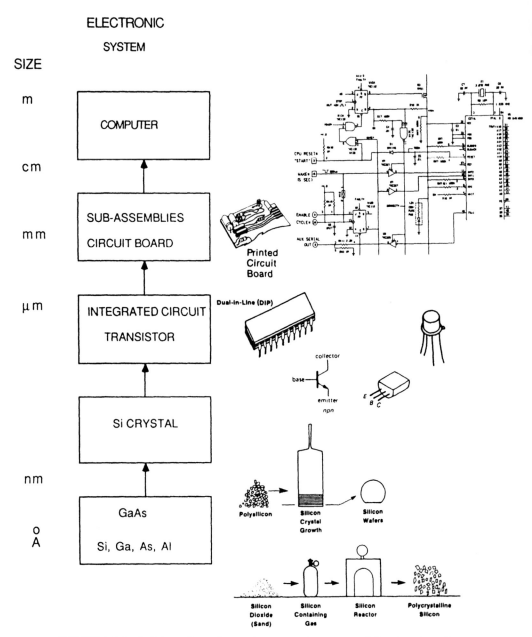

FIGURE 1.1.3. Buildup of an electronic system (computer) following the scheme in Figure 1.1.1.

1.1. INTRODUCTION

the computer. In all cases, the system structures and interactions become more complex as one proceeds from the lowest boxes, containing simple atoms and molecules (the molecular level), to more complex molecules, to macromolecules, and so on up to the integrated chemical systems and larger entities. A feeling of the relative size of the various features of interest in these systems can be obtained from Figure 1.1.4.

The same concept of a hierarchical organization also holds for large structures and machines, such as buildings and automobiles. While this book does not deal with structural materials, the same concepts apply. Indeed, many advanced materials, such as composites and alloys, can be considered integrated chemical systems, whose behavior (strength, flexibility, stability) depends on the microstructure and the interactions between the components.

Chemists most frequently study reactions or synthesize species at the molecular level. Synthetic methods at this level have attained a highly developed state, and very complicated molecules, such as chlorophyll and large metal clusters, can be constructed. Such synthetic reactions are usually carried out by a series of reactions in homogeneous media, specifically, in the liquid or gas phase. The modes of reaction and bonding and the nature of intermolecular interactions at this level are well understood. The synthetic methods for larger molecules (polymers and macromolecules) are similarly highly developed, and a large number of polymers and copolymers with different chemical properties and structural characteristics can be produced at will. Again, these usually arise from reactions of one or more types of monomers (e.g., reactants A and B) that polymerize in a homogeneous medium to produce polymers of the form . . . AAAA. . . , . . . BBBB. . . , . . . AABBAA. . . , . . . ABABAB. . . , and so on. However, when one wants to synthesize a polymer chain made of a number of different monomeric units (W, X, Y, Z) in a specific sequence, for example, . . . WXWYZX. . . , one must turn to synthesis at an interface. Thus polypeptides (1) and oligonucleotides (2) of known sequence are synthesized by a solid-phase method that involves attachment of the first monomer to a resin bead, followed by repeated treatments with different linking agents, monomers, and protective agents to grow the desired polymer structure. Polymers are of immense technological importance and macromolecules are involved in many biological processes, the details of reaction mechanisms and the nature of molecular interactions at this level are not as developed as those at the molecular level.

When we proceed to the next level, that of the "integrated chemical system" (3) (also sometimes called the *nanostructure*, microsystem, or

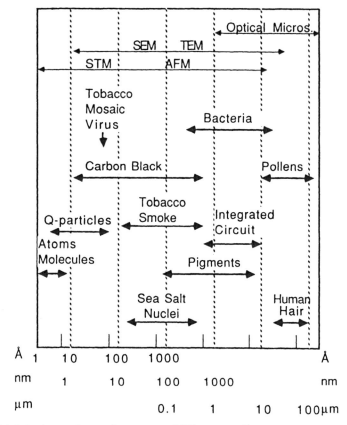

FIGURE 1.1.4. Approximate size ranges of different small structures. Integrated-circuit features are those for smallest fabricated structures as of 1991. The top lines show the microscopic techniques that can be used for imaging at the size range: STM, scanning tunneling; AFM, atomic force; SEM, scanning electron; TEM, transmission electron. [Taken, in part, from G. W. Kreichbaum and P. Kleinschmit, *Angew. Chem. Intl. Ed. Engl.*, **28**, 1416 (1989).]

mesomeric level), we are concerned with structures that contain a number of different components and several phases. The operation at this level frequently requires consideration of interfacial processes and mechanisms of mass and charge transport. Although such systems can be building blocks to even larger, more complex ones that involve a number of different integrated chemical systems, they are also of interest in their own right in connection with the construction of smaller devices, such as ones that can carry out complicated, multistep reactions or processes or that can act as analytical sensors. For example, as discussed in more detail below, many biological systems involve a number of different enzymes held in a membrane matrix. Systems for the utilization of solar energy to drive useful chemical reactions require several components that provide different functions: light absorption, charge separation, catalysis, and redox chemistry. The different components in this system are assembled on a suitable support in a structure that will produce the desired reactions when irradiated.

This monograph is concerned primarily with such systems. *We define these integrated chemical systems (ICSs) as heterogeneous, multiphase systems involving several different components (e.g., semiconductors, polymers, catalysts, membranes) designed and arranged for specific functions or to carry out specific reactions or processes. Often the different components will be organized structurally and will show synergistic effects. Usually it is the interaction of the components of the ICS that determines its properties.* Rather arbitrarily we limit the sizes of the structural elements of an ICS to the scale of molecular dimensions to a few micrometers. Thus the integrated circuits that are used in electronic devices are ICSs. Note that integrated circuits are constructed mainly by chemical fabrication techniques. Chemical reactions are not intentionally carried out with integrated circuits, but undesired reactions, such as electrochemically induced corrosion and current flow-induced metal migration, can occur in these systems. These processes become of greater importance as the size scale of the circuit is decreased. Integrated circuits are also of interest in connection with ICSs, because closely related devices have been used as analytical sensors and because some of the techniques used in the fabrication of integrated circuits are appropriate for the construction of ICSs as well, as discussed in Chapter 2.

1.2. EXAMPLES OF INTEGRATED CHEMICAL SYSTEMS

The character and applications of ICSs can be understood better by considering several examples of different types in more detail. Thus we discuss here, rather briefly, two biological organelles that are truly sophisticated ICSs, the chloroplast and the mitochondrion. We also describe several man-made ICSs, with examples of heterogeneous catalysts, photoelectrochemical systems, chemical microsensors, and a system employed in instant color photography.

1.2.1. Biological Integrated Chemical Systems

Living organisms depend on ICSs to carry out many functions, such as photosynthesis and oxidative phosphorylation. Consider a typical cell of a higher green plant (Fig. 1.2.1). Many of the organelles contained in it are ICSs. We consider here two of these, the chloroplast and the mitochondrion.

The chloroplast, shown in more detail in Figure 1.2.2, carries out the conversion of solar energy to chemical energy in the form of glucose and ATP (adenosine triphosphate) (4). The process of light absorption occurs in the grana, which are stacks of thylakoid vesicles containing light-absorbing centers and various proteins and enzymes. Note that the size of the membrane stacks in the chloroplast are of the order of 1 μm; the individual subunits within the membrane are of macromolecular size.

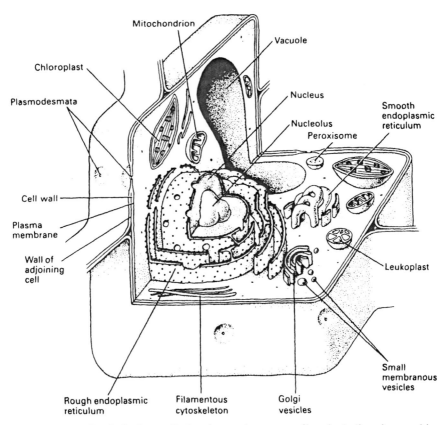

FIGURE 1.2.1. Typical plant cell showing various organelles, including the two biological integrated chemical systems discussed, the chloroplast and the mitochondrion. [From J. Darnell, H. Lodish, and D. Baltimore, *Molecular Cell Biology.* Copyright (c) 1986 Scientific American Books, Inc. Reprinted with permission of W. H. Freeman and Company.]

1.2. EXAMPLES OF INTEGRATED CHEMICAL SYSTEMS

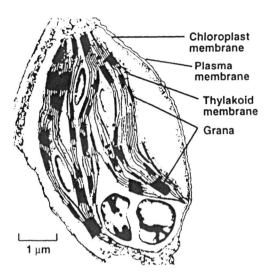

FIGURE 1.2.2. Drawing of a typical chloroplast form, showing stacks of thylakoid membranes that contain the components of the photosynthetic apparatus.

A more detailed schematic diagram of the thylakoid membrane and its various components is shown in Figure 1.2.3. The currently accepted model of photosynthesis in green plants involves two photosystems, designated I and II. The light absorbing center in photosystem II (PS II) is called P680 and involves a pair of chlorophyll molecules. When a photon of light is absorbed, an electron is transferred to the primary acceptor, Q_1, while an electron is taken from (i.e., a hole is transferred to) the water-splitting enzyme, Z. The hole ultimately causes the oxidation of water to O_2. The electron is passed, via a group of plastoquinone molecules, PQ, to proteins Fe$-$S, cytochrome f, and plastocyanin, which carry the electrons to the photosystem I (PS I) reaction-center protein. A photon of light absorbed by the P700 chlorophyll of photosystem I results in the transfer of an electron to the primary acceptor A_1, while the hole produced is filled by the electron coming from photosystem II. The electron in A_1 is passed via secondary acceptors, Fe$-$S centers, and soluble ferrodoxin to a flavoprotein that carries out the reduction of NADP$^+$ (nicotinamide adenine dinucleotide phosphate) to NADPH (reduced NADP). In the process of passage of the electrons from PS II to PS I via the PQ pool, protons are pumped across the thylakoid membrane. The increase in hydrogen ion concentration inside the membrane results in the conversion of ADP (adenosine diphosphate) to ATP. The reactions that are carried out in the photosynthetic process

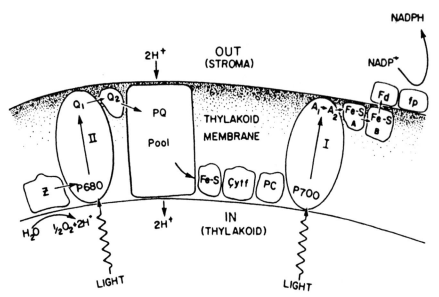

FIGURE 1.2.3. Model of the thylakoid membrane with the various components of the system. Structures I and II are the reaction center proteins of photosystems I and II. Z, the Mn-containing water-splitting enzyme; P680 and Q_1, the primary donor and acceptor species in II; Q_1, Q_2, PQ, plastoquinone molecules; Fe—S, iron-sulfur protein; Cyt f and PC, electron carriers between I and II; P700 and A_1, the primary donor and acceptor species in I; Fe—S_A, Fe—S_B, secondary acceptors; Fd, soluble ferredoxin Fe—S protein; f_p, flavoprotein that carries out reduction of $NADP^+$. [Reprinted with permission from J. R. Bolton, in *Inorganic Chemistry: Towards the 21st Century*, M. H. Chisholm, ed., *ACS Symposium Series 211*, American Chemical Society, Washington, DC, 1983. Copyright 1983 American Chemical Society.]

can be summarized as

$$2H_2O + 2\ NADP^+ \xrightarrow{h\nu} 2\ H^+ + 2\ NADPH + O_2 \quad (1.2.1)$$

$$H^+ + ADP^{3-} + P_i \rightarrow ATP^{4-} + H_2O \quad (1.2.2)$$

where P_i represents inorganic phosphate. The products are later converted, in a series of dark reactions in another system, to glucose:

$$6CO_2 + 18\ ATP^{4-} + 12\ NADPH + 12\ H_2O \rightarrow$$
$$6H^+ + C_6H_{12}O_6 + 18\ ADP^{3-} + 18\ P_i + 12\ NADP^+$$

$$(1.2.3)$$

1.2. EXAMPLES OF INTEGRATED CHEMICAL SYSTEMS

The various components in the photosynthetic system are organized in a particular arrangement so that electrons can be passed from one component to another in a specific order and direction (*vectorial charge transfer*). The operation depends on all of the components being oriented in a special arrangement within the membrane. The membrane environment is also important; if enzyme Z is removed from the membrane, it cannot oxidize water. If the primary electron acceptors of PS I or PS II are blocked, the photogenerated electrons and holes recombine, and the efficiency of the photosynthetic process is greatly diminished. Thus the chloroplast is a highly developed and advanced ICS containing a membrane support, electron-transfer catalysts, charge-transferring redox couples, and photosensitive centers all organized in a particular way to carry out the process of photosynthesis.

The mitochondrion (Fig. 1.2.4) is an organelle that exists in almost all organisms and is involved in the utilization of pyruvate and oxygen to produce ATP. As with the chloroplast, a membrane structure con-

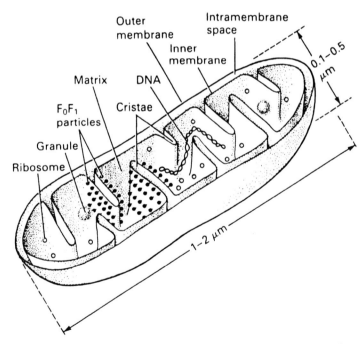

FIGURE 1.2.4. Diagram of a mitochondrion. [From J. Darnell, H. Lodish, and D. Baltimore, *Molecular Cell Biology*. Copyright (c) 1986 Scientific American Books, Inc. Reprinted with permission of W. H. Freeman and Company.]

taining imbedded enzymes is employed to carry out a series of electron-transfer and proton-transfer reactions. The different enzymes are located near one another in an interactive way to provide for vectorial charge transfer, and the membrane provides a specific environment for the enzymes that allows a coupling of the electron transfer to proton transfer across the membranes. A schematic diagram of the membrane of the mitochondrion and the major reactions are shown in Figures 1.2.5 and 1.2.6. In steps preceding those shown in Figure 1.2.5, glycolysis occurs in the cytoplasm, resulting in the conversion of glucose to pyruvate and the production of some NADH and ATP:

$$\underset{\text{(glucose)}}{C_6H_{12}O_6} + ADP^{3-} + 2P_i + 2NAD^+ \rightarrow$$

$$2\underset{\text{(pyruvate)}}{C_3H_3O_3^-} + 2H^+ + 2\,NADH + 2ATP^{4-} \quad (1.2.4)$$

The conversion of pyruvate to CO_2 in the mitochondrion occurs with the formation of NADH. The oxidation of NADH and the reduction of O_2 occurs in a separate series of reactions at the inner mitochondrial membrane, as shown in Figure 1.2.6. The overall reaction is just the reverse of Eq. (1.2.1):

$$2\,H^+ + 2\,NADPH + O_2 \rightarrow 2H_2O + 2\,NADP^+ \quad (1.2.5)$$

The net result of the reaction is the pumping of protons across the inner membrane to produce the necessary pH change to cause the hydrogen ion concentration increase needed to drive conversion of ADP to ATP [Eq. (1.2.3)].

We find analogies of both of these biological ICSs in artificial systems for the conversion of radiant energy to chemical energy (in a photoelectrochemical system) and in the oxidation of an organic "fuel" to produce electrical energy (in a fuel cell involving heterogeneous catalysts). Throughout this book we will use biological systems as guides in the design of artificial ICSs. After all, nature has had the experience of millions of years of evolution in the development of optimal structures and arrangements to carry out desired functions. However, the ICS need not mimic directly or try to reproduce the actual biological system. Biological systems need to operate under special constraints and have taken a certain design path based on evolutionary origins and naturally available materials. Attempting to design artificial systems following too closely to biological ones would certainly severely limit us. Recall, for example, that there are no biological systems in which electrons flow

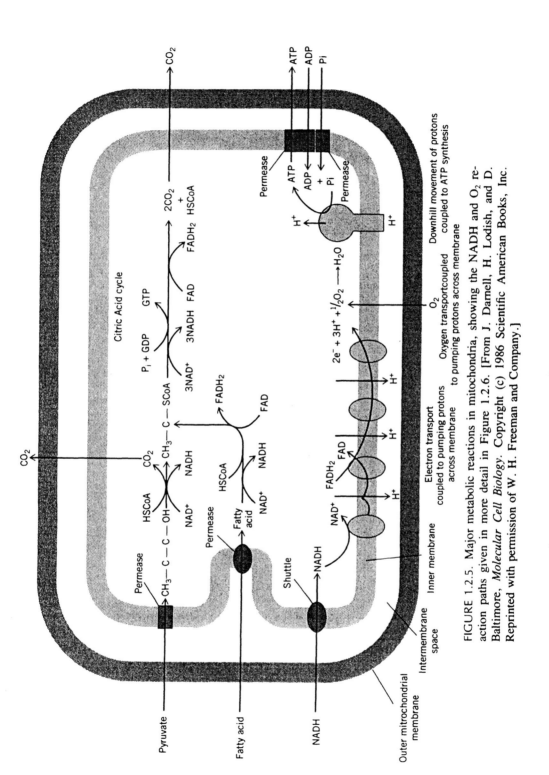

FIGURE 1.2.5. Major metabolic reactions in mitochondria, showing the NADH and O_2 reaction paths given in more detail in Figure 1.2.6. [From J. Darnell, H. Lodish, and D. Baltimore, *Molecular Cell Biology*. Copyright (c) 1986 Scientific American Books, Inc. Reprinted with permission of W. H. Freeman and Company.]

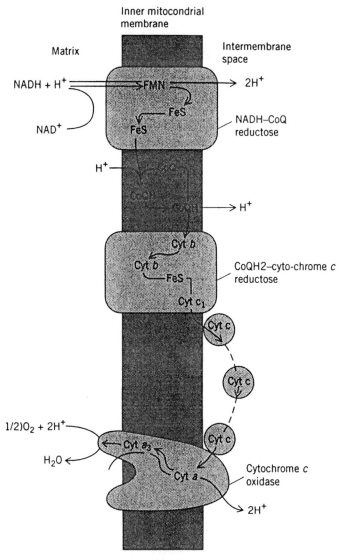

FIGURE 1.2.6. Diagrammatic representation of inner mitochondrial membrane and pathways for electron and proton transfer. [From J. Darnell, H. Lodish, and D. Baltimore, *Molecular Cell Biology*. Copyright (c) 1986 Scientific American Books, Inc. Reprinted with permission of W. H. Freeman and Company.]

through metal wires or that use semiconductors like silicon. Rather, biological systems will be considered as analogs and guides to the principles for designing the ICS. One must, however, stand in awe at the exquisite organization and structure in such biological systems. Moreover, such systems are self-organizing and often self-repairing. The ar-

1.2.2. Heterogeneous Catalysts

Some heterogeneous catalysts are ICSs that use several different components on a support to carry out a particular reaction. An example of such a system is shown schematically in Figure 1.2.7 (5). In this system, SiO_2 (in the form of particles or a glass surface) is coated with the polymer abbreviated PQ^{2+}. This polymer is formed by reduction of the monomer **I**

$$(MeO)_3Si\text{—}N^+\text{=}\text{=}N^+\text{—}Si(OMe)_3$$

I

to form a layer of $PQ^{2+}\ 2X^-$ (where X^- represents a halide ion, such as Br^-). Platinum particles are incorporated in the polymer layer by immersing it in a solution containing $PtCl_6^{2-}$, where the following ion-exchange process occurs:

$$SiO_2/PQ^{2+}, 2X^- + PtCl_6^{2-} \rightarrow SiO_2/PQ^{2+}, PtCl_6^{2-} + 2X^- \quad (1.2.6)$$

Treatment of the $SiO_2/PQ^{2+}, PtCl_6^{2-}$ with H_2 results in the formation of Pt particles within the polymer matrix. This system is capable of

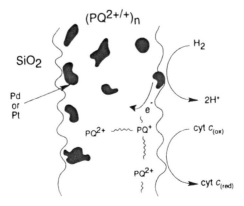

FIGURE 1.2.7. A heterogeneous catalytic ICS involving a silica support with a redox polymer ($PQ^{2+/+}$) and metal (Pt or Pd) particles that is capable of mediating the reaction between dissolved H_2 and horse heart cytochrome c (see Ref. 5).

catalyzing a redox reaction, such as the reduction of the large biological enzyme horse heart cytochrome c with H_2, by providing separate sites for the oxidation and reduction half-reactions. The H_2 diffuses into the polymer layer, where it is oxidized at the Pt particles. The polymer near the Pt particles is then reduced from the PQ^{2+} to the PQ^+ form. Electron transfer through the polymer matrix produces PQ^+ at the interface with the liquid, where reduction of Cyt c occurs. This ICS comprises a silica support, a redox couple mediator, and a Pt catalyst and is capable of effecting a reaction (between hydrogen and Cyt c) that would not occur in the absence of the catalyst.

Many other heterogeneous catalysts and electrocatalytsts can be considered as ICSs. For example, the dimensionally stable anode, introduced by Henri B. Beer and developed by Vittorio de Nora for use in chloralkali cells for the evolution of chlorine, employs a titanium substrate with a mixture of ruthenium oxide (RuO_2) and titanium oxide (TiO_2) as well as other components. The mixture and structure is adjusted to give a high surface area for the electrode reaction with good catalytic properties to promote the anodic evolution of chlorine while maintaining high stability to avoid the degradation of the electrode surface during generation of the strongly oxidizing species.

1.2.3. Photoelectrochemical Systems

Many systems have been developed for the conversion of radiant energy (e.g., sunlight) to electrical or chemical energy. For example, a semiconductor material can absorb photons to create electron–hole pairs, which are then utilized in oxidation and reduction reactions to generate desired products (Fig. 1.2.8). The production of hydrogen, which is sometimes proposed as a potential fuel, by reduction of H^+ or H_2O is often desired. However, this reduction is a kinetically difficult process and usually requires the use of a hydrogen evolution catalyst for efficient generation of H_2 at a potential near the thermodynamic one. Moreover, unless the photogenerated electrons and holes are captured in rapid chemical reactions, they tend to recombine and lower the efficiency of the photoreaction. A number of systems based on semiconductors have been described to carry out photoreactions; the overall reaction can be represented as

$$2H^+ + \text{Red} \xrightarrow[\text{(semiconductor)}]{h\nu} H_2 + \text{Ox} \quad (1.2.7)$$

where Red represents a "sacrificial (electron) donor," that is, a molecule that is irreversibly oxidized by the photogenerated holes to the reaction

1.2. EXAMPLES OF INTEGRATED CHEMICAL SYSTEMS

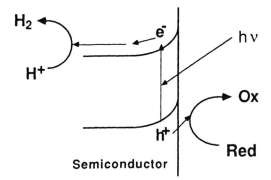

FIGURE 1.2.8. Schematic representation of light-driven reaction on an n-type semiconductor, here showing the reduction of H^+ to hydrogen and the oxidation of a sacrificial electron donor, Red. The incoming photon generates the electron (e^-)–hole (h^+) pair.

product, Ox. These kinds of systems are discussed in more detail in Chapter 6. One scheme, based on a viologen/Pt catalyst system similar to the heterogeneous catalysis scheme described earlier, is shown in Figure 1.2.9. Here light is absorbed by the GaAs (gallium arsenide) semiconductor. Direct formation of hydrogen at the GaAs surface by the photogenerated electrons, produced with an energy near the conduction band edge of the GaAs, is very slow. In other words, GaAs as

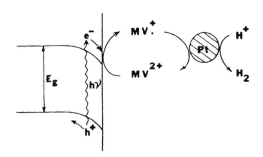

FIGURE 1.2.9. Photoproduction of hydrogen on p-type GaAs with methyl viologen and colloidal Pt in solution. [Reprinted with permission from F-R. Fan, B. Reichman, and A. J. Bard, *J. Am Chem. Soc.*, **102**, 1488 (1980). Copyright 1980 American Chemical Society.]

an electrode material shows a high hydrogen overpotential. However, electron transfer to a one-electron redox couple oxidant, such as MV^{2+}, occurs readily. Thus the reaction can be promoted by having MV^{2+} in solution or coating the electrode with a viologen-bearing polymer (**II**) containing Pt particles that act as

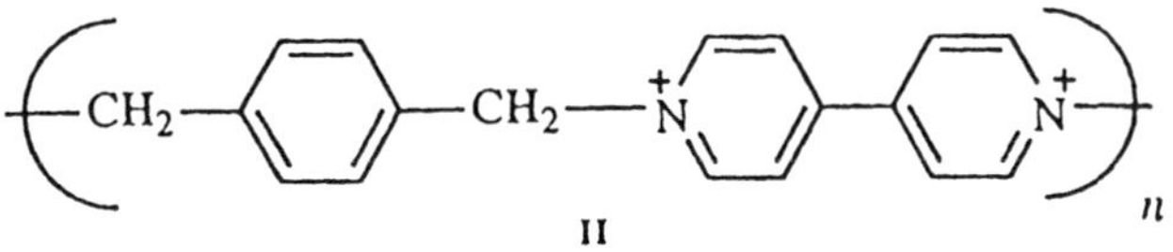

$$\left(-CH_2-\underset{}{\bigcirc}-CH_2-\overset{+}{N}\underset{}{\bigcirc}-\underset{}{\bigcirc}\overset{+}{N}-\right)_n$$

II

catalysts (6, 7). A similar system, with a Si substrate and polymer **I**, described earlier, has also been discussed (8).

A different system involving small semiconductor particles supported in a polymer matrix is shown in Figure 1.2.10 (3, 9). Here CdS is precipitated in a Nafion film to form the light-absorbing

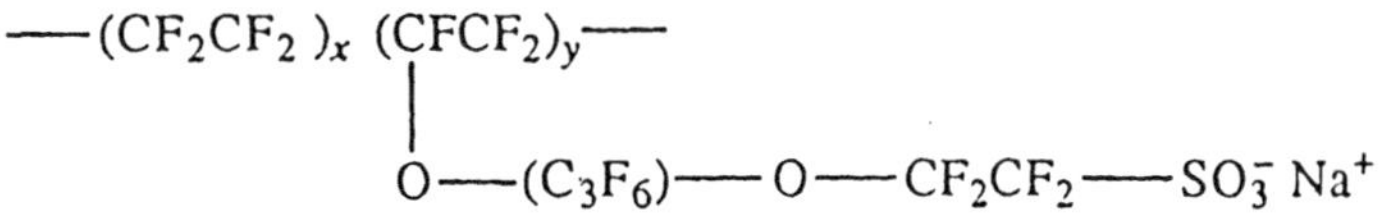

$$—(CF_2CF_2)_x\,(CFCF_2)_y—$$
$$|$$
$$O—(C_3F_6)—O—CF_2CF_2—SO_3^-\,Na^+$$

Nafion

semiconductor matrix. Nafion is a Teflon-like polymer with attached sulfonate groups on the polymer backbone. Associated with these $—SO_3^-$ groups are cations, so that Nafion is a polyelectrolyte. To produce the ICS, Cd^{2+} is first introduced in an ion-exchange process by soaking a film of Nafion in a solution containing Cd^{2+}:

$$2\,P\text{--}SO_3^-Na^+ + Cd^{2+} \rightarrow (P\text{--}SO_3^-)_2\,Cd^{2+} + 2\,Na^+ \quad (1.2.8)$$

where $P\text{--}SO_3^-$ represents the Nafion sulfonates.

Treatment of this film with H_2S forms CdS within the polymer film:

$$(P\text{--}SO_3^-)_2\,Cd^{2+} + H_2S \rightarrow 2\,P\text{--}SO_3^-H^+ + CdS \quad (1.2.9)$$

The actual structure and particle size of the CdS formed depend on the precipitation conditions. Methyl viologen can then be introduced into the polymer layer by ion exchange

$$P\text{--}SO_3^-H^+ + MV^+ \rightarrow P\text{--}SO_3^-MV^+ + H^+ \quad (1.2.10)$$

and introduction of Pt into the polymer film as described earlier. Irradiation of the film in the presence of sulfide ion in solution leads to the

1.2. EXAMPLES OF INTEGRATED CHEMICAL SYSTEMS

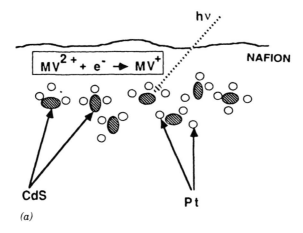

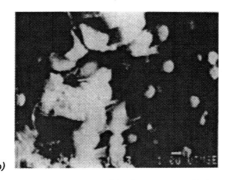

FIGURE 1.2.10. ICS for photoproduction of hydrogen, consisting of particles of CdS and Pt contained in a Nafion matrix. (a) schematic diagram of system; (b) scanning electron micrograph of cross section of Nafion membrane showing CdS particles (before Pt introduction). Calibrating bar is 1 μm. [Reprinted with permission from M. Krishnan, J. R. White, M. A. Fox, and A. J. Bard, *J. Am. Chem. Soc.*, **105**, 7002 (1983). Copyright 1983 American Chemical Society.]

production of hydrogen with concomitant oxidation of sulfide ion, which serves in this scheme as the sacrificial donor, Red [Eq. (1.2.7)]. Here we have a good example of an ICS. The polymer film serves as a support as well as the proper ion-exchange medium for assembly of the system. The CdS particles behave as the light-capturing agents in which absorbed photons are transduced into electron–hole pairs. MV^{2+} serves as the electron relay, and Pt particles serve as catalysts. This system can be considered as a very rough analog of the photosynthetic scheme discussed in Section 1.2.1.

1.2.4. Instant Color Photographic Film

Another ICS that involves a photosensitive system, but which operates in a very different manner, is photographic film used in instant photography. This ICS, which allows the photographic production of an image and its instant development to form a color image, involves a multilayer construction with a number of interfaces and components. As an example the Polaroid SX-70 system is discussed briefly. More detailed descriptions of this and other systems and the chemistry involved are available (10, 11). Schematic cross sections of the SX-70 film before and after development are shown in Figure 1.2.11. The substrate and the transparent protective layer are both polyesters. The multilayer negative, shown in more detail in Figure 1.2.12, contains three light-sen-

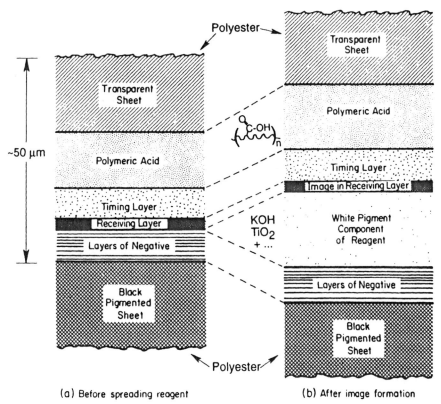

FIGURE 1.2.11. Schematic cross sections of SX-70 instant photography film (a) before and (b) after development. Details of the layers of negative are shown in Figure 1.1.12. [Reprinted with permission from E. H. Land, H. G. Rogers, and V. K. Walworth, in *Neblette's Handbook of Photography and Reprography*, 7th ed., J. M. Sturge, ed., Van Nostrand Reinhold, New York, 1977. Copyright 1977 Van Nostrand Reinhold.]

1.2. EXAMPLES OF INTEGRATED CHEMICAL SYSTEMS

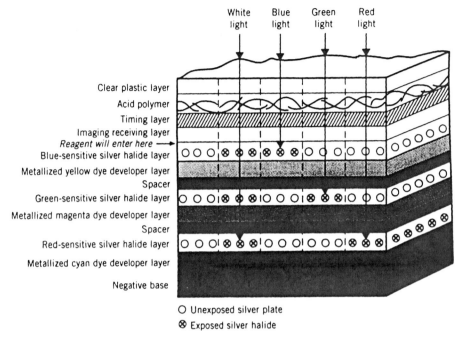

FIGURE 1.2.12. Schematic cross section of negative layers in SX-70 film during exposure. Color components are metallized dye developers: molecules capable of acting as both photographic developers and image dyes. Above each dye developer is an appropriately sensitized emulsion, which during processing controls the reactions of that dye developer. The blue-sensitive emulsion controls the yellow dye developer; the green-sensitized emulsion controls the magenta dye developer; and the red-sensitized emulsion controls the cyan dye developer. The vertical arrows represent incident light during exposure. The horizontal black arrow at the left of the initial integral structure indicates the interface which will be cleaved and entered by the reagent as it is spread during processing. [Reprinted with permission from E. H. Land, H. G. Rogers, and V. K. Walworth, in *Neblette's Handbook of Photography and Reprography*, 7th ed., J. M. Sturge, ed., Van Nostrand Reinhold, New York, 1977. Copyright 1977 Van Nostrand Reinhold.]

sitive silver halide layers and, below each, layers containing metal-ion-based dyes, which are attached to a hydroquinone developer, whose structures are shown in Figure 1.2.13. The movement of these dyes into the image-receiving layer after exposure and development forms the final color photograph.

The principles of the initial exposure step largely follow conventional photography based on sensitized silver halide layers. Absorption of a photon into a silver halide (e.g., AgBr) center leads to reduction of the AgBr with oxidation of a nearby hydroquinone molecule, called the

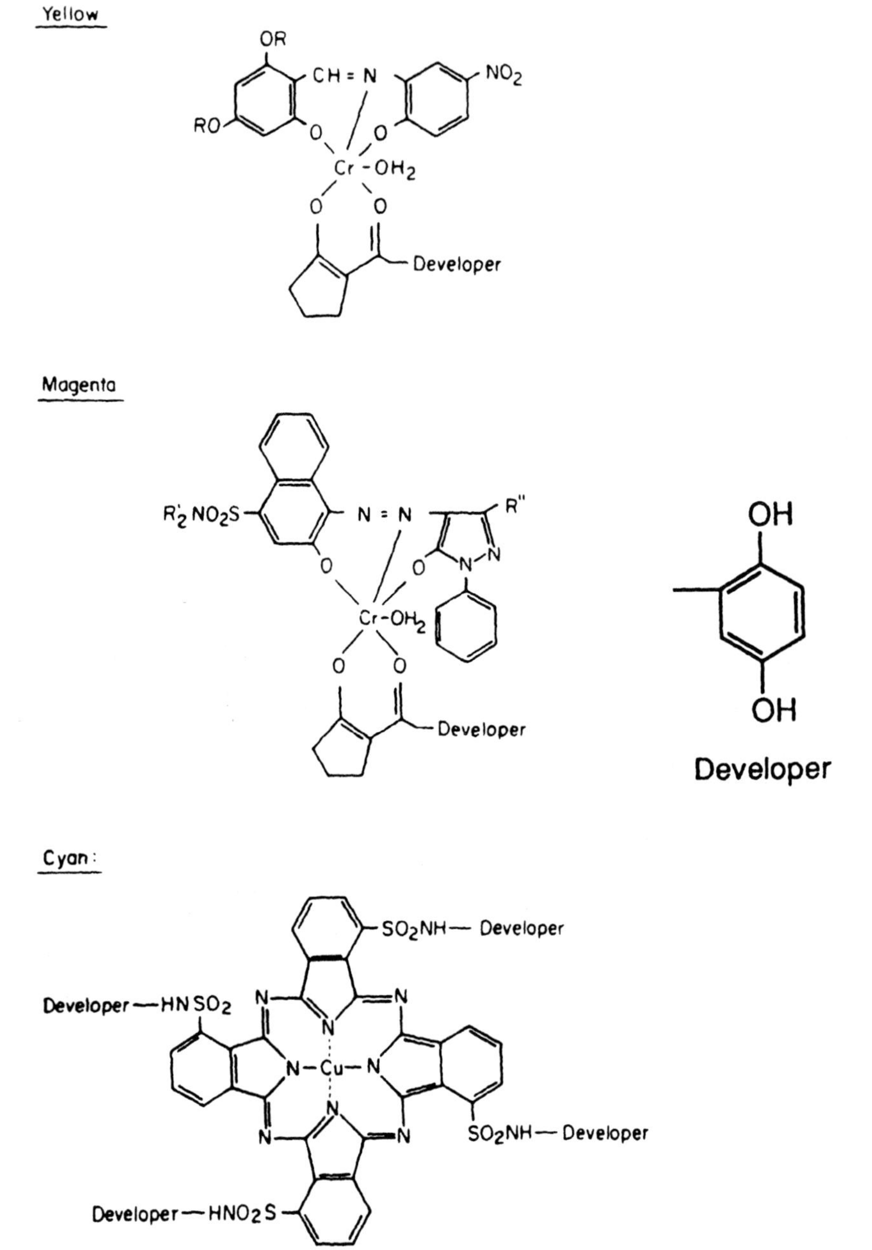

FIGURE 1.2.13. Metallized dye developers of types used in the SX-70 instant photography process. [Reprinted with permission from E. H. Land, H. G. Rogers, and V. K. Walworth, in *Neblette's Handbook of Photography and Reprography*, 7th ed., J. M. Sturge, ed., Van Nostrand Reinhold, New York, 1977. Copyright 1977 Van Nostrand Reinhold.]

1.2. EXAMPLES OF INTEGRATED CHEMICAL SYSTEMS

auxiliary developer:

$$Ag^+Br^- + h\nu \rightarrow (AgBr)^* \quad (1.2.11)$$

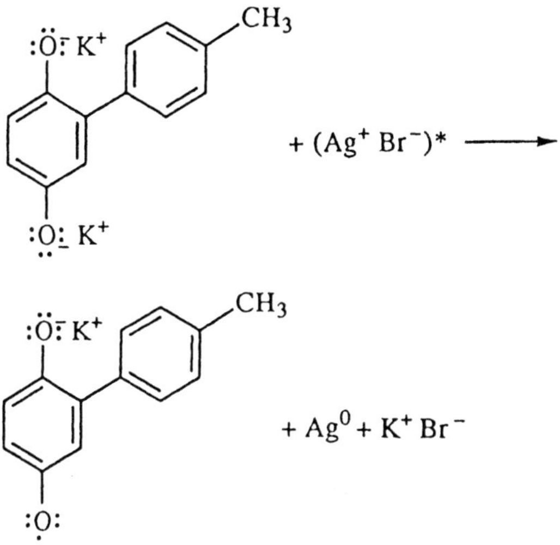

[hydroquinone dipotassium salt with tolyl substituent] $+ (Ag^+ Br^-)^* \longrightarrow$

$\quad$ (1.2.12)

[semiquinone/quinone product] $+ Ag^0 + K^+Br^-$

The oxidized auxiliary developer in turn can oxidize the hydroquinone attached to the dye molecule:

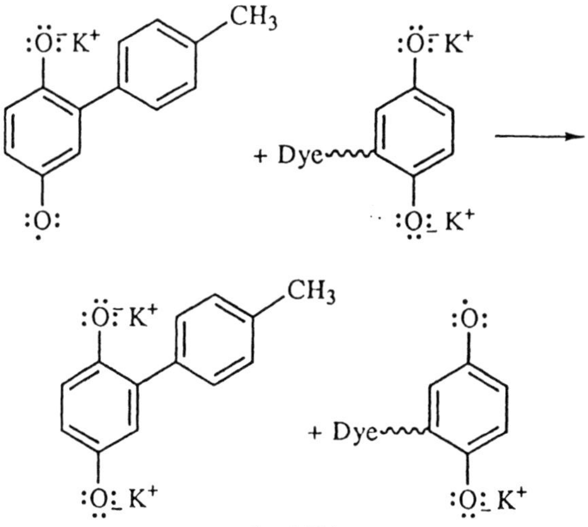

[oxidized biphenyl derivative] $+$ Dye~~~[hydroquinone-K⁺ salt] $\longrightarrow$

$\quad$ (1.2.13)

[biphenyl hydroquinone dipotassium salt] $+$ Dye~~~[quinone]

(insoluble)

Thus, near the sites that have been exposed to light of the appropriate wavelength, the dye molecules contain (oxidized) quinone centers, while at unexposed sites, the dye molecules contain reduced (hydroquinone) centers. During the development stage, a strongly alkaline (KOH) reagent is injected just below the image-receiving layer (Fig. 1.2.12). The oxidized dye molecules are insoluble in KOH, while the reduced molecules dissolve and move into the image-receiving layer, where they become immobilized:

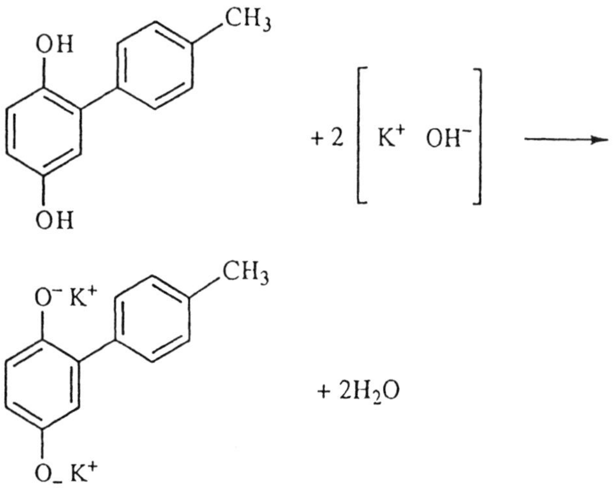

$$\text{[hydroquinone-tolyl compound with 2 OH groups]} + 2\left[K^+\ OH^-\right] \longrightarrow$$

$$\text{[same compound with 2 }O^-K^+\text{ groups]} + 2H_2O \tag{1.2.14}$$

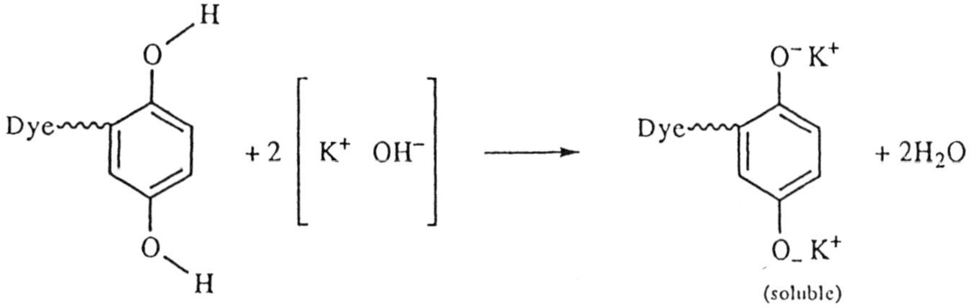

$$\text{Dye}-\text{[hydroquinone with 2 OH groups]} + 2\left[K^+\ OH^-\right] \longrightarrow \text{Dye}-\text{[hydroquinone with 2 }O^-K^+\text{ groups]}\ (\text{soluble}) + 2H_2O \tag{1.2.15}$$

The development process is shown schematically in Figure 1.2.14. The polymeric acid layer serves to neutralize the excess alkali and stop the development process; the timing layer, with a controlled permeability, determines the length of the development process (i.e., how long the system remains strongly alkaline before OH^- neutralization by the polymeric acid layer occurs).

The instant color photographic film is somewhat different from the previously described ICS, because in operation it undergoes an irreversible change in structure. Most ICSs we will consider are designed to operate without irreversible chemical or physical changes. However, the photographic ICS does involve a multicomponent system with submicrometer features and many interfaces. The action of the light demon-

1.2. EXAMPLES OF INTEGRATED CHEMICAL SYSTEMS

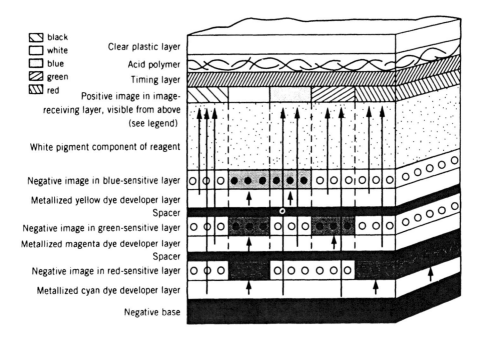

• Developed silver

FIGURE 1.2.14. Schematic cross section of negative layers in SX-70 film after development with the white pigment of the reagent shown as a new inner layer. During processing, the alkali of the reagent dissolves the dye developers, and each diffuses toward the overlying emulsion layer, as indicated by the arrows. In the fully exposed regions of each emulsion, reduction of silver halide results in oxidation and immobilization of the associated dye developer, which therefore remains in the negative. In unexposed regions, each dye developer transfers through the emulsion, through the layers above it and through the pigment to the image-receiving layer, where it may be viewed from above against the white pigment. Thus, in the region exposed to red light, the cyan dye developer is concealed from view; the yellow and magenta dye developers form the visible image, which absorbs blue light and green light but reflects red light and therefore appears red. Similarly, green exposure results in immobilization of the magenta dye developer beneath the pigment and transfer of yellow and cyan dye developers to form an image that appears green. Blue exposure results in immobilization of the yellow dye developer beneath the pigment and transfer of magenta and cyan dye developers to form an image that appears blue. In regions of white light exposure, all three dye developers remain in the negative. Where there is no exposure, all three transfer to produce an image that appears black. The transparent polymeric acid layer effects pH reduction by capturing alkali ions. The timing layer determines the time of initiation of pH reduction as controlled by its rate of permeability of alkali ions. [Reprinted with permission from E. H. Land, H. G. Rogers, and V. K. Walworth, in *Neblette's Handbook of Photography and Reprography*, 7th ed., J. M. Sturge, ed., Van Nostrand Reinhold, New York, 1977. Copyright 1977 Van Nostrand Reinhold.]

strates the use of radiation-induced processes to form structures within the system. It also contains many of the components of other types of ICSs, such as polymeric support and protective layers, sensitizers, and redox and acid–base couples.

1.2.5. Microsensors

Many different types of sensors have been constructed for the analysis of species contained in solution or in the gas phase (12–14) (see also Section 7.3). For example, electrochemical and many electronic sensors are based on changes in electric current induced by interactions of molecules with electrodes. Optical devices (called *optrodes*) are based on molecule-induced changes in absorbance or luminescence. Several different types of electrochemical sensors are shown in Figure 1.2.15 (12). These often employ integrated-circuit fabrication technology to form metal electrodes and other structures on silicon or SiO_2 layers. A critical factor in the operation of such devices is the nature of the chemically sensitive layer that covers the surface or interconnects the electrodes and gives the device the required chemical sensitivity and selectivity. Consider the chemiresistor shown in Figure 1.2.15. Interdigitated gold electrodes on an insulating SiO_2 layer are connected by a battery or ac (alternating-current) source, and the current flowing between the electrodes is used to provide information about the conductivity of the interelectrode medium. This measurement can only provide molecule specific information if the connecting medium shows high selectivity for certain species. The principles of FET and CHEMFET technology are illustrated in Figure 1.2.16. In one type of solid-state field-effect transitor (FET) (15), the current that flows between the source (S) and the drain (D) (in the example, produced by diffusion of an n-type dopant into the p-type Si substrate to form a highly doped n^+-type zone) is controlled by the electrical nature of the p-Si beneath the gate (G) electrode. With an uncharged gate, the S/G and G/D regions both behave as n/p junctions and essentially no current flows between them. If the gate region is made positive, an inversion layer forms within the p-type layer between source and drain and a current flows between them. (The principles of semiconductors and junctions are discussed in Chapter 6.) Thus, the current flow between S and D is a measure of the electrical charge on the gate electrode. In a CHEMFET (16), the gate region is covered by a suitable chemically sensitive layer and its potential is controlled with respect to a reference electrode in the solution. If the gate responds selectively and reproducibly to a solution species, the drain current can be used to measure the concentration of solution species. Note again typical characteristics of an ICS in these devices—multiphase

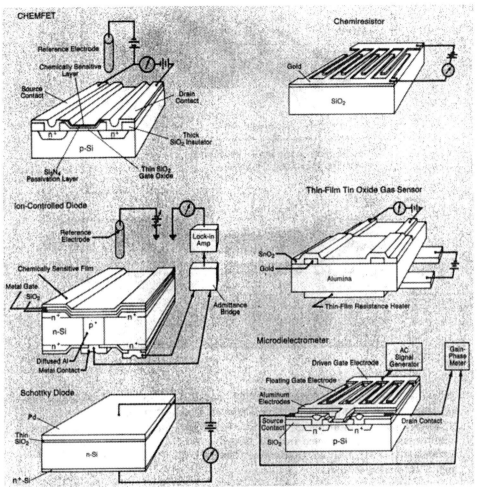

FIGURE 1.2.15. Different types of electrochemical or electronic sensors. [Reprinted with permission from H. Wohltjen, *Anal. Chem.*, **56**, 87A (1984). Copyright 1984 American Chemical Society.]

construction; a support; micrometer-size structural features; conductive, insulating, and semiconductive regions; and chemically sensitive layers.

Enzyme electrodes (17) (Fig. 1.2.17) are another type of electrochemical sensor that is an ICS. These are based on the modification of an electrode surface by attachment of an enzyme, and usually other components, to produce a system that causes a selective reaction that is detected as a potential or current change. For example, the system shown in Figure 1.2.17 is a detector for the amino acid arginine and requires both a piece of liver tissue and the enzyme urease immobilized on a pH detector (glass–reference electrode pair). A slice of bovine liver tissue is held within a suspension of urease near the electrode surface with a

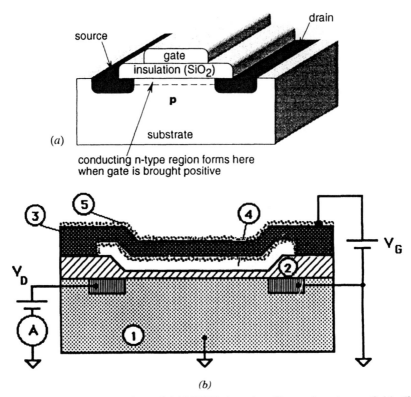

FIGURE 1.2.16. (a) An n-channel MOSFET (metal oxide semiconductor field-effect transistor). (b) A CHEMFET based on a suspended gate FET (see also Fig. 1.2.15): 1, Si substrate; 2, insulator; 3, suspended metal gate; 4, gate gap; 5, chemically sensitive layer. [(b) Reprinted from M. Josowicz and J. Janata, in *Chemical Sensor Technology*, T. Seiyama, ed., Kodansha, Scientific Ltd., Tokyo, 1988, Vol. 1. Copyright 1988 Kodansha Ltd.]

dialysis membrane. Arginine that diffuses into the mixture is converted to urea at the liver tissue:

$$\text{Arginine} \xrightarrow{\text{liver tissue}} \text{ornithine} + \text{urea} \qquad (1.2.16)$$

The urea is hydrolyzed by the urease:

$$\text{Urea} \rightarrow CO_2 + 2\, NH_3 \qquad (1.2.17)$$

The ammonia diffuses through the gas permeable membrane into the electrolyte held at the pH electrode surface, where it causes an increase in pH. The different components of the system thus contribute specific reactions and controlled fluxes that produce an electrical potential response proportional to arginine level.

1.2. EXAMPLES OF INTEGRATED CHEMICAL SYSTEMS

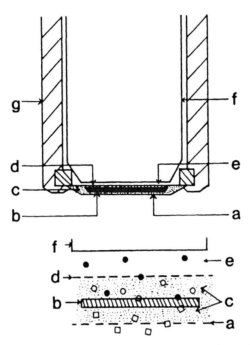

FIGURE 1.2.17. Schematic diagram of liver tissue-enzyme electrode for the detection of arginine: a, dialysis membrane; b, bovine liver tissue slice; c, urease enzyme suspension; d, gas-permeable membrane; e, internal electrolyte; f, glass-reference combination pH electrode; g, plastic electrode body. □, arginine molecules; ○, urea intermediate; ●, ammonia product. [Reprinted with permission from G. A. Rechnitz, *Science*, **214**, 287 (1981). Copyright 1981 by the AAAS.]

Optical sensor ICSs can be constructed with a similar strategy (18). The system shown in Figure 1.2.18 is a fluorescence sensor for glucose (19). The end of a bifurcated fiberoptic is modified as shown with the reagent, canavalin A (R—), immobilized on sepharose and coated on the walls of a hollow fiber. Dextran labeled with the fluorescent dye fluorescein acts as a ligand (L) for canavalin A and forms a complex with it. This complex on the walls of the hollow fiber is outside the excitation and emission light paths, so that no emission signal appears when all of the ligand, L, is attached to R—. The hollow fiber is permeable to glucose, which can diffuse in from the sample solution, but it is impermeable to any free dextran-labeled fluorescein inside the hollow fiber. Glucose that diffuses through the hollow fiber displaces some L from the R—L complex on the fiber walls via the following reaction:

$$\underset{\text{(glucose)}}{A} + \underset{\substack{\text{(canavalin A with}\\\text{dextran-fluorescein)}}}{R-L} \rightarrow R-A + \underset{\substack{\text{(dextran-}\\\text{fluorescein)}}}{L} \quad (1.2.18)$$

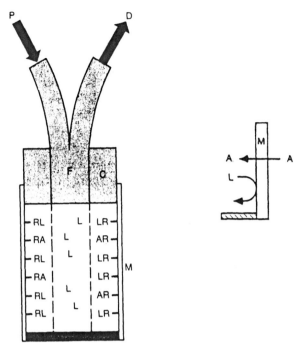

FIGURE 1.2.18. An optical sensor (optrode) for glucose based on fluoresecence detection. F is a bifurcated fiberoptic; light enters at P and fluorescence is detected at D. C is the cladding around the fiber bundle. R is the reagent containing canvalin A immobilized on sepharose and coated on the walls of the hollow fiber, M. L is the competing ligand, dextran labeled with fluorescein, and A is glucose. (See Refs. 18a and 19.) [Reprinted with permission from W. R. Seitz, *Anal. Chem.*, **56**, 16A (1984). Copyright 1984 American Chemical Society.]

The liberated ligand diffuses into the light path, where it emits light under excitation and signals the glucose level. This ICS thus employs a silica substrate as a light path, a hollow fiber and sepharose as a support, and an immobilized biological material as a selective reactant.

1.2.6. Molecular Electronic Devices

Integrated chemical systems have been proposed as electronic devices that can be used to control signals (20, 21). A typical device is shown in Figure 1.2.19. In this system, interdigitated metal (Pt or Au) bands of micrometer dimension are deposited by photolithographic techniques on an insulating (Si_3N_4 or SiO_2) substrate. These are covered with a thin polymeric film whose electronic conductivity is a function of its extent of oxidation. For example, poly-3-methylthiophene is an electrical con-

1.2. EXAMPLES OF INTEGRATED CHEMICAL SYSTEMS

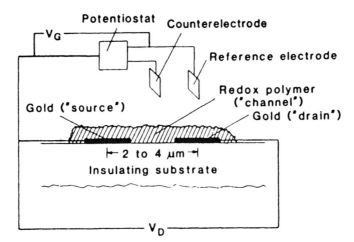

FIGURE 1.2.19. An electrochemically based transistor, where current between gold "source" and "drain" electrodes is determined by the conductivity, and hence the state of oxidation, of the redox polymer (e.g., poly-3-methylthiophene). (See Refs. 21 and 22.) [Reprinted with permission from M. S. Wrighton, *Science*, **231**, 32 (1986). Copyright 1986 by the AAAS.]

ductor in the oxidized form, but is an insulator in the reduced form. The current flow between an adjacent pair of electrodes (that behave in an analogous way to the source and drain in a solid-state FET) is controlled by the state of oxidation of the polymer (the "channel" or gate). Oxidation and reduction of the polymer can be accomplished electrochemically by making the source electrode, which is in contact with the polymer, part of an electrochemical cell and adjusting its potential (V_G) with respect to an external reference electrode. When V_G is made sufficiently positive, the polymer will be oxidized and current will flow between the gold band electrodes. When V_G is brought to a value where the polymer is reduced, no current will flow. Thus, V_G controls the current flow between the electrodes, just as V_G in a solid-state FET controls the current flow between source and drain. One must point out that the response time of this electrochemical FET is much slower than that of a solid-state device, but the principle of a chemically based FET is important. For example, the state of oxidation of the polymer layer can also be varied by contacting it with different solution components, for example, O_2 and H_2. The response of such a device to these gases is shown in Figure 1.2.20 (22). Similar responses would be shown for other oxidizing and reducing gases, so that, as with many other chemical sensors, selectivity will require specific chemical coatings.

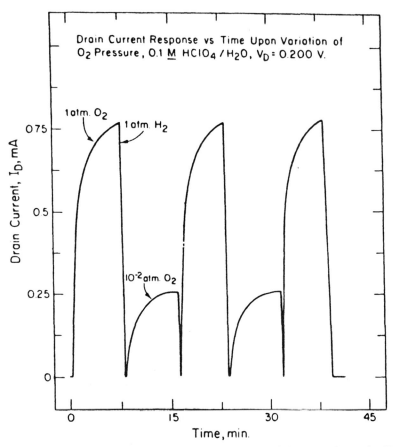

FIGURE 1.2.20. Response of electrochemical transistor of the type shown in Figure 1.2.19 to H_2 and O_2, which react with the poly-3-methylthiophene layer to change its conductivity. [Reprinted with permission from J. W. Thackeray and M. S. Wrighton, *J. Phys. Chem.*, **90**, 6674 (1986). Copyright 1986 American Chemical Society.]

1.3. CONCLUSIONS

The examples discussed here are representative of many of the ICSs that have been described. Others, such as liquid chromatographic stationary phases or other systems for chemical separations or multiphase systems that can be employed in synthesis, might also be mentioned. In Chapters 2 and 3, the general principles of the construction and characterization of ICS are discussed, and in Chapters 4, 5, and 6, some of the electrochemical and photoelectrochemical ICSs that have been developed are described in some depth. Finally, future ICSs are proposed in Chapter 7. As implied by the definition of ICS given above, the emphasis here is on systems that are employed to control reactions, to act as sensors, or which in general behave as devices. The role of the ICS for structural purposes might also be considered. For example, a large structure, such as a skyscraper, is built of different components. Here, the ICS level might be that of the composite materials, ceramics and reinforced plastics, that are sometimes used. Although consideration of materials and their properties is beyond the scope of this work, some of the principles of construction and the methods of characterization described in the following chapters may be of interest in connection with such structural ICSs.

REFERENCES AND NOTE

1. (a) R. B. Merrifield, L. D. Vizioli, and H. G. Boman, *Biochemistry,* **21,** 5020 (1982); (b) R. B. Merrifield, *Angew. Chem. Intl. Ed. Engl.,* **24,** 799 (1985); (c) A. Marglin and R. B. Merrifield, *Ann. Rev. Biochem.,* **39,** 841 (1970).
2. (a) D. M. J. Gait, *Oligonucleotide Synthesis, a Practical Approach,* IRL Press, Oxford, 1984; (b) K. Itakura and A. D. Riggs, *Science,* **209,** 1401 (1980); (c) J. W. Engels and E. Uhlmann, *Angew. Chem. Int. Ed. Engl.,* **28,** 716 (1989).
3. We coined this name in connection with systems designed to carry out photoelectrochemical reactions: (a) A. J. Bard, F-R. Fan, G. A. Hope, and R. G. Keil, *ACS Symposium Series 211,* American Chemical Society, Washington, DC, 1983, p. 93; (b) M. Krishnan, J. R. White, M. A. Fox, and A. J. Bard, *J. Am. Chem. Soc.,* **105,** 7002 (1983).
4. J. Darnell, H. Lodish, and D. Baltimore, *Molecular Cell Biology,* Scientific American Books, New York, 1986.
5. D. C. Bookbinder, N. S. Lewis, and M. S. Wrighton, *J. Am. Chem. Soc.,* **103,** 7656 (1981).
6. F-R. Fan, B. Reichman, and A. J. Bard, *J. Am. Chem. Soc.,* **102,** 1488 (1980).
7. H. D. Abruña and A. J. Bard, *J. Am. Chem. Soc.,* **103,** 6898 (1981).
8. D. C. Bookbinder, J. A. Bruce, R. N. Dominey, N. S. Lewis, and M. S. Wrighton, *Proc. Natl. Acad. Sci. USA,* **77,** 6280 (1980).
9. A. W-H. Mau, C-B. Huang, N. Kakuta, A. J. Bard, A. Campion, M. A. Fox, J. M. White, and S. E. Webber, *J. Am. Chem. Soc.,* **106,** 6537 (1984).
10. E. H. Land, H. G. Rogers, and V. K. Walworth, in *Neblette's Handbook of Photography and Reprography,* 7th ed., J. M. Sturge, ed., Van Nostrand-Reinhold, New York, 1977, p. 256.
11. V. K. Walworth, *Kirk–Othmer Encyclopedia of Chemical Technology,* 3rd ed., Wiley, New York, Vol. 6, p. 646.
12. H. Wohltjen, *Anal. Chem.,* **56,** 87A (1984).
13. R. C. Hughes, A. J. Ricco, M. A. Butler, and S. J. Martin, *Science,* **254,** 74 (1991).
14. J. Janata, *Principles of Chemical Sensors,* Plenum, New York, 1989.

15. P. Horowitz and W. Hill, *The Art of Electronics*, Cambridge University Press, Cambridge, UK, 1989, p. 113.
16. M. Josowicz and J. Janata, in *Chemical Sensor Technology*, T. Seiyama, ed., Kodansha, Ltd., Tokyo, 1988, Vol. 1, p. 153.
17. (a) G. A. Rechnitz, *Anal. Chem.*, **54,** 1194A (1982); (b) *Science*, **214,** 287 (1981).
18. (a) W. R. Sietz, *Anal. Chem.*, **56,** 16A (1984); (b) *CRC Crit. Rev. Anal. Chem.*, **19,** 135 (1988).
19. J. S. Schultz, U.S. Patent 4344438, 1982.
20. C. E. D. Chidsey and R. W. Murray, *Science*, **231,** 25 (1986).
21. M. S. Wrighton, *Science*, **231,** 32 (1986).
22. J. W. Thackeray and M. S. Wrighton, *J. Phys. Chem.*, **90,** 6674 (1986).

Chapter 2
CONSTRUCTION OF INTEGRATED CHEMICAL SYSTEMS

Our ability to arrange atoms lies at the foundation of technology.
—K. Eric Drexler
Engines of Creation
Anchor Press/Doubleday
Garden City, New York, 1986

2.1. INTRODUCTION

We examine here the materials and techniques available for the construction of integrated chemical systems. The above quotation and recent discussions of "nanotechnology," "designed interfaces," and "tailored electrode surfaces" suggest that *molecular engineering* of interfaces and materials is at hand. If this is so, then we are still at the rather early stages and are just learning how to produce materials and structures with the desired physical and chemical properties, especially structures containing several components prepared with very high (nanometer-range) resolution. If we were truly expert molecular engineers, we could produce on demand a compound that would specifically interact with one form of DNA (e.g., that of a cancer cell or an invading organism), and not with the host cells, or would form a structure like that of the mitochondrion or chloroplast. It is probably safer at this time to call our construction efforts *molecular carpentry, plumbing,* or *electrical wiring,* rather than *engineering*. This chapter represents, then, a first attempt at producing a "handbook of molecular engineering" and lists some of the different general classes of materials that have been employed in ICS and some of the means of assembling these. Further details on these topics will be given in Chapters 4 and 6 on modified electrodes and photoelectrochemical systems.

2.2. MATERIALS OF CONSTRUCTION

A typical ICS consists of a number of components, each serving a different function. A list of these, with some examples, includes the following:

1. *Supports*—classified by (e.g.) electrical properties (insulators, semiconductors, conductors) or optical properties. Typical supports: metals, polymers, bilayer membranes, clays, zeolites.
2. *Catalysts*—metals, metal oxides, organometallic compounds, enzymes, modified enzymes, catalytic antibodies.
3. *Charge carriers and mediators*—electron carriers (redox couples, electronically conductive polymers); proton and other ion carriers (acid–base couples, polyelectrolytes, ionically conductive polymers).
4. *Linking and coupling agents*—silanes, cyanuric chloride.
5. *Photosensitive centers*—inorganic semiconductors, pigments, organic and organometallic sensitizers.
6. *Electroactive centers*—redox couples.
7. *Chemically sensitive centers*—biological receptors (antibodies, enzymes), metal-selective ligands, shape-selective centers.

Note that a given component can serve more than one function in an ICS. For example, the support material can also serve as a charge-transport medium, such as the polymer Nafion. Similarly, a given material can play different roles in different applications. For example, a semiconductor might be used as a support, a catalyst, or a photoactive center. A more detailed consideration of these functions and materials follows.

2.2.1. Supports

The support material serves to hold the various components in the proper arrangement. It can also serve as a template to produce the desired structure in the ICS. For example, the pore size of a support material such as porous Vycor (a porous silica glass) or a zeolite or pillared clay can control the size of a material synthesized within it. The support may

2.2. MATERIALS OF CONSTRUCTION

also control, in the final ICS, the rate of transport of different components within or through the system, that is, function as a membrane.

The electronic conductivity of the support may be of importance. This can be varied over more than 20 orders of magnitude, as shown in Figure 2.2.1. Electronic insulators, such as SiO_2, Si_3N_4, and Al_2O_3, often serve as supports in integrated circuits and other devices. Other insulators, such as diatomaceous earth (kieselguhr), magnesia, cellulose, and starch,

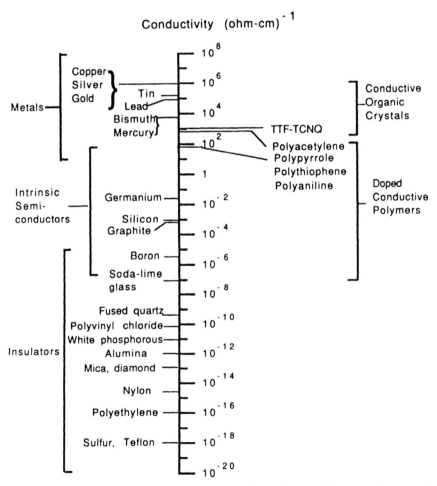

FIGURE 2.2.1. Approximate electrical conductivities of materials. Actual observed conductivities depend on doping levels and impurities. The electronic conductivities are for room temperature in units of $ohm^{-1} cm^{-1}$ (or siemens (S) per centimeter). The classification on the left of metals, semiconductor, or insulator is only approximate. [Adapted with permission from D. O. Cowan and F. M. Wiygul, *Chem. Eng. News*, July 21, 1986, p. 29. Copyright 1986 American Chemical Society.]

are used as chromatographic supports. Electronic conductors, such as graphite, metals, and conductive organic crystals (e.g., TTF^+TCNQ^-), are used as substrates for electrodes (see Section 4.2). Semiconductors can also be supports either as a bulk material (e.g., the Si substrate in an FET) or deposited as a thin film on an insulating or conducting support (e.g., SnO_2 deposited on glass; TiO_2 formed on a Ti substrate). In general, the structural form of the support, in addition to its composition, is important in its behavior. For example, materials like Al_2O_3 and TiO_2 can be prepared in many forms, such as powders, high-surface-area aerogels produced by a sol–gel process, and porous films produced by anodization of the metals.

Polymers of various types can be used as supports (Fig. 2.2.2). These may be relatively inert, such as polyethylene or poly(vinylstyrene), or may be polyelectrolytes or ion-exchange materials, such as

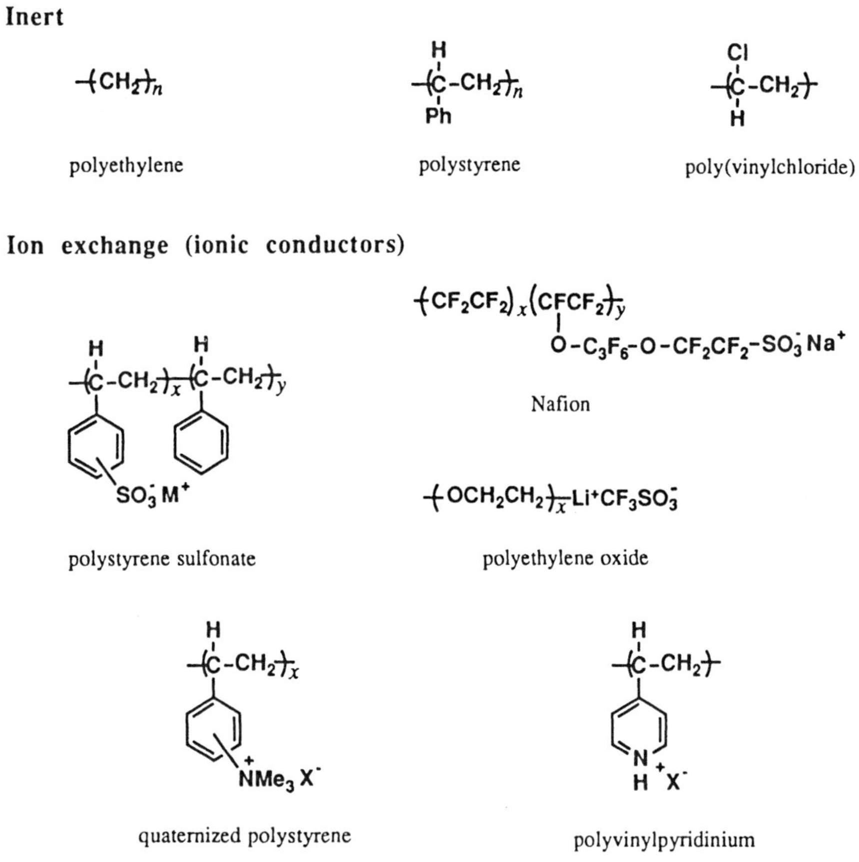

Inert

$-(CH_2)_n-$

polyethylene

$-(\overset{H}{\underset{Ph}{C}}-CH_2)_n-$

polystyrene

$-(\overset{Cl}{\underset{H}{C}}-CH_2)-$

poly(vinylchloride)

Ion exchange (ionic conductors)

$-(\overset{H}{C}-CH_2)_x(\overset{H}{C}-CH_2)_y-$ (with phenyl and phenyl-$SO_3^- M^+$ groups)

polystyrene sulfonate

$-(CF_2CF_2)_x(\overset{}{C}FCF_2)_y-$
$O-C_3F_6-O-CF_2CF_2-SO_3^- Na^+$

Nafion

$-(OCH_2CH_2)_x- Li^+ CF_3SO_3^-$

polyethylene oxide

$-(\overset{H}{C}-CH_2)_x-$ (with phenyl-$\overset{+}{N}Me_3 X^-$)

quaternized polystyrene

$-(\overset{H}{C}-CH_2)-$ (with pyridinium-$\overset{+}{N}H$ X^-)

polyvinylpyridinium

FIGURE 2.2.2. Some examples of polymers used as supports. Electronically conductive polymers are shown in Figure 2.2.4.

2.2. MATERIALS OF CONSTRUCTION

poly(vinylstyrene sulfonate) or Nafion. Ion-exchange polymers are especially useful because they allow the transport of ions through the matrix (as illustrated in Fig. 2.2.3). They can also serve as templates for construction of a system via incorporation of a suitable ion followed by subsequent modification via a chemical reaction, for example, the Nafion/CdS system discussed in Section 1.2.3. Several polymeric materials are also electronic conductors (Fig. 2.2.4) (1-4). Polymers often serve as membranes that control the flux of species into and through the system.

Membranes can also be produced by the spontaneous assembly of surface-active materials, such as lecithin or long-chain fatty acids. Surfactants can also be formed into organized layers on a surface, lipid bilayers, micelles, and vesicles (5, 6). These different arrangements are shown in Figure 2.2.5, and the formation of organized assemblies are discussed below.

Structured inorganic materials, such as clays and zeolites (7–11) (Fig. 2.2.6), are very useful support materials, because they have known and rigid structures, often with pores or channels of definite size, that can be used to give the ICS a desired orientation and configuration. They are also ion-exchange materials, usually cation exchangers. Other inorganic materials that can serve as supports include layered dichalcogenides (e.g., WS_2, SnS_2), zirconium organophosphonates and organophosphates, and vanadium hydrogen phosphate (12, 13).

Composite supports are also possible. For example, conductive polymers and semiconductors can be formed on or within clays and zeolites. Finally, the support should not simply be considered as an inert structure that holds the components in a particular arrangement and perhaps allows for an increase in surface area. The nature of the support can be important in the reactivity and mode of action of the components, since the support provides a unique environment for catalysts and reactants (14). This principle applies, for example, in enzyme catalysis in membranes.

2.2.2. Catalysts

In considering catalysts for ICS one can draw on the extensive literature in the field of heterogeneous catalysis (14–16). Catalysts often promote electron and proton transfers. Metals, such as Pt and Pd, and metal oxides, like RuO_2, can serve to accelerate the transfer of electrons to or from a desired species. For example, the reduction of protons to hydrogen or the reduction of oxygen to water is promoted by Pt. The oxidation of water to oxygen is catalyzed by several Ru compounds, including RuO_2. In these electron-transfer reactions, the catalyst particle often

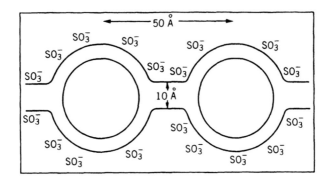

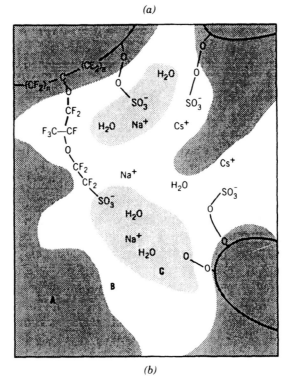

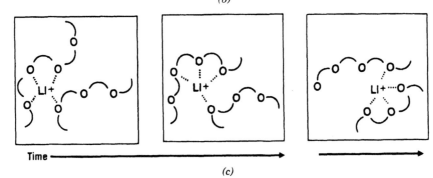

2.2. MATERIALS OF CONSTRUCTION

serves as a "microelectrode" that mediates both the oxidation and reduction reactions. An example of this mediation is the reaction of reduced methyl viologen (MV^+) with protons to form H_2 described in Section 1.2.3. The direct reaction between MV^+ and H^+ is slow, because MV^+ is oxidized in a one-electron reaction ($MV^+ - e \rightarrow MV^{2+}$), while two electrons are needed to generate hydrogen ($2H^+ + 2e \rightarrow H_2$). Since the simultaneous collision of two MV^+ and an H^+ is very improbable, and the reaction between a single MV^+ and H^+ is insufficiently energetic to produce a free hydrogen atom, the homogeneous reaction is slow. At Pt particles, both reactions can occur at the metal surface, exchanging electrons with the Pt. In this case, the Pt also serves to assist proton reduction through the strong adsorption of hydrogen atoms. In multi-electron-transfer reactions, the catalyst may serve to store the needed number of charges to effect the desired reaction, as in the four-electron-transfer reaction from O_2 to H_2O. In choosing a catalyst it is important that it be chemically stable under the reaction conditions. Noble-metal catalysts, like Pt and Ru, are popular because they are difficult to oxidize and dissolve.

Catalysts also serve to speed up a given bond-breaking or bond-forming reaction, often with high selectivity. Well-known examples include the catalytic cracking of hydrocarbons at zeolites and over alkaline-earth oxides. Since many of the useful metal catalysts are expensive, a great deal of effort has gone into the synthesis of other materials that can play an equivalent role. Organometallic species can serve as catalysts for many different types of reactions in homogeneous solution. These can be immobilized into an ICS to play a similar role.

Enzymes isolated from biological materials are often employed in chemical systems such as chemical sensors. These generally have the

---

FIGURE 2.2.3. Proposed modes of ion transport through ionic conductors. (a) Model of Nafion proposed by Gierke. The large circles represent clusters of inverted micelles, with sulfonate groups and water in the interior within a hydrophobic fluorocarbon matrix. Cations move from cluster to cluster through channels. [Reprinted with permission from K. A. Mauritz, C. J. Hora, and A. J. Hopfinger, in *Ions in Polymers*, A. Eisenberg, ed., American Chemical Society Advances in Chemistry Series, No. 187 (1980). Copyright 1980 American Chemical Society.] (b) More detailed model of Nafion (three-region model): where A is fluorocarbon, B is interfacial zone, and C are ionic clusters. [Reprinted with permission from H. L. Yeager and A. Steck, *J. Electrochem. Soc.*, **128**, 1880 (1981). Copyright 1981 The Electrochemical Society.] (c) Model of Li^+ conduction in polyethylene oxide in which local motion of the polymer chains assists movement of coordinated ion through solid. [Reprinted with permission from D. F. Shriver and G. C. Farrington, *Chem. Eng. News*, May 20, 1985, p. 42. Copyright 1985 American Chemical Society.]

polymer	structure	typical methods of doping	typical conductivity (ohm-cm)$^{-1}$
polyacetylene		electrochemical, chemical (AsF_5, I_2, Li, K)	500 (2000 for highly-oriented films)
polyphenylene	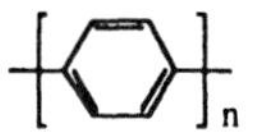	chemical (AsF_5, Li, K)	500
poly(phenylene sulfide)	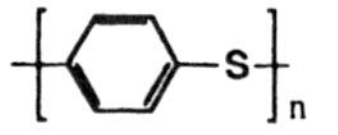	chemical (AsF_5)	1
polypyrrole	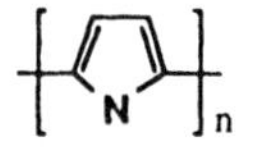	electrochemical	600
polythiophene	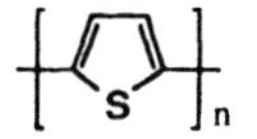	electrochemical	100
poly(phenylquinone)	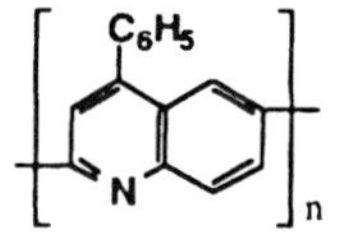	electrochemical, chemical (sodium naphthalide)	50
polyaniline	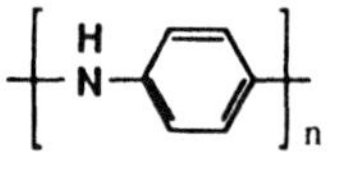	electrochemical	500

FIGURE 2.2.4. Electronically conductive polymers.

FIGURE 2.2.5. Some structures produced from surfactants. (*a*) spherical micelle: (1) schematic; (2) computer-generated model of dodecanoate, where black represents head group, white the hydrocarbon tail and stippled the terminal methyl. [Reprinted with permission from *Nature* (K. A. Dill et al., 1984, 309, 42) Copyright 1984 Macmillan Magazines Limited.]. (*b*) Schematic representations of (1) spherical vesicle; (2) multicompartment liposome. (*c*) (1) lamellar micelle; (2) bilayer lipid membrane.

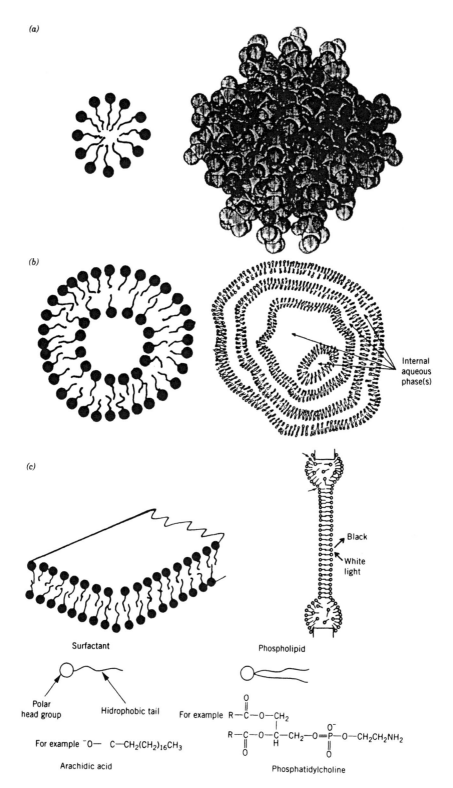

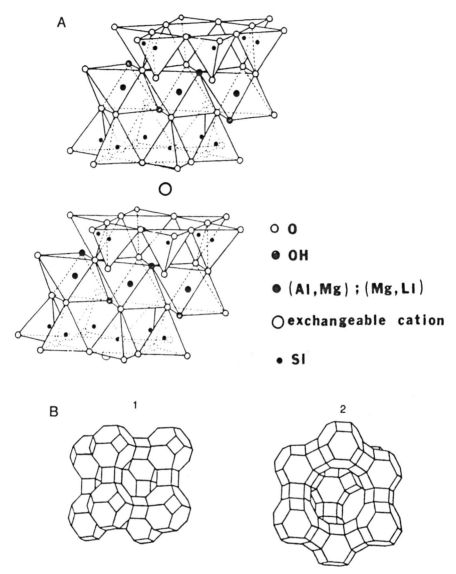

FIGURE 2.2.6. Schematic representations of typical silicate structures that can be used as supports. (A) 2:1 layer smectic clay (e.g., montmorillonite) consisting of sheets of 1 AlO_6 octathedron to 2 SiO_4 tetrahedra. The sheets have an overall negative charge that is balanced by exchangeable cations (e.g., Na^+). [Reprinted with permission from P. K. Ghosh and A. J. Bard, *J. Phys. Chem.*, **88**, 5519 (1984). Copyright 1984 American Chemical Society.] (B) Zeolite frameworks: (1) Faujasite (Zeolites X and Y); (2) Zeolite A. [Reprinted with permission from W. Hölderich, and M. Hesse, F. Näumann, *Angew. Chem. Intl. Ed. Engl.*, **27**, 226 (1988). Copyright 1988 VCH Verlagsgesellschaft.]

2.2. MATERIALS OF CONSTRUCTION

advantage of high selectivity, but often are of limited stability. These can sometimes be immobilized and be modified structurally to change their behavior from that in the natural state. There has been recent interest in antibodies that are expressed for a species with a structure similar to that assumed for the transition state of a desired chemical reaction. Such catalytic antibodies can behave like enzymes, such as in cleaving peptides or esters (17).

2.2.3. Charge Carriers and Mediators

In some ICSs, electrons or ions are transferred between species or across membranes. This charge transfer can be facilitated by appropriate mediators. For example, for electron transfer across an ion-exchange membrane, suitable redox couples, Ox/Red, can be used; favorites are $MV^{2+/+}$, $Ru(bpy)_3^{3+/2+}$ (bpy = 2, 2'-bypyridine), $Ru(NH_3)_6^{3+/2+}$, and $Fc^{0/+}$ (Fc = ferrocene or a substituted ferrocene). These are often also used as mediators for charge transfer to enzymes or semiconductors. The desired characteristics of a redox couple for such a role include high stability of both oxidized and reduced forms, good solubility in the medium, a standard potential in the desired range, and reasonably rapid electron-transfer kinetics (Table 2.2.1). For mediation of enzymes (oxidoreductases), the mediator must also be able to access the prosthetic group of the enzyme. Acid–base couples similarly are used to mediate or promote proton transfers. Electron transport through a system can also be accomplished by electronic conductors formed within the support. For example, metallic particles, conductive polymers, or organic metals can be formed within a polymer or zeolite matrix.

2.2.4. Linking and Coupling Agents

Different types of interactions are used in ICS assembly. Coupling agents are a "molecular glue" for interconnecting components or attaching them to substrate surfaces. In many cases an ICS can be assembled without the explicit need to bind the components together with covalent bonds, such as by making use of strong adsorptive, ion exchange, or hydrophobic interactions. However, more stable systems result from the purposeful linking and attachment of components via strong covalent bonds, such as silyl $(-O-Si-)$, ester $[-O-(C=O)-]$, or amide $[-NH-(C=O)-]$ linkages. Several linking agents and their relevant reactions are shown in Fig. 2.2.7. In all cases, a substituent of interest, indicated as Z, is shown being linked to another component with a free $-OH$ or $-COOH$ group. These groups could represent the

TABLE 2.2.1. Selected Oxidation–Reduction Mediators in Aqueous Solution (pH 7)[a]

Mediator	Potential (V vs. NHE)
Ru(bpy)$_3^{3+/2+}$	1.27
Br$_2$/Br$^-$	1.09
Fe(phen)$_3^{3+/2+}$	1.07
IrCl$_6^{2-/3-}$	1.00
Ru(CN)$_6^{3-/4-}$	0.86
Os(bpy)$_3^{3+/2+}$	0.84
Mo(CN)$_8^{3-/4-}$	0.77
1,1-Dicarboxylic acid ferrocene	0.64
W(CN)$_8^{3-/4-}$	0.49
Co(phen)$_3^{3+/2+}$	0.38
Fe(CN)$_6^{3-/4-}$	0.36
1,4-Benzoquinone/hydroquinone	0.28
N,N,N',N'-Tetramethyl-p-phenylenediamine (TMPD)	0.27
1,2-Naphthoquinone/hydroquinone	0.14
Anthraquinone-2-sulfonate/hydroquinone	−0.22
Methyl viologen (+2/+)	−0.45
4,4'-Dimethyl-1,1'-trimethylene-2,2'-bipyridyl	−0.69

[a] More extensive listings of mediator compounds, especially for use with biological systems in aqueous solutions, are given in: W. Clark, *Oxidation-Reduction Potentials of Organic Systems*, Williams & Wilkins, Baltimore, 1960; M. L. Fultz and R. A. Durst, *Anal. Chim. Acta*, **140**, 1 (1982); R. Szentrimay, P. Yeh, and T. Kuwana, in *Electrochemical Studies of Biological Systems*, D. Sawyer, ed., American Chemical Society, Washington, DC, 1977, p. 143; P. N. Bartlett, P. Tebbutt, and R. G. Whitaker, *Prog. React. Kinet.*, **16**, 55 (1991); A. J. Bard, R. Parsons, and J. Jordan, *Standard Potentials in Aqueous Solution*, Marcel Dekker, New York, 1985.

surface of oxidized metal, semiconductor, or carbon. Silanes of various types have been used extensively (18). Cyanuric chloride and dicyclohexyl carbodiimide (DCC) are representative of a number of possible organic coupling agents. Inorganic species can also be used as linking agents. For example, trinuclear acetatohydroxoiron(III) nitrate [Fe$_3$(OCOCH$_3$)$_7$OH·2H$_2$O] NO$_3$ will, on heating, attach via a dehydration reaction to aluminosilicates (e.g., clays and zeolites); zirconyl chloride and hydroxy–aluminum ions can play a similar role. These are used, for example, to form pillars between the sheets of clays to form a more rigid structure, but can probably also serve to attach other components to them. Covalent bonds to a surface can sometimes form spontaneously. For example, the interaction of S with Au, Hg, or Ag is so

2.2. MATERIALS OF CONSTRUCTION

Silanes

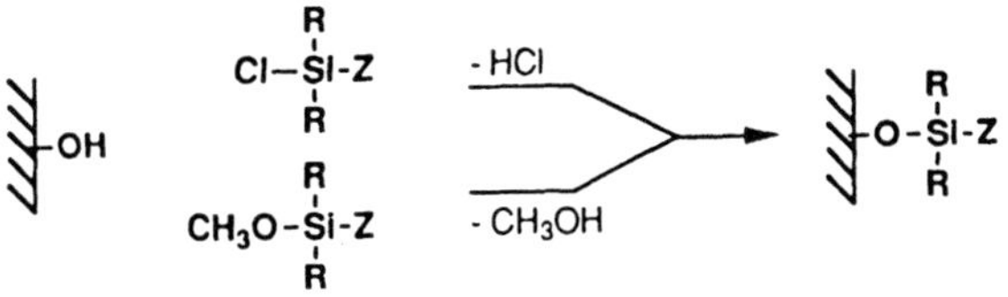

$$\text{surface-OH} + \begin{matrix} \text{Cl-Si(R)}_2\text{-Z} \\ \text{CH}_3\text{O-Si(R)}_2\text{-Z} \end{matrix} \xrightarrow[-\text{CH}_3\text{OH}]{-\text{HCl}} \text{surface-O-Si(R)}_2\text{-Z} \quad (1)$$

For example,

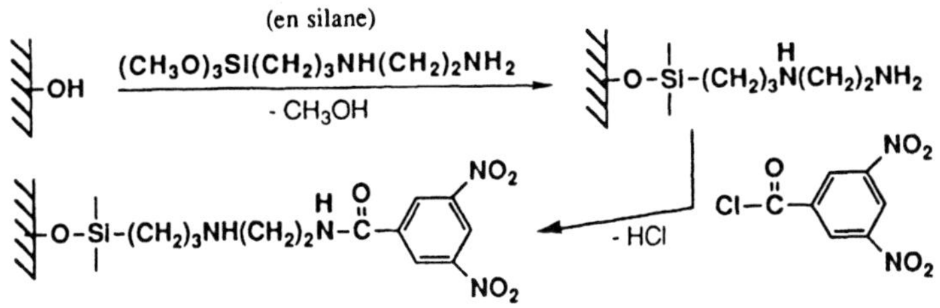

$$\text{surface-OH} \xrightarrow[-\text{CH}_3\text{OH}]{(\text{CH}_3\text{O})_3\text{Si(CH}_2)_3\text{NH(CH}_2)_2\text{NH}_2 \text{ (en silane)}} \text{surface-O-Si-(CH}_2)_3\text{N(H)(CH}_2)_2\text{NH}_2$$

$$\xrightarrow[-\text{HCl}]{\text{Cl-C(=O)-C}_6\text{H}_3(\text{NO}_2)_2} \text{surface-O-Si-(CH}_2)_3\text{NH(CH}_2)_2\text{N(H)-C(=O)-C}_6\text{H}_3(\text{NO}_2)_2 \quad (2)$$

Dicyclohexyl carbodiimide (DCC)

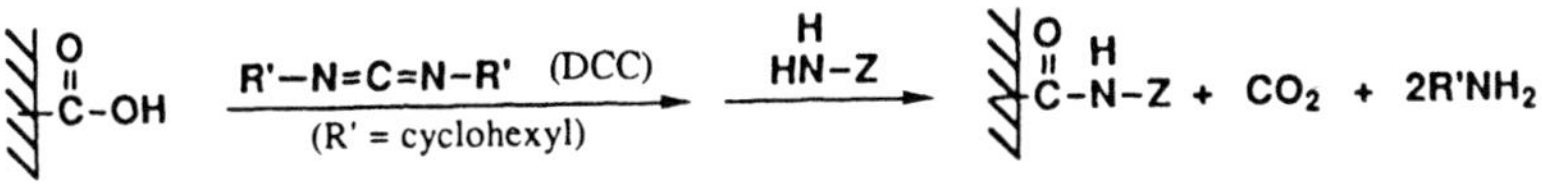

$$\text{surface-C(=O)-OH} \xrightarrow[\text{(R' = cyclohexyl)}]{\text{R'-N=C=N-R' (DCC)}} \xrightarrow{\text{HN(H)-Z}} \text{surface-C(=O)-N(H)-Z} + \text{CO}_2 + 2\text{R'NH}_2 \quad (3)$$

For example,

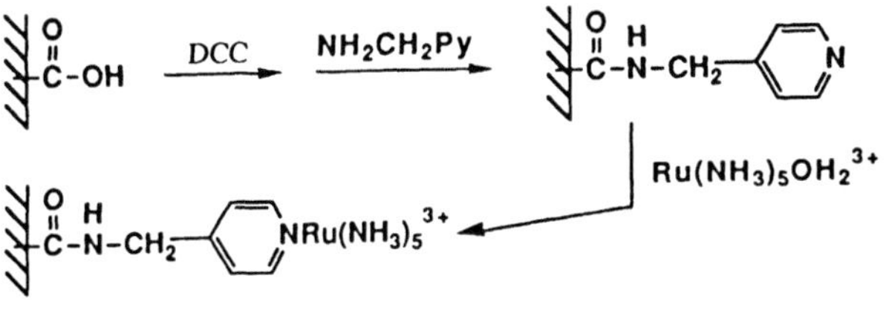

$$\text{surface-C(=O)-OH} \xrightarrow[]{\text{DCC}} \xrightarrow{\text{NH}_2\text{CH}_2\text{Py}} \text{surface-C(=O)-N(H)-CH}_2\text{-Py} \quad (4)$$

$$\xrightarrow{\text{Ru(NH}_3)_5\text{OH}_2{}^{3+}} \text{surface-C(=O)-N(H)-CH}_2\text{-PyRu(NH}_3)_5{}^{3+}$$

Cyanuric Chloride

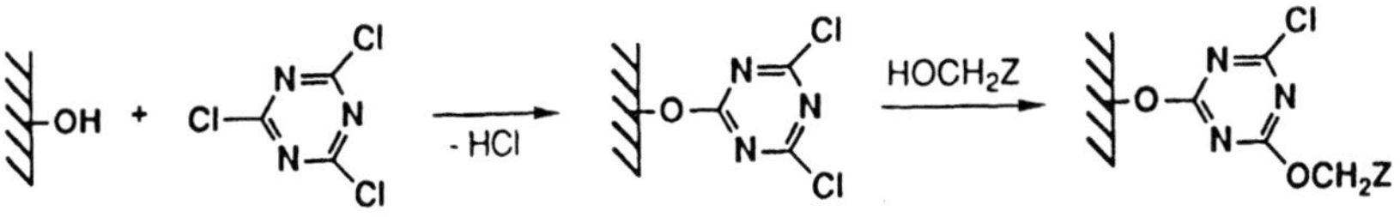

$$\text{surface-OH} + \text{Cl-C}_3\text{N}_3\text{Cl}_2 \xrightarrow{-\text{HCl}} \text{surface-O-C}_3\text{N}_3\text{Cl}_2 \xrightarrow{\text{HOCH}_2\text{Z}} \text{surface-O-C}_3\text{N}_3(\text{Cl})(\text{OCH}_2\text{Z}) \quad (5)$$

FIGURE 2.2.7. Representative coupling reagents and typical reactions for their use in attachment to a support surface. Z represents a group (e.g., a ferrocene or viologen) that is being immobilized to a surface. During reactions of silanes with a surface, multiple bonds to the surface may form, or crosslinking between adjacent silanes can occur.

strong that exposure of the surfaces of these metals to species like RSH or RSSR leads to derivatization of the surface with RS molecules.

2.2.5. Photosensitive Centers

Species that absorb light are needed in ICSs that are used to carry out photochemical reactions. These behave simply as antennas for transporting the excitation energy through the system or as actual transducers for conversion of the energy of the absorbed photon to that of an electron–hole (e^-h^+) pair and can be isolated molecules that absorb at the proper wavelength, for example, aromatic hydrocarbons like 9,10-diphenylanthracene, or $Ru(bpy)_3^{2+}$, or larger particles of pigment. Semiconductor particles, such as TiO_2 and CdS, are efficient light absorbers and transducers, as discussed in more detail in Chapter 6.

2.2.6. Chemically Sensitive and Electroactive Centers

In the design of analytical sensors or surfaces that are selective for a particular reaction, molecules that show some specificity in their interaction with a target species are required. In general, chemically sensitive centers play a role in catalysis, transport, and recognition and are usually based on sites with the size, shape, and interacting groups or environments specific to a particular molecule. The general aim in this area is to form sites that are capable of *molecular recognition*, for example, by the design of receptor sites capable of binding a substrate strongly and selectively. A number of different approaches have been taken (19–21), and several representative structures that have been employed in molecular recognition are shown in Figure 2.2.8. The general principle is suggested in *A*, where a cavity of the appropriate size and shape and containing several interacting groups is shown. Examples of several designed molecules, such as the cavitands (cryptands and spherands) (20, 21) are shown. Flexible molecules that feature a cleft have also been used (19). Cyclodextrins are naturally occurring, rigid, bucket-shaped molecules with hydrophobic interiors and hydrophilic exteriors (22). The size of the interior of the bucket varies with the particular cyclodextrin molecule (Fig. 2.2.9). The molecule can be attached to a carbon surface by the following reactions (23):

$$\sim C-(CO)-OH \xrightarrow{SO_2Cl_2} \sim C-(CO)-Cl \xrightarrow{-CD-OH}$$
$$\sim C-(CO)-O-CD \qquad (2.2.1)$$

2.2. MATERIALS OF CONSTRUCTION

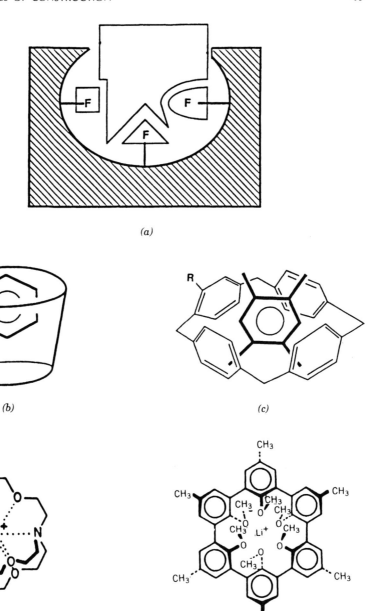

FIGURE 2.2.8. Molecules that show some degree of molecular recognition. (*a*) Schematic diagram of an active site showing cavity with three interacting functional groups (F). (*b*) Cyclodextrin containing incorporated benzene molecule (see also Fig. 2.2.9). (*c*) Cyclophane complexing durene [reprinted with permission from J. Rebek, Jr., *Science*, **235**, 1478 (1987). Copyright 1987 by the AAAS.] (*d*) [2.2.2] cryptand complexing K^+; (*e*) a spherand complexing Li^+.

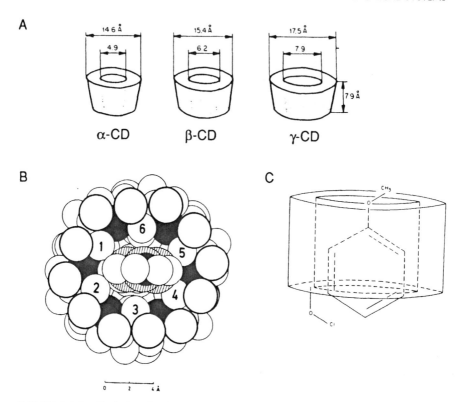

FIGURE 2.2.9. Cyclodextrin (CD). (A) Sizes of different cyclodextrins (α, β, γ). [Reprinted with permission from J. Szejtli, in *Inclusion Compounds*, J. L. Atwood, J. E. D. Davies, and D. D. MacNicol, eds., Academic Press, London, 1984, Vol. 3, Chapter 11, p. 332. Copyright 1984 Academic Press]. (B) Space-filling model of the complex between α-CD and aromatic molecule. [Reprinted with permission from W. Saenger, in *Inclusion Compounds*, J. L. Atwood, J. E. D. Davies, and D. D. MacNicol, eds., Academic Press, London, 1984, Vol. 2, Chapter 8, p. 248. Copyright 1984 Academic Press.]. (C) Structural diagram of the complex between a CD and anisole. [Reprinted with permission from R. Breslow, in *Inclusion Compounds*, J. L. Atwood, J. E. D. Davies, and D. D. MacNicol, eds., Academic Press, London, 1984, Vol. 3, Chapter 14, p. 478. Copyright 1984 Academic Press.]

where ~C represents the carbon surface and **CD** is cyclodextrin. Because of the size of the interior and its environment, aromatic molecules of the size to fit in the cage (e.g., substituted phenyls) will complex with it. This provides a degree of molecular recognition and orientation. Macrocyclic ligands, such as the crown ethers and the cryptands, also are used for this purpose. Similarly, many ligands, through chelation effects, can be used as specific sites for metal ions. Of course, biological systems contain a wide range of sites, such as those in enzymes, anti-

2.2. MATERIALS OF CONSTRUCTION

bodies, pheromone receptors, and neurotransmitter receptors, that show highly specific interactions and can often distinguish between optical isomers of the same molecule. One can also envision synthetic polymers and inorganic structures that serve this function through specific structural features and functional groups. For example, pillared clays and zeolites have interior spaces and channels of a size that allow small molecules to penetrate to the interior while excluding larger ones.

Electroactive centers are provided on a surface by tethering redox groups of the type described in Section 2.2.3 to a surface. They provide recognition, in a less selective way than the sites described above, through their specific redox potential. It is possible to couple a receptor site, such as a chelating agent, with a redox center, for example, for the complexation and selective reduction or oxidation of a metal ion.

2.2.7. Other Components

One can envision other ICS components that could serve specific functions. For example, pressure selectivity could be accomplished by incorporation of piezoelectric materials, such as $BaTiO_3$ or poly(vinylidene fluoride). These convert a mechanical stress into an electrical signal that could then be detected by another component. Alternatively, a piezoelectric material could be used to convert an electrical signal into mechanical motion, as, for instance, in producing the synthetic equivalent of muscle fibers or flagella. Polymers that change dimension on oxidation and reduction (see Chapter 4), or when triggered by temperature or chemical changes (24), could also serve this function. Components sensitive to temperature or magnetic field might also be useful in certain applications. It would by useful if ICSs, especially portable or mobile ones, could be powered by integral power sources. Probably the most likely ones would be electrical, such as, miniature batteries or capacitors. These might be constructed with the same materials as larger ones by scaling down current methods of fabrication. The disadvantage of such sources is the need for replenishment of the active materials (for primary cells) or recharging. Another possibility would be a miniature fuel cell. An interesting approach to fabrication of such a device is shown in Figure 2.2.10 (25). In this thin-film fuel cell two Pt electrodes are separated by a thin membrane of pseudoboehmite ($Al_2O_3 \cdot 2.1H_2O$) or Nafion. Although the whole device is bathed in a mixture of H_2 and O_2, for some reason the presence of the membrane causes a differential reaction rate at the two electrodes, so that H_2 oxidation occurs preferentially at the outer electrode and O_2 reduction occurs at the inner electrode. Open-circuit voltages of up to 0.95 V and power densities of

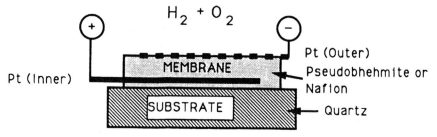

FIGURE 2.2.10. Schematic cross section of miniature fuel cell based on differential reaction of hydrogen and oxygen at inner and porous outer electrodes. [Reprinted from C. K. Dyer, *Nature*, **343**, 547 (1990). Copyright 1990 Macmillan Magazines Limited.]

1–5 mW/cm^2 were reported. An alternative would be a miniature fuel cell built along more conventional lines, with the fuel and oxygen streams separated and fed independently to membrane (e.g., Nafion)-separated electrodes.

2.3. PRINCIPLES OF CONSTRUCTION

2.3.1. Means of Attachment

The various components in an ICS, as in most other chemical systems, can be interconnected by different types of interactions with greatly differing energies. Of course, this division of modes of bonding into different groups is strictly formal and is useful only as a general classification. In many, if not most, cases, an intermolecular interaction involves several different modes operating simultaneously. For example, the formation of a bond between a metal ion and a ligand can involve both covalent and electrostatic forces. Similarly, when Ru(bpy)$_3^{2+}$ is taken up by a Nafion film, both electrostatic and hydrophobic interactions must be considered. The different classes of interactions, listed in roughly the order of decreasing energy of interaction (per bond or unit interaction) are as follow:

Covalent attachment—carried out by directly linking groups of interest or via linking agents, as described above (text and Fig. 2.2.7). Other examples are given in Chapter 4.

Electrostatic interactions—occur via ion exchange, such as the replacement of one ion by another in a polyelectrolytic material, via ion pairing, or via donor–acceptor interactions. This mode is important in ion-exchange polymers, such as poly(styrene sulfonate) and Nafion, and also in clays and zeolites.

Hydrogen bonding—which is especially important in biological systems. Although the energy of an individual hydrogen bond is small, a given intermolecular interaction may be very strong, because it involves the formation of a large number of hydrogen bonds between the species.

Hydrophobic interactions—occur because of the tendency of fatty or lipophilic groups to interact with one another in an aqueous, or more generally, a polar, environment. These interactions are involved in the tendency of surface-active substances to form monolayers at a water surface and are of importance in the formation of lipid bilayers, membranes, micelles, and vesicles (Fig. 2.2.5).

Van der Waals forces—are quite weak, but may be involved in the physical adsorption of species to solid surfaces.

Physical entrapment—can hold species inside pores of supports. For example, micelles containing a species held via a hydrophobic interaction can be entrapped inside a polyacrylamide matrix.

2.4. SOME PRINCIPLES OF ASSEMBLY

Many of the techniques used to assemble ICSs are different than the chemical synthetic methods used in homogeneous media. Clearly the structure of a system and the spatial resolution (Fig. 2.4.1) that is attained in its assembly in all dimensions are important. In general, there are techniques available to produce high resolution in thickness (the Z direction), even down to monolayer and molecular dimensions. As discussed below, methods of deposition of thin films of nanometer to micrometer dimensions are well developed. However, techniques to produce features of similar size in the lateral (X, Y) direction are only just emerging. There are two different approaches to fabrication of ICSs: (1) from the top down, such as by deposition of layers of material and then controlled removal to produce the desired pattern or arrangement (as in integrated-circuit chip fabrication); and (2) from the bottom up, such as by typical chemical synthetic procedures on a defined surface (as in peptide or DNA synthesis). The former methods depend heavily on techniques like photolithography and surface etching with microtips, while the synthetic methods often make use of templates to provide the desired spatial arrangement and resolution. Both approaches are discussed below. For fabrication of actual devices for practical applications,

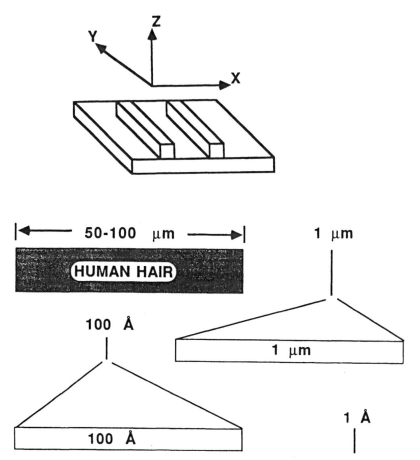

FIGURE 2.4.1. Relative size scales of interest in integrated chemical systems (see also Fig. 1.1.4) and directional notation used.

the ability to mass-produce systems at a relatively low unit cost will be an important factor in selecting the appropriate approach.

2.4.1. Self-Organizing Systems

Some systems spontaneously assemble into an organized arrangement under the proper conditions. Indeed, self-assembly is central in biosynthesis and the formation of biological ICSs. Biological self-organization from the biocomponents can be made to occur *in vitro* in certain cases. For example, the tobacco mosaic virus can be assembled from the single strand of viral RNA and the protein subunits (Fig. 2.4.2), and under the proper conditions, some denatured proteins will refold into the native

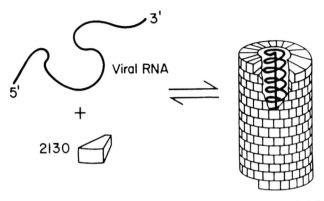

Tobacco Mosaic Virus

FIGURE 2.4.2. Self-assembly of tobacco mosaic virus (TMV) from a single strand of viral RNA and 2130 protein subunits. These were obtained by dissociation of TMV into its component parts (26). A helical structure forms that is 3000 Å in length, 180 Å in diameter, with a helical pitch of 23 Å and a central hole of 40 Å. [Reprinted with permission from J. S. Lindsey, *New J. Chem.*, **15**, 153 (1991). Copyright 1991 Gauthier-Villars Publishers.]

form with the formation of several disulfide and many hydrogen bonds (26). The idea of using self-assembly in the fabrication of artificial ICSs, such as "molecular electronic devices," is certainly an attractive one; efforts in this area have been reviewed (27). A number of examples of artificial self-assembling systems are known; these, along with biological examples, are given in Table 2.4.1. In systems that self-assemble, the attractive energy involved in intermolecular interactions must compensate for the entropic tendencies to produce randomization.

One of the more widely studied chemical approaches to self-assembly is probably that involving surface active substances that form organized assemblies at the air–water interface. These can be manipulated with a Langmuir film balance, and monolayer and multilayer organized films (Langmuir–Blodgett or LB films) can be deposited on solid substrates (28, 29). Molecules that form assembled monolayers typically contain a polar head group (e.g., —COOH, —OH) with a long-chain hydrocarbon or fluorocarbon tail (Fig. 2.4.3). When these are added to water, such as a solution in chloroform (the *spreading solvent*), they form a monolayer at the air–water interface, with the polar head groups pointing toward the water and the hydrocarbon tails pointing toward the air. When the surfactant molecules are close together, attractive hydrophobic interactions can occur between the long chains. Monolayer films of such surfactants are formed, studied, and manipulated with the film balance

TABLE 2.4.1. Examples of Self-Organizing Systems in Biological and Artificial Systems

Biological Systems

DNA (double-helix formation) membranes
Proteins (including disulfide bond formation and hydrogen bonding)
Viruses (e.g., tobacco mosaic virus reconstitution)
Protein assemblies (microtubules, viral coats, hemoglobin)
Ribosomes

Artificial Systems

Crystals
Monolayers of surfactants at air–water interface
Langmuir–Blodgett monolayers and multilayers on solids
Self-assembled monolayers (e.g., thiols on Au, silanes on SiO_2)
Lipid bilayer membranes
Liquid crystals
J- and H-dye aggregates
Topochemical polymerization
Topochemical reactions in or on ordered solids (mica, clays, zeolites)

Source: Taken, in part, from J. S. Lindsey, *New J. Chem.*, **15**, 153 (1991).

(Fig. 2.4.4). A number of different film balances are commercially available, including computer-controlled ones. The water or aqueous solution is contained in a Teflon trough (4), which is filled with substrate solution just above the trough rim. The film is contained between two barriers, one that is fixed and connected to a torsion wire (6) and one that can be moved across the surface of the trough by an external drive (the sweep barrier) (5). The force exerted by the film against the barrier connected to the torsion wire can be measured to yield the surface pressure. One records a curve of surface pressure (in millinewtons per meter or, equivalently, dynes per centimeter) against area as the movable barrier is swept toward the fixed one (Fig. 2.4.5). Because the area between the two barriers and the amount of fatty acid that has been added are both known, the area axis can be given in terms of square angstroms of surface area per molecule of fatty acid. The highly idealized isotherm can be interpreted as follows. When the molecules on the surface are far apart, only a small force is required to move them together (1). As they approach under higher pressure, a transition occurs to what is sometimes described as a ''liquid-like state'' (2, 2') and on further compression to a close-packed monolayer (3). This compact monolayer undergoes little change in area on compression, and extrapolation of the

2.4. SOME PRINCIPLES OF ASSEMBLY

Typical surfactant molecule that forms monolayers:

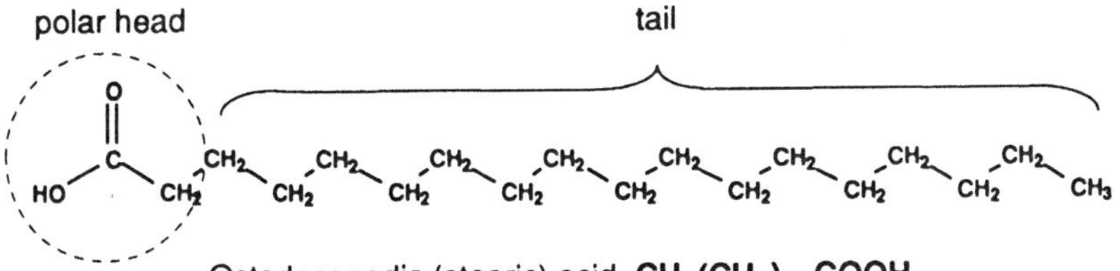

Octadecanodic (stearic) acid, $CH_3(CH_2)_{16}COOH$

Surfactants that form monolayers or micelles:

<u>cationic</u>

hexadecyltrimethyl ammonium bromide (CTAB), $CH_3(CH_2)_{15}N^+(CH_3)_3\ Br^-$

dioctadecyldimethyl ammonium chloride (DODAC), $[CH_3(CH_2)_{17}]_2N^+(CH_3)_2\ Cl^-$

<u>anionic</u>

sodium dodecylsulfate (SDS) [or sodium lauryl sulfate (NaLS)], $CH_3(CH_2)_{11}OSO_3^-\ Na^+$

sodium oleate, $CH_3(CH_2)_7CH=CH(CH_2)_7COO^-\ Na^+$

<u>neutral</u>

1-(1,1-dimethyl-3,3-dimethylbutane)-4-polyoxyethylene(9.5)-benzene or polyoxyethlene-t-octylphenol (TX-100), $(CH_3)_3CCH_2C(CH_3)_2\ C_6H_4(OCH_2CH_2)_{9-10}OH$

polyoxyethylene(6) alcohol, $CH_3(CH_2)_7(OCH_2CH_2)_6OH$

FIGURE 2.4.3. Typical surfactant molecules. For a more extensive listing, see J. H. Fendler, *Membrane Mimetic Chemistry*, Wiley, New York, 1982 and W. L. Hinze, *Ann. Chim.*, **77**, 167 (1987).

isotherm in this region to zero force yields the area occupied by a molecule in monolayer film (in the example, about 20 $Å^2$/molecule). Further movement of the sweep barrier toward the fixed one will ultimately result in collapse of the film (4). Once formed on the solution surface, the film can be transferred to an appropriate solid phase, such as a metal

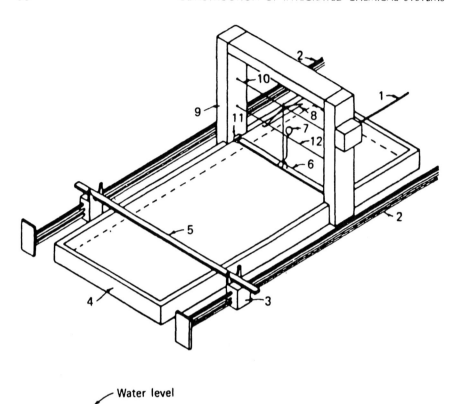

FIGURE 2.4.4. Typical modern film balance. (1) torsion wire control; (2) sweep control; (3) sweep holder; (4) trough; (5) moving barrier (sweep); (6) stationary barrier (float); (7) mirror; (8) calibration arm; (9) head; (10) main torsion wire; (11) barriers around float; (12) wire for mirror. Inset shows water (subphase) level in trough. Many different trough arrangements with alternative methods of measuring force on float are available. [Reprinted with permission from J. H. Fendler, *Membrane Mimetic Chemistry*, Wiley, New York, 1982. Copyright 1982 John Wiley & Sons.]

electrode or glass slide, by pulling the solid through the film–solution interface in a controlled manner while maintaining the surface pressure by movement of the sweep barrier. Devices to control the lifting of the solid phase under constant pressure (not shown in Fig. 2.4.4) are available. Multiple dips of the solid phase can be used to build up films containing several layers of the same or different substances (Fig. 2.4.6). The final result is an organized layer (the LB film) on a surface. This film can be used as a substrate to incorporate other molecules (e.g., electroactive or photosensitive ones) to form the desired system. Alter-

2.4. SOME PRINCIPLES OF ASSEMBLY

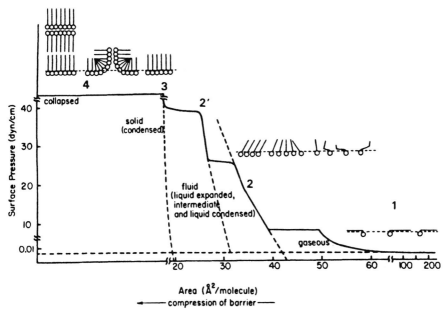

FIGURE 2.4.5. Highly idealized schematic representation of surface pressure (Π)/surface area (A) isotherm. Actual Π-A isotherms are considerably less distinct and usually do not show all of the transitions indicated here; see, for example, Refs. 5, 30, and 31. Note the breaks in the X and Y axes of the graph. [Reprinted with permission from J. H. Fendler, *Membrane Mimetic Chemistry*, Wiley, New York, 1982. Copyright 1982 John Wiley & Sons.]

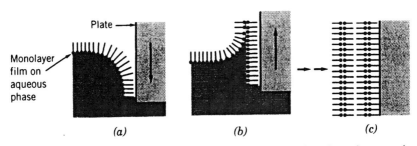

FIGURE 2.4.6. Buildup of a multilayer LB film by dipping a plate through a monolayer film at the air–water interface: (*a*) as plate with a hydrophobic surface is pushed through the monolayer, a film forms on the plate in the orientation shown; (*b*) as the plate is withdrawn, a second layer forms, in a head-to-head arrangement; (*c*) multiple dips of the plate result in a multilayer film.

natively the surfactant molecules themselves may have photoactive or electroactive tail groups attached. The formation of mixed films, involving two or more surfactants, is also possible. If the film preparation procedures are carried out under scrupulously clean conditions, and the films are transferred to clean and very smooth substrates, relatively uniform films can be prepared. However, even under the best conditions, the film structures contain defects. LB films also tend to be rather fragile. However, it is possible to introduce polymerizable groups within the monolayer-forming molecules and then induce polymerization after monolayer formation. A similar strategy has been employed to produce polymerized micelles and vesicles formed from surface-active materials.

Self-assembling films can also be formed on a surface without the use of a film balance (30). For example, if the head group of the surface-active molecule is one (e.g., a thiolate) that will interact strongly with the solid surface (e.g., gold), a monolayer film will self-assemble on the solid surface, when it is immersed in a solution of the surfactant for a sufficient time. An alternative approach is to bind a long-chain silane to a glass surface using the type of linkage chemistry shown in Fig. 2.2.7 (31). Monolayers, which are anchored to the substrate, should be more stable than LB films.

We might stress the important role of the spatial arrangement of the components in an ICS on the properties. An assembled film should show physical properties (e.g., conductivity) different from those of the same or similar material in a random arrangement. An example of this phenomenon is the behavior of films of sexithiophenes (6T) (see Fig. 7.4.1) (32). Conjugated thiophene oligomers and polymers (Fig. 2.2.4) form organic conductors and semiconductors. The structure of films of 6T depend on the nature and location of substituent groups. For example, substitution of an n-hexyl group in a β position ($\beta\beta'$DH6T) disrupts packing of the 6T chains, while substitution at the ends (α, ωDH6T) leads to better organization because of interactions among the hydrocarbon chains. The conductivity and carrier mobility in films of these molecules increases in the order α, ωDH6T > 6T >> $\beta\beta'$DH6T (32).

Liquid crystalline materials also show a tendency to self-organize. By proper treatment of a substrate to induce a desired orientation, multilayers of oriented liquid crystalline materials can be fabricated. An example of a device based on such a molecule, a liquid crystal (LC) porphyrin, is shown in Figure 7.4.2 (33). An organized solid film of this material was produced by melting to the LC phase, filling the cell, and allowing the material to cool. Different LC materials form the basis of the widely used LC displays.

Surface-active substances (e.g., phospholipids) can form free-stand-

2.4 SOME PRINCIPLES OF ASSEMBLY

ing membranes made up of a pair of monolayers, called *bilayer lipid membranes* (BLMs). These are produced by painting a solution of a lipid, such as lecithin, across a small hole in a Teflon sheet separating two solutions (Fig. 2.4.7). There have been extensive studies of the transport of ions and molecules across such a BLM film, with and without an electric field applied across the membrane. The potential that is established across the BLM when the solution compositions are changed is also of interest, as in the design of sensors. Different substances (carriers) can be added to the BLM to alter its permeability, or pore-forming substances can be introduced to produce selective transport of particular ions or molecules across it.

Inorganic multilayers that are analogous to LB films have been prepared. A particularly interesting example involves the synthesis of multilayers of zirconium 1,10-decanediylbis(phosphonate) on Si and Au (34). These are formed as indicated in the following scheme:[1]

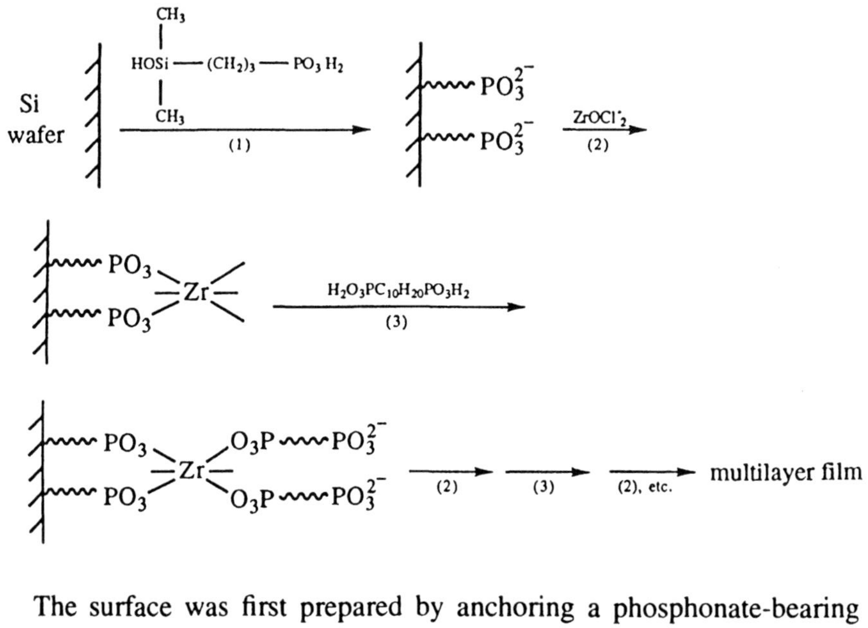

$$\text{Si wafer} \xrightarrow[(1)]{\text{HOSi}-(\text{CH}_2)_3-\text{PO}_3\text{H}_2, \text{CH}_3, \text{CH}_3} \left\{\begin{array}{c}\text{PO}_3^{2-}\\ \text{PO}_3^{2-}\end{array}\right\} \xrightarrow[(2)]{\text{ZrOCl}_2}$$

$$\left\{\begin{array}{c}\text{PO}_3\\ \text{PO}_3\end{array}\right\}\text{Zr} \xrightarrow[(3)]{\text{H}_2\text{O}_3\text{PC}_{10}\text{H}_{20}\text{PO}_3\text{H}_2}$$

$$\left\{\begin{array}{c}\text{PO}_3\\ \text{PO}_3\end{array}\right\}\text{Zr}\left\{\begin{array}{c}\text{O}_3\text{P}\sim\text{PO}_3^{2-}\\ \text{O}_3\text{P}\sim\text{PO}_3^{2-}\end{array}\right\} \xrightarrow[(2)]{} \xrightarrow[(3)]{} \xrightarrow[(2), \text{etc.}]{} \text{multilayer film}$$

The surface was first prepared by anchoring a phosphonate-bearing molecule to it, such as through silane linkage chemistry. A multilayer film (up to eight layers) was then built up by alternate exposures of the surface to solutions of $ZrOCl_2$ and 1,10-decanediylbis(phosphonate), as shown.

[1]Reprinted from H. Lee, L. J. Kepley, H.-G. Hong, and T. E. Mallouk, *J. Am. Chem. Soc.*, **110**, 618 (1988). Copyright 1988 American Chemical Society.

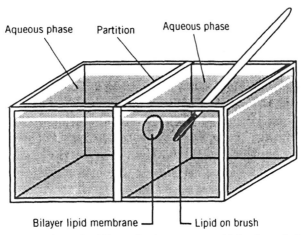

FIGURE 2.4.7. Formation of a bilayer lipid membrane (BLM) by painting a lipid solution across a pinhole (≤ 2 mm diameter) in a partition separating two solutions. Electrodes (e.g., Ag/AgCl) can be introduced into the solutions to apply an electric field across the BLM.

As indicated in Table 2.4.1, organized structures can also be produced in many other ways, for example, as a solid grown under controlled conditions. For example, crystals can be formed by controlled growth in the solution or vapor phase or by electrocrystallization on an electrode surface. Solids can also be formed by the controlled mixing of two different solutions to produce rings or layers (e.g., Liesegang ring formation) (35). For example, when a polymer membrane [e.g., poly(vinylalcohol)] separates a solution of $AgNO_3$ from one of KI or NaBr, the interdiffusion of the two solutions will produce a series of layers of AgI or AgBr within the film (Fig. 2.4.8) (36). These structures form because of the effects of the relative rates of precipitation, nucleation, and diffusion and represent a spatially oscillating reaction. Metal films within a polymer layer can similarly be formed by interdiffusion of a metal-ion solution and a reducing agent. For example, a film of Pt will form in (and on) a Nafion membrane, when it separates a solution of K_2PtCl_6 from one containing hydrazine (37). A similar strategy, i.e., interdiffusion from two solutions into a membrane, can be used to form a film of semiconductor particles (e.g., CdS in Nafion) (38). An example of a silver layer formed within a polyimide film on an electrode surface is shown in Figure 2.4.9 (39). In this example, Ag^+ diffused into the film from the solution and reducing agent (reduced polyimide) was generated electrochemically within the film. By suitable masking techniques, two-dimensional metal structures can be formed within the films, as shown in Figure 2.4.9.

Computer simulations have shown how structures, even rather com-

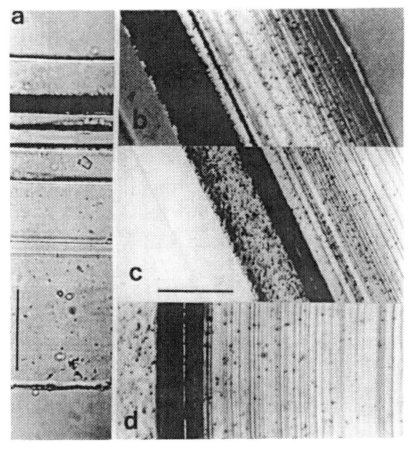

FIGURE 2.4.8. Photomicrographs of silver halide layers precipitated within a poly(vinylalcohol) film (film cross sections are shown). (*a*) AgI; scale bar, 0.05 mm; (*b*) and (*c*) AgBr, in polarized light with crossed (*b*) and uncrossed (*c*) polarizers, scale bar, 0.05 mm; (*d*) same sample at higher magnification, scale bar, 0.025 mm. [Reprinted with permission from K. F. Mueller, *Science*, **225**, 1021 (1984). Copyright 1984 by the AAAS.]

plex ones, can arise spontaneously in a system with simple components, when the system evolves with time following certain rules.[2] Consider, for example, that all snowflakes form by the conversion of water vapor to ice of the same crystalline form. Yet each snowflake shows a different overall pattern. The same principles may apply to biological systems

---

[2]This, in a sense, is a principle of evolution. "The appearance of order does not necessarily involve design; it can arise spontaneously by natural processes." (R. Hanbury Brown, *The Wisdom of Science*, Cambridge University Press, New York, 1986).

(a)

(b)

2.4. SOME PRINCIPLES OF ASSEMBLY

(e.g., the growth of algae) or chemical systems (e.g., electrodeposits of dendritic metal or conductive polymer films).

Studies of the formation of such structures form the basis of complex-systems theory and computations involving a class of mathematical systems known as *cellular* automata have been carried out (40). A cellular automaton is made up of a group of identical components with each component evolving with time according to a simple set of rules. Structures arise from an initial distribution of components in a set of boxes a_i that can have one of several values; for example, in a three-level system, each box can have a value of 0, 1, or 2. An arbitrary initial distribution evolves by following a given rule for how the value in box a_i in the next time increment, written as a_i', depends on its value and that of the neighboring boxes, a_{i+1} and a_{i-1}. For example, in the three-level system, the rule for evolution can be:

If $(a_{i-1} + a_i + a_{a+1}) =$	6	5	4	3	2	1	0
Then a_i' goes to	1	2	0	1	1	2	0

In this manner a complex pattern will be generated with time, whose form depends on the initial distribution. If the rule for evolution is changed, different patterns result. Typical patterns are shown in Ref. 40 and can be obtained with the computer program Mathematica. These theoretical constructions from cellular automata may provide a guide for the experimental construction of ICSs and indeed may be the basis by which certain biological structures form, such as by distribution of components in a self-organized membrane that reacts with time to build a given configuration.

2.4.2. Template Synthesis

Another approach to obtaining a desired structure is to use a naturally occurring or synthesized material as a substrate or template (41). For

---

FIGURE 2.4.9. (*a*) Transmission electron micrograph of a polyimide film containing an electrogenerated silver layer. Cross section of film (1000 Å thick) between interfaces A and B is shown. The silver metal is seen as a diffuse region of particles and as the dark band of thickness t_i, near the midpoint of the film. Scale bar, 1.0 μm. (*b*) Methods of depositing controlled two-dimensional structures in the film by controlling cathode area (imaged cathode) or masking the solution side of the film. [Reprinted with permission from S. Mazur and S. Reich, *J. Phys. Chem.*, **90**, 1365 (1986). Copyright 1986 American Chemical Society.]

example, clays and zeolites occur as very small particles with a known and well-defined structure (Fig. 2.2.6). Cations introduced into these by an ion-exchange process will often be located at defined sites within the crystal. For example, there is evidence that some ions that enter between the aluminosilicate sheets in a clay particle do so by expanding the distance between the sheets. The next ions of the same species will then tend to go into this same layer rather than into unoccupied layers of the same particle. This effect can lead to segregation, where different species are located together at different places within the same particle. Similarly, the limited channel diameters in zeolites can lead to selective introduction of some species within the channels or cages of the zeolite while others can only exchange on the external surfaces. A number of devices based on zeolite templates have been proposed (41). These same principles should be applicable to intercalation or topochemical reactions in other materials, including layered compounds, such as graphite or transition-metal dichalcogenides (42).

Another approach involves the synthesis of a molecule on a suitable template. Replication of DNA is the prototypical example of such a process (43). In this reaction, the two strands of DNA from the double helix separate, followed by pairing of the strand bases to the complementary deoxyribonucledoside triphosphates via hydrogen bonds. The triphosphates are then joined together to form a new DNA strand by the enzyme DNA polymerase. An analogous process can be carried out *in vitro* in the polymerase chain reaction. Related techniques of synthetic self-assembly have been carried out (27). For example, template polymerizations follow the scheme shown below, in which a template chain directs the assembly of monomers, but is not contained in the final daughter product chain (44):[3]

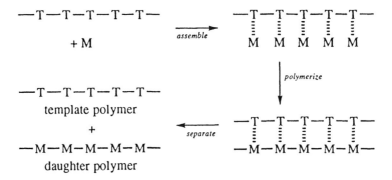

[3]Reprinted from J. S. Lindsey, *New. J. Chem.*, 15, 153 (1991). Copyright 1991 Gauthier-Villars Publishers.

2.4. SOME PRINCIPLES OF ASSEMBLY

Analogous replication schemes in inorganic systems have also been described (45). For example, it has been claimed that by using particles of the clay montmorillonite in a "breeding solution" containing Na^+, K^+, Mg^{2+}, Al^{3+}, and $Si(OH)_4$, new silicate layers form by intercalating synthesis within the layers of the original parent clay particle. Ultimately, new clay particles are formed that are reported to maintain the structural characteristics of the parent particles over a number of generations.

A number of other potential template materials have been used (41). These include less structured systems, like micelles and LB films, pores in membrane filters, and domains in block copolymers. For example, Nucleopore polycarbonate membranes (~ 10 μm thick) are commercially available as filters, with cylindrical pores with diameters of 30 nm–10 μm. These pores can be used as templates for the synthesis of fibers and tubules of polymers and metals (46). Tubules of the conducting polymer polypyrrole can be produced by using the membrane to separate a solution of pyrrole monomer from a solution of ferric ion, which can oxidize the monomer and initiate the polymerization within the pores. This method not only produces a monodisperse array of polypyrrole tubules, but the conductivity of the tubules so produced is reported to be higher than that of the same bulk-polymerized material because of orientational effects of the pore walls. A similar approach can be used to form metal (e.g., Au) tubules electrochemically inside the pores of an inorganic (Al_2O_3, Anapore) membrane (47).

Another template approach is the use of block copolymers to grow small particles of semiconductors and metals (48). In a typical example, a functionalized phosphine-containing diblock copolymer of two components, A and B [where A = racemic 2-exo-3-endo-bis(diphenylphosphino)bicyclo[2.2.1]heptene; B = methyltetracyclododecene], of composition $A_{60}B_{300}$, was prepared by ring-opening metathesis polymerization. Such a polymer contains spherical domains because of self-organization of the polymer material (as also occurs in Nafion; see Fig. 2.2.3). Silver ion can be incorporated into these domains; this converts to small (< 100 Å) particles of silver on heating. The size and shape of the domains in such copolymers is controlled by nature of the components, the length of each block, and the degree of polymerization. By an analogous approach ZnS and CdS nanoclusters were produced (49). This and other template approaches are very useful in the preparation of particles of semiconductors for microheterogeneous photoelectrochemical systems, as discussed in Section 6.4.2.

A drawback of most template methods, especially those using support materials like zeolites and clays, is that the structure one obtains is not controllable in the same way that structures on an integrated-circuit chip are. Thus templates often produce randomly distributed arrays (e.g., of particles in a polymer film or fibrils in a membrane), even when the sizes of the components in the array are very uniform. In constructing multicomponent integrated systems, one must often plan a series of synthesis steps to give a desired structure, such as to first synthesize a semiconductor particle, and then photodeposit a metal nearby. Alternatively, one can synthesize the template itself to yield the desired arrangement of components.

2.5. TECHNIQUES OF ASSEMBLY

Typical techniques that can be used for the controlled deposition of thin films and the construction of molecular assemblies include many that are familiar for the fabrication of semiconductor devices (50): spin coating and photolithography, vacuum evaporation and sputtering, molecular-beam epitaxy, plasma deposition and etching, chemical vapor deposition, and liquid-phase etching and deposition. In addition to these, surface chemical synthetic techniques and electrochemical techniques (deposition and photodeposition, etching and photoetching, polymerization) can be used. While it is beyond the scope of this work to discuss the instrumentation and methodology of these methods in detail, a brief description of each with some typical applications will be given.

2.5.1. Spin Coating and Photolithography

In constructing an ICS, one sometimes would like to form a thin uniform film (e.g., of a polymer) on a support surface. This can sometimes be accomplished by "drop coating," or simply dropping a small amount of solution of the dissolved polymer on the substrate and allowing it to evaporate. However, the uneven evaporation around a drop usually produces a very uneven film; films that are thicker on the outside (ring-shaped films) are often produced. Spin coating is thus often used to form more uniform films (Fig. 2.5.1A). The substrate, such as a slice of Si or a metal foil, is placed on a horizontal stage that can be rotated at a high velocity [e.g., 2500 rev/min (rpm)]. Sometimes the stage is heated. A given volume of a solution of polymer in a volatile solvent is dropped on the substrate while it is spinning. The solution spins from the surface forming a uniform layer of polymer as the solvent evaporates. This technique has been used to form polymer and clay layers on electrode

2.5. TECHNIQUES OF ASSEMBLY

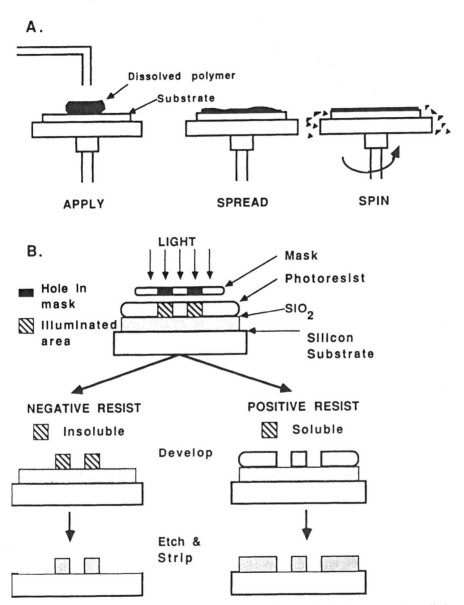

FIGURE 2.5.1. (*A*) Spin coating. The solution containing dissolved polymer is applied to substrate and then spun to remove excess material and form film. Dynamic spin coating, where the material is applied to a spinning substrate is also possible. (*B*) Schematic representation of photolithography for etching an SiO_2 layer on Si (e.g., in the manufacture of a solid-state device) with positive and negative photoresists.

surfaces and is also employed to form photosensitive polymer layers (*photoresists*) in photolithography.

The basic principles of photolithography are illustrated in Figure 2.5.1B. Two types of photoresists are used. Positive photoresists are insoluble polymers that become soluble in a developer solution on irradiation, while negative photoresists are initially soluble but become insoluble on irradiation. After a thin layer of photoresist is placed on a substrate by spin coating, it is irradiated through a mask that contains the desired pattern. With a positive photoresist, wherever the light strikes, the polymer becomes more soluble. After exposure, the film is developed by immersion in a developer solution that dissolves the exposed polymer areas. Typical chemistry involving a Novolac positive photoresist (a phenol formaldehyde polymer) is given in Figure 2.5.2, where the unexposed resin is insoluble, but exposed portions will dissolve in strong base. The developed film thus has exposed areas where it has been irradiated. Metal patterns can now be deposited on the substrate by vacuum evaporation, dopants can be diffused into the substrate, or the substrate can be oxidized or etched at the exposed areas. Following the desired treatment, the remaining resist layer can be stripped away

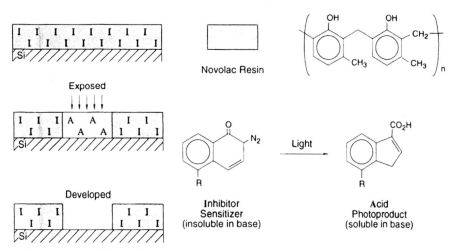

FIGURE 2.5.2. Schematic representation of chemistry in positive photoresists. Photolysis of sensitizer (inhibitor) causes conversion to an acid form in the areas exposed to light. These can then be selectively dissolved (developed) in a solution of aqueous base (e.g., tetramethylammonium hydroxide). In a negative photoresist (e.g., a polyisoprene), the action of light on the sensitizer causes crosslinking in the polymer, rendering the illuminated area less soluble. The resist is developed in an organic solvent, like xylene.

2.5. TECHNIQUES OF ASSEMBLY

with a suitable solvent or by using an oxygen plasma to leave the structured substrate. Additional structuring can now be carried out by new deposition and photolithographic steps. The procedures with negative photoresists are analogous to those just described. The structural resolution attainable by photolithography using visible light is of the order of 0.8–1 μm. Higher resolution is attainable with UV, electron or ion beams, and X-ray sources that may utilize other photoresist materials, such as poly(methylmethacrylate) (51).

High-resolution lithography can also be done with organized monolayer films, either LB or self-assembled, of the type discussed in the last section. These films form quite uniform and complete monolayers on smooth substrates, although there is considerable evidence that they contain defects of various types, for example, because of impurities like dust particles, boundaries in the different film domains, or defects on the substrate. Lithography of these films usually involves scratching through them, such as with the tip of a scanning tunneling microscope (see Section 2.5.7). One example of lithography of a self-assembled monolayer is shown in Figure 2.5.3 (52). A monolayer of ω-mercaptohexadecanoic acid [$HS(CH_2)_{15}COOH$] was first assembled on a gold film. The surface of such a film is hydrophilic, because the carboxylate groups are exposed. A line (0.1–1 μm) was then scratched in the film with a scalpel or carbon fiber, to expose gold. Then a different surfactant with a hydrophobic terminal group, [$CH_3(CH_2)_{11}S]_2$, was adsorbed on the exposed gold areas. This technique could be used to form patterns on the surface with different wettability (hydrophilic vs. hydrophobic) with features in the 0.1–1-μm scale. An alternative approach is the deposition of an alkanethiolate on a gold surface in a desired pattern, such as by placing it on the surface with a pen or stamp (53). This layer protects the gold from attack from an aerated cyanide solution, so that immersion of a patterned gold surface in such a solution leads to etching of features in the exposed gold. Note that such mechanical lithography, even if carried out with an automated instrument, is quite slow compared to photolithography, and it will probably be less useful for mass production of devices. However, it could be used to fabricate photolithography masks (e.g., for X-ray methods) and other unique structures.

2.5.2. Vacuum Evaporation and Sputtering

These techniques are used to deposit metals and semiconductors. The principles are illustrated in Figure 2.5.4. In vacuum evaporation, the substrate is contained in high vacuum in a bell jar. A metal source (e.g.,

Au, Ag) is heated electrically or by an electron beam and the atoms evaporate from the surface and transfer to the substrate. In sputtering, the source is in the form of a target, and after evacuation, an inert gas like Ar is introduced into the vacuum chamber. An RF plasma (see below) is formed, and the target is bombarded with a beam of ions (Ar^+). Clusters of atoms broken from the target surface deposit on the substrate. Targets may be metals (e.g., Pt, Ti), semiconductors, or insulators. It is also possible to carry out reactive sputtering where a small amount of a reactant, such as oxygen, is introduced along with the inert gas, so that the metal cluster (e.g., Ti) reacts on its way to the target to form a layer of the reaction product (e.g., TiO_2). The surface of the substrate material is often cleaned before deposition by sputtering it with a beam of ions, thus etching away any layers of impurities. The thickness

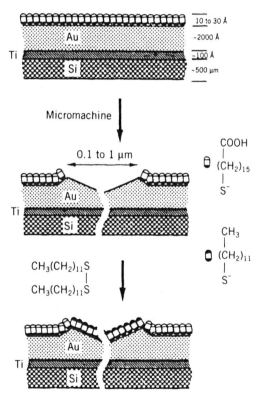

FIGURE 2.5.3. Schematic illustration of the formation of 0.1-1-μm hydrophobic lines in a hydrophilic surface with micromachining and SAMs (self-assembled monolayers). [Reprinted from N. L. Abbott, J. P. Folkers, and G. M. Whitesides, *Science*, **257**, 1380 (1992). Copyright 1992 by the AAAS.]

2.5. TECHNIQUES OF ASSEMBLY

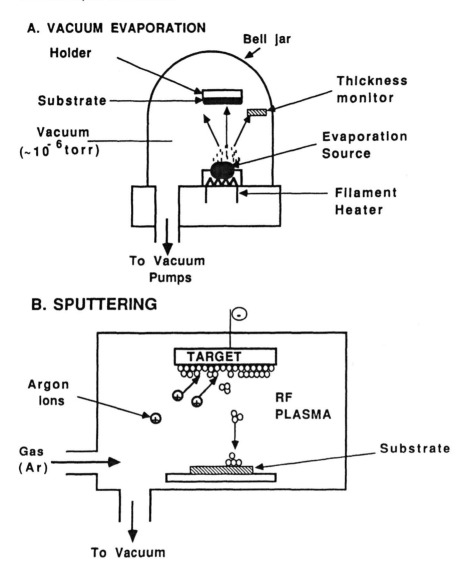

FIGURE 2.5.4. (A) Vacuum evaporation apparatus. (B) Sputtering apparatus.

of an evaporated or sputtered layer can be controlled by observation of a thickness gauge placed in the chamber. This consists of a piezoelectric crystal on which a film forms during the deposition step producing a change in the crystal's resonant frequency.

The strength of adhesion of films on substrates depends on the cleanliness of the substrate and on specific interactions between substrate and film. Sometimes better adhesion can be obtained by coating the substrate with an intermediate layer of a different material. For example, Au and Pt do not adhere well to glass. However, a more adherent film can be produced by first coating the glass with Cr before noble-metal deposition.

2.5.3. Plasma Processing

Plasma deposition and plasma etching can be used to deposit materials and pattern surfaces. A gas containing the desired components is subjected to a radiofrequency (RF) field (Fig. 2.5.5A). The RF energy causes dissociation of the components in the gas to form a plasma: a mixture of positive ions and electrons. Species in the plasma strike the substrate and can either deposit films or etch the surface.

In plasma etching, a gas like CF_4, when introduced into the chamber, will produce F atoms that can etch substrates like Si, SiO_2, Ti, and W. Under controlled conditions, the photoresist layer will not be etched, so that this dry etching can be used in place of liquid etching in photolithography. The etching chamber is frequently designed to cause the reactant species to move only in a direction perpendicular to the surface to be etched, so that the etching process is anisotropic (Fig. 2.5.5B).

A typical example of a plasma deposition process is the formation of amorphous hydrogenated silicon from silane (SiH_4) (Fig. 2.5.5C). In this method, SiH_4 and other gases added for doping or alloying are introduced into the reactor chamber at a pressure of 0.1–1 torr. Decomposition occurs in the RF field and an amorphous Si:H film forms on the substrate. It is also possible to deposit polymer layers on substrates by introduction of a monomer (e.g., vinylferrocene) into the plasma; layers of poly(vinylferrocene) have been deposited on electrode surfaces by this technique.

Related etching techniques, called *ion-beam milling* and *reactive-ion etching*, utilize a directed or focused ion beam to remove material from a surface by both physical (sputtering) and chemical (reactive) means (49, 54). Focused ion beams can also be used for deposition. The apparatus involves an ion-beam column in a vacuum chamber with a fine capillary tube to feed a gas whose composition determines the species deposited (Fig. 2.5.6). For example, for the deposition of gold, dimethylgold hexafluoroacetylacetonate might be used. Focused ion beams for milling and deposition allow the fabrication of structures with very high resolution with feature sizes down to about 50 nm.

A. PLASMA ETCHING

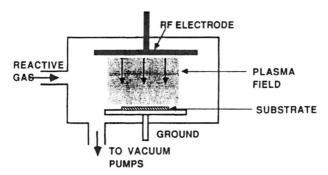

B. TYPES OF ETCHING

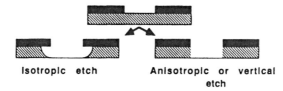

C. PLASMA DEPOSITION

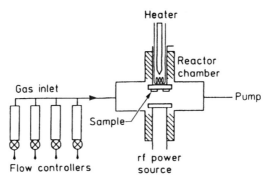

FIGURE 2.5.5. (*A*) An RF diode plasma reactor. Radiofrequency field can alternatively be coupled into the reactor inductively from a coil outside the chamber. (*B*) Differences in isotropic etching, with undercutting below photoresist layer and anisotropic etching. (*C*). Plasma reactor for deposition of amorphous Si:H. [Reprinted from R. A. Street, *Hydrogenated Amorphous Silicon*, Cambridge University Press, New York, 1991. Copyright 1991 Cambridge University Press].

2.5.4. Chemical Vapor Deposition (CVD)

A CVD process involves a gas-phase chemical reaction that occurs at the surface of the substrate to form a film of the desired composition. The apparatus for CVD (Fig. 2.5.7) involves a means of controlling the

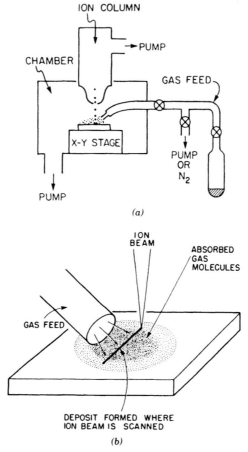

FIGURE 2.5.6. (a) Schematic diagram of apparatus for focused ion-beam deposition. Ion beams are typically Ar^+ or Ga^+ of 20–500 keV energy. Both the ion column and the vacuum chamber are separately pumped. (b) Close-up of the deposition area, where the reactive gas is fed to the area where deposition is desired via a capillary tube. [Reprinted with permission from J. Melngailis, "Focused Ion Beam Induced Deposition—A Review," in *Electron-Beam, X-Ray, and Ion-Beam Submicrometer Lithographies for Manufacturing*, M. Peckerar, ed., SPIE Proceedings, Vol. 1465, 1991, p. 36.]

flow of the reactant gases into the reaction chamber and a method of introducing energy at or near the substrate to cause the desired reaction to occur. In the simplest CVD systems, reactant gases, sometimes diluted with an inert carrier gas, mix above a heated substrate. The reaction of the gaseous components form a thin film on the substrate surface of defined composition. CVD films of insulators (e.g., SiO_2 and Si_3N_4), semiconductors (Si, GaAs, ZnSe), or conductors (Au, Pt, $MoSi_2$) can

2.5. TECHNIQUES OF ASSEMBLY

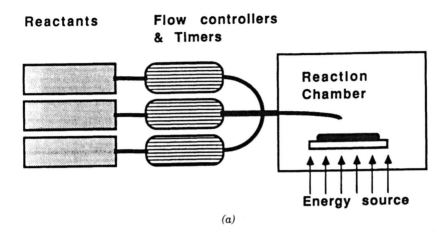

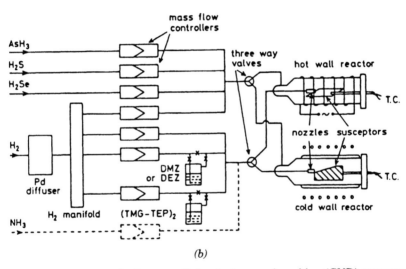

FIGURE 2.5.7. (a) Schematic diagram of chemical vapor deposition (CVD) apparatus. (b) Apparatus for MOCVD of ZnSe. DMZ = dimethylzinc; DEZ = diethylzinc; (TMG-TEP)$_2$ = trimethylgallium triethylphosphine adduct; T.C. = thermocouple. The hot wall reactor is heated with an external resistance heater, while in the cold wall reactor, the substrate was heated inductively. [Reprinted with permission from G. Fan and J. O. Williams, *J. Chem. Soc. Faraday Trans. 1*, **83**, 323 (1987). Copyright 1987 The Royal Society of Chemistry.]

be prepared. Typical CVD reactions are listed in Table 2.5.1. When the reactant vapor contains a metalloorganic compound, the technique is called *metal-organic chemical vapor deposition* (MOCVD). The apparatus for CVD can be quite elaborate and involve close control of gas flows and temperatures (Fig. 2.5.7B). However, a rather simple appa-

TABLE 2.5.1. Typical CVD Reactions to Produce Different Films

Reactants	Products
$SiCl_4 + 2\ H_2$ →	$Si + 4\ HCl$
$SiH_4 + O_2$ →	$SiO_2 + 2\ H_2$
$3\ SiH_4 + 4\ NH_3$ →	$Si_3N_4 + 12\ H_2$
$2\ Al(OC_3H_7)_3$ →	$Al_2O_3 + H_2O + C_xH_y$
$Ti(OC_3H_7)_4 + 2\ H_2O$ →	$TiO_2 + 4\ C_3H_7OH$
$Zn(C_2H_5)_2 + H_2Se$ →	$ZnSe + 2\ C_2H_6$

ratus can also be employed. For example, a film of TiO_2 can be deposited on a metal substrate by flowing together a stream of titanium isopropoxide with water vapor through an inverted funnel held over the substrate heated by a hot plate (Fig. 2.5.8A). Although most CVD reactors operate at or near atmospheric pressure, low-pressure CVD systems (LPCVD) (0.1–100 mtorr) are also used. It is also possible to promote the CVD reactions by input of RF energy (plasma-enhanced CVD) or UV light (photochemical CVD).

A related technique is *spray pyrolysis or spray deposition* in which a liquid containing the reactants is sprayed onto the heated substrate. For example, a film of CdS can be obtained by spraying an aqueous solution of $CdCl_2$ and thiourea through a nozzle onto a heated substrate (Fig. 2.5.8B). This approach is also frequently used to prepare conductive films of doped SnO_2 (from a mixture of $SnCl_4$ and HF) on glass substrates.

2.5.5. Molecular-Beam Epitaxy

Molecular-beam epitaxy (MBE) is a method of producing thin crystalline films of metals and semiconductors, of thickness down to atomic dimensions, as well as artificially layered structures, by means of controlled atomic or molecular beams. In MBE (Fig. 2.5.9) the substrate is contained in an ultrahigh vacuum. Beams of different substances (e.g., Ga, As, In, Al) are formed by heating these elements. The flow of these beams into the chamber is controlled by shutters over the beams, so that one or more beams can flow at a given time at a precisely controlled rate toward a heated crystalline substrate. A film grows on the substrate of a composition determined by the relative flow rates of the different beams. The film is said to be *epitaxial* if the crystallinity and orientation of the deposited layer are determined by that of the substrate. The im-

2.5. TECHNIQUES OF ASSEMBLY

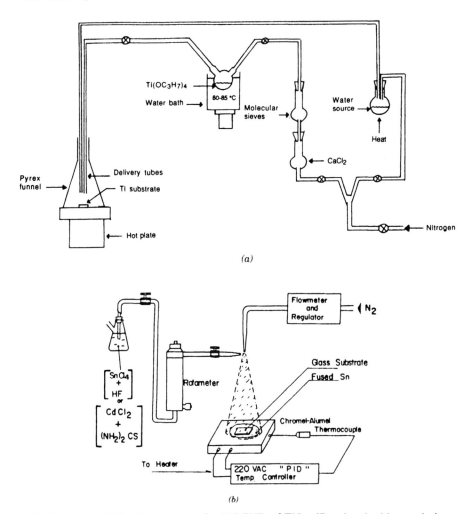

FIGURE 2.5.8. (a) Simple apparatus for MOCVD of TiO_2. [Reprinted with permission from K. L. Hardee and A. J. Bard, *J. Electrochem. Soc.*, **122**, 739 (1975). Copyright 1975 The Electrochemical Society.] (b) Simple apparatus for spray pyrolytic deposition of CdS or SnO_2. [Reprinted with permission from J. G. Ibanez, O. Solorza, and E. Goimez-del-Campo, *J. Chem. Ed.*, **68**, 872 (1991). Copyright 1991 Division of Chemical Education, American Chemical Society.]

pinging species of the molecular beams react at the substrate surface with the resulting atoms migrating to lattice sites of the substrate to form a relatively defect-free film with a precise orientation with respect to the substrate or underlying layer. When the layers consist of thin films of dissimilar materials deposited on top of one another, the process is called *heteroepitaxy*. It is possible to deposit individual atomic layers (e.g., of

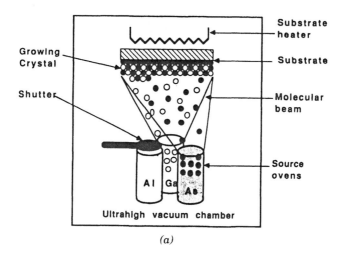

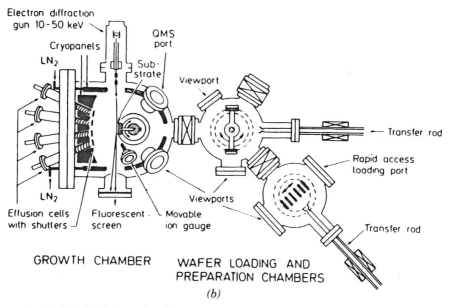

FIGURE 2.5.9. (a) Schematic of apparatus for molecular-beam epitaxy. (b) Cross section of UHV (ultrahigh-vacuum) system used for MBE with sample loading and preparation chambers and showing electron gun and fluorescent screen for monitoring of deposition process by reflection high-energy electron diffraction (RHEED). [Reprinted with permission from K. Ploog, *Angew. Chem. Intl. Ed. Engl.*, **27**, 593 (1988). Copyright 1988 VCH Verlagsgesellschaft.]

2.5. TECHNIQUES OF ASSEMBLY

the lattice-matched materials $Al_xGa_{1-x}As$ and GaAs) by MBE with very high accuracy and to prepare multiple layers with periodic modulation of chemical composition (*superlattices*) with thicknesses equal to or less than that of the spacings in the elementary unit cell of the bulk material. When the layer thicknesses are of the dimensions of the order of angstroms, the properties of the layers, such as their band gaps (as determined via their optical absorption and emission spectra) are different from those of the bulk materials. This difference arises because the number of molecules across the layer thickness is too small to form bands characteristic of the bulk material. Such layers are known as *quantum wells*. Similar effects are seen with very small particles, as discussed in Chapter 6. A transmission electron microscope picture of InGaAs quantum wells separated by thicker layers of InP that were prepared by MBE is shown in Figure 2.5.10. The figure also shows photoluminescence spectra of the wells of different dimensions. Note that these emissions are at shorter wavelengths (i.e., at higher energies) than those corresponding to bulk InGaAs.

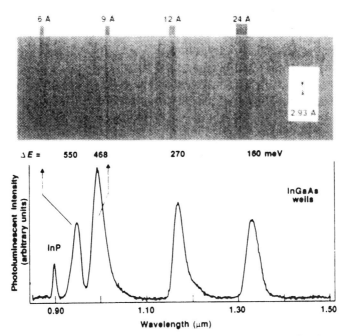

FIGURE 2.5.10. Transmission electron microscope (TEM) image of InGaAs quantum wells of indicated thicknesses separated by 150-Å InP barriers. Below are photoluminescence spectra at 6K of the quantum wells. Energy difference ΔE is measured from the energy gap of bulk InGaAs. [Reprinted with permission from V. Narayanamurti, *Science*, **235**, 1023 (1987). Copyright 1987 by the AAAS.]

2.5.6. Electrochemical Methods

A number of different electrochemical approaches to formation of surface structures are possible. Thin films and submonolayers of metals and semiconductors can be deposited. Metals and semiconductors can be etched, the latter sometimes requiring simultaneous illumination (photoetching). Polymer films can be deposited, either from solutions of the polymer or by polymerization of a monomer on an electrode. Controlled thin films of oxides and other materials can be prepared by anodizing the substrate metal in contact with the appropriate electrolyte solution. A brief overview of some of the electrochemical techniques that have been employed follows.

Anodization. When a metal electrode is made the anode in an electrochemical cell, a film often forms on the metal surface. For example, passage of current through a titanium anode in contact with a solution of phosphoric acid will form a film of TiO_2 on the surface. Oxide films on W, Al, Ta, Si, and many metals can be produced by this procedure. The film is often an insulator or semiconductor whose thickness and porosity can be controlled by the solution conditions, the potential applied to the anode, and the time of electrolysis (Fig. 2.5.11) (55). Anodization produces unique structures that cannot be obtained by chemical means. For example, electrochemical dissolution of Si in HF can lead to a very porous surface structure which will, unlike bulk Si, photoluminesce in the visible region (56). This has been attributed to the formation of very small (quantum-size) structures on the Si surface, although the formation of amorphous hydrogenated Si or other Si compounds has also been proposed to account for the luminescence.

Films other than oxides can be produced by appropriate choice of electrolyte. Thus sulfide and selenide layers can be produced by anodization on Cd and Zn. Electrochemical anodization is also often used to clean electrode surfaces and prepare them for modification. For example, Pt and C electrodes subjected to anodization will form thin oxidized surface films that can couple to linking agents to form modified surface layers. Alternatively, anodic treatment can have a pronounced effect on the behavior of an electrode material. Highly oriented pyrolytic graphite (HOPG) is a layered material with a very smooth surface. The fresh smooth surface obtained by peeling off the top layer of HOPG is hydrophobic, as measured by the large contact angle with water, and electrode reactions are generally sluggish at this fresh basal plane surface. Anodization results in the breakup of the surface with exposure of edge planes and formation of oxide sites. The surface becomes more hydro-

2.5. TECHNIQUES OF ASSEMBLY

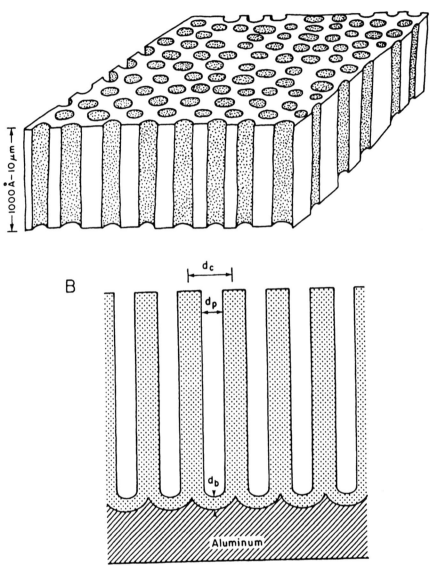

FIGURE 2.5.11. Structure of porous aluminum oxide film prepared by anodization of Al foil. Schematic representation of (A) pore structure and (B) cross section. The Al_2O_3 film forms by electron or ion transport driven by the electric field through a barrier layer (thickness, d_b) of Al_2O_3. [Reprinted with permission from C. J. Miller and M. Majda, *J. Electroanal. Chem.*, **207**, 49 (1986). Copyright 1986 Elsevier.]

philic and electrode reactions now occur more readily. Similarly, untreated Pt is usually covered with a layer of organic impurities that can hinder its behavior as an electrode. On cyclic anodization and cathodization, resulting in oxidation of surface species and a reorganization of the surface, much better and reproducible electrochemical responses are observed.

Electrodeposition. Different types of thin films can be deposited on electrode surfaces. The electrodeposition of metals and alloys is an old and highly developed field. Under highly controlled conditions, very thin (~nanometers) films and modulated structures of the superlattice type can be prepared. An example is given in Figure 2.5.12, where alternating films of Ni and Ni—P of 2-4 nm thickness were deposited on a Cu disk substrate (57). This was accomplished by rotating the substrate at 12-20 revs/min so that it alternately contacted two different plating baths. Deposition of single films of metals (e.g., Au, Pd, Ni, Cu) of controlled thickness involves a well-developed technology and

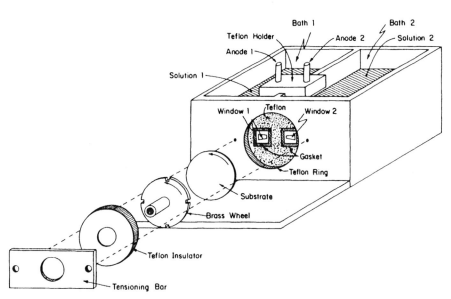

FIGURE 2.5.12. Plating apparatus for alternate deposition of Ni and Ni—P on a rotating substrate. The Ni bath was $NiCl_2$ in a boric acid solution, and the Ni—P bath contained $NiCl_2$ in a phosphoric/phosphorus acid medium. The inset shows a schematic cross section of the structure produced with 2-4-nm alternating layers with a total thickness of the order of 1 μm. [Reprinted with permission from L. M. Goldman, B. Blanpain, and F. A. Spaepen, *J. Appl. Phys.*, **60**, 1374 (1986). Copyright 1986 American Institute of Physics.]

2.5. TECHNIQUES OF ASSEMBLY

less elaborate apparatus. Semiconductor films can also be produced on an inert substrate. For example, CdSe can be deposited on a metal substrate by reduction from a bath containing Cd^{2+} and Se.

Three-dimensional structures on surfaces can be produced by electrodeposition through suitably prepared masks in a process similar to photolithography. A typical example is illustrated in Figure 2.5.13 (58). A

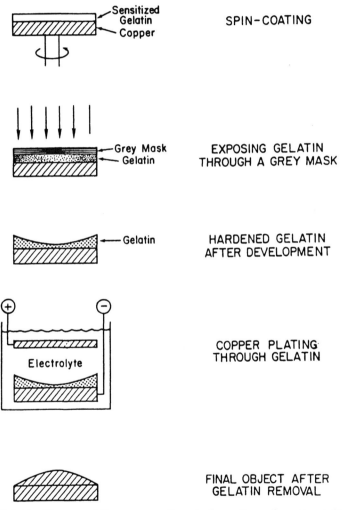

FIGURE 2.5.13. Diagram of the process of producing a three-dimensional Cu electrodeposit by plating through a gelatin layer that has been modified by light exposure and chemical development. [Reprinted with permission from J. C. Angus, U. Landau, S. H. Liao, and M. C. Yang, *J. Electrochem. Soc.*, **133**, 1152 (1986). Copyright 1986 The Electrochemical Society.]

sensitized gelatin coating was first spin-coated on the copper substrate. This was exposed to light through a mask (or photographic negative); crosslinking of the gelatin occurs in the exposed areas. The gelatin was then developed in water or water/isopropanol to wash away the unexposed areas and form a hardened gelatin layer in the exposed ones. The thickness of the hardened gelatin, and hence the rate of diffusion of solution metal ions through it to the substrate, depends on the degree of exposure to the light—the greater the exposure, the thicker the gelatin. Thus copper electrodeposition through the developed gelatin film occurred more rapidly at the less exposed sites to produce a three-dimensional metal structure on the substrate surface with a thickness and configuration governed by the exposure time, the mask, and the deposition time.

Controlled monolayers or submonolayers can also be produced by *underpotential deposition* (UPD) (59). UPD involves the deposition of an element (A) on a substrate material of a different element (B) at lower potentials than those needed to deposit A on a substrate of A. UPD is usually carried out with a less noble metal (e.g., Cu, Pb) deposited on a more noble one (e.g., Ag, Au, Pt). The difference in potential for deposition of A on B compared to that for A on A, the underpotential, is related to the free energy of adsorption of A on B, or by the difference in work functions of the two metals. An alternative thermodynamic explanation can also be given for UPD. For example, the deposition of a monolayer of a metal (e.g., Cu) on an inert substrate (e.g., Pt) occurs at less negative potentials (more easily) than it does on bulk copper. From the Nernst equation for the $Cu^{2+} + 2e = Cu$ half-reaction:

$$E = E° + \left(\frac{RT}{2F}\right) \ln \frac{a_{Cu^{2+}}}{a_{Cu}} \qquad (2.5.1)$$

the potential to start the deposition of Cu on bulk copper from a solution of Cu^{2+} satisfies eq. (2.5.1) with $a_{Cu} = 1$. The activity of Cu when it is deposited on an inert substrate is much smaller than unity and only increases toward 1 as a monolayer is formed. Thus the deposition of up to a monolayer of Cu occurs at more positive potentials.

A current–potential (i–E) curve for the deposition of a metal on an inert substrate via UPD shows a "prepeak" for the deposition of a monolayer of metal at potentials several tenths of a volt positive of the main metal deposition wave (Fig. 2.5.14). By controlling the potential and the time of deposition, one can deposit fractions of monolayers of metals. As shown in Figure 2.5.14, a much sharper peak is obtained with a single-crystal substrate on which the deposited metal forms an

2.5. TECHNIQUES OF ASSEMBLY

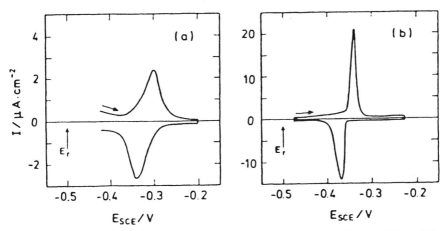

FIGURE 2.5.14. Current–potential curves showing the underpotential deposition of Pb on (a) polycrystalline Ag and (b) Ag(111) single crystal from a solution 0.2 mM Pb^{2+} and 0.5 M NaClO$_4$ (pH 2) at a scan rate of 1 mV/s. Deposition and stripping of bulk Pb (on top of the UPD Pb layer) occurs at potentials more negative than those shown. [Reprinted with permission from D. M. Kolb, in *Advances in Electrochemistry and Electrochemical Engineering*, H. Gerischer and C. W. Tobias, eds., Wiley, New York, 1978, Vol. 11, p. 125. Copyright 1978 John Wiley & Sons.]

epitaxial layer. Recently there has been interest in the deposition of multilayers of two (or more) elements via UPD in a method named *ECALE* (electrochemical atomic-layer epitaxy) (60). Consider the deposition of CdTe by this approach. First a layer of Te is deposited via UPD on a gold substrate. The Te-containing solution is flushed from the cell and replaced by one containing Cd^{2+}. Cd is then deposited on the Te layer. By alternating the solutions, multilayers can be built up, with the deposition conditions and surface reactions controlled by the potential of the substrate.

Etching and Photoetching. Anodization of a metal or semiconductor in an appropriate electrolyte solution will cause it to dissolve. When combined with photolithographic techniques, where intact photoresist layers inhibit anodic dissolution, metal structures can be prepared by selective dissolution. With semiconductors (e.g., n-GaAs), conditions can be found where, at a given potential, dissolution occurs on a semiconductor immersed in a solution only where the material is irradiated (see Chapter 6). In this case structures can be formed on a semiconductor surface by photoelectrochemical etching, where the semiconductor is held at a given potential in contact with the chosen etchant and irradiated through a mask.

Polymerization. The deposition of polymer films electrochemically can be accomplished from solutions of the polymer, where oxidized or reduced forms of the polymer are insoluble and precipitate on the electrode surface. An alternative approach involves immersion of the electrode in a solution of a monomer whose polymerization can be induced electrochemically. These methods are described in more detail in Chapter 4.

2.5.7. High-Resolution Deposition and Etching

The scanning tunneling microscope (STM) (61) and scanning electrochemical microscope (SECM) (62), which are used for the examination of surface topography with high resolution, can also be used for fabrication purposes. The principles of the STM are shown in Figure 2.5.15. A small tip, usually tungsten or Pt/Ir, which is formed by etching a wire to a sharp point, is scanned near the surface of the substrate. This scanning can be carried out in the X and Y directions with subangstrom-scale resolution by the use of piezoelectric elements as voltage-to-distance transducers. When the tip is brought to within a few angstroms from the surface, a tunneling current can flow between substrate and tip. The tip is scanned across the surface, with a feedback loop maintaining the tunneling current at a constant level ($\sim$ nanoamperes) by moving the tip up and down in the Z direction. A plot of the voltage of the Z piezoelectric versus X or Y distance shows the surface electronic distribution and topography.

However, the tip interacts in a variety of ways with the substrate surface, and these interactions can be used to modify the substrate (61c, 63). Lines, holes, and mounds of angstrom dimensions can be formed on a surface with the STM. For example, with the introduction of a suitable reactant in the gas phase the high electric field between tip and substrate can cause plasma-processing type reactions. Even single atoms can be moved around and fixed on a surface with the STM (64). More frequently the tip of the STM [or atomic force microscope (AFM)] is used to scratch ("micromachine") features in films of organized monolayers or polymeric resists. For example, LB films (65) and self-assembled monolayers of alkanethiols (66) can be etched with the STM tip.

The SECM involves a similar arrangement, with the sample immersed in a solution and the tip serving as the working electrode in an electrochemical cell containing additional auxiliary and reference electrodes (62). The current flow in the SECM is governed by a faradaic process at the tip, specifically, a redox process involving some solution species. The presence of a sample substrate, either a conductor or an insulator, will affect the flow of current at the tip, and this effect is used to image

2.5. TECHNIQUES OF ASSEMBLY

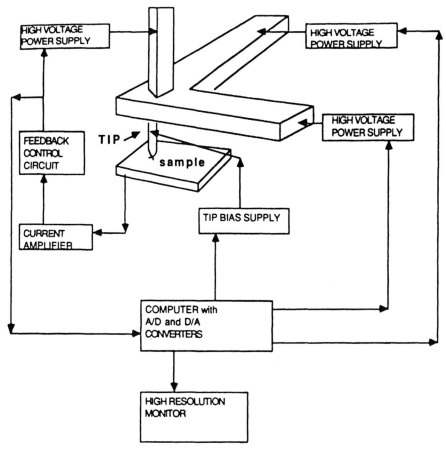

FIGURE 2.5.15. Schematic diagram of scanning tunneling microscope. In many microscopes the piezoelectric element tripod scanner shown here is replaced by a piezotube scanner.

surfaces. However, the SECM can also be used for deposition and etching with high resolution in the X, Y, and Z dimensions (67). For example, lines of silver, about 200 nm wide, can be deposited by Ag^+ reduction at the tip. Metal and semiconductor surfaces can also be etched by generation of a suitable etchant at the tip, such as Br_2 produced at the tip will etch holes in GaAs (68). Representative examples of SECM fabrication are shown in Figure 2.5.16. This arrangement is basically a high-resolution form of electroplating and electrochemical machining, where metals are deposited or anodically dissolved with a shape that is determined by the shape and spacing of the counter electrode. This approach should also be useful for other types of electrosynthetic reactions, such as the deposition of conductive polymers.

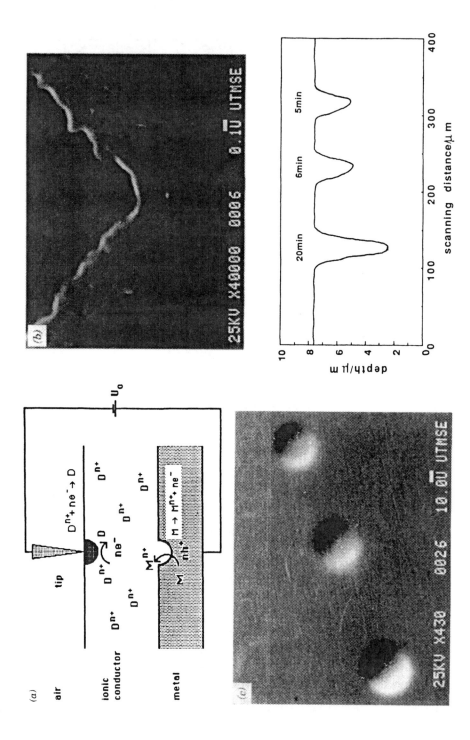

FIGURE 2.5.16. (a) Schematic diagram of deposition and etching of metals in an ionic conductor film with the SECM. Deposition of metal D occurs at the tip electrode and localized etching of metal, M, takes place on the substrate. [Reprinted with permission from D. H. Craston, C. W. Lin, and A. J. Bard, *J. Electrochem. Soc.*, **135**, 785 (1988). Copyright 1988 The Electrochemical Society.] (b) SEM picture of a pattern of silver lines deposited in a Nafion film by SECM by the tip reaction $Ag^+ + e \rightarrow Ag$. [Reprinted with permission from D. H. Craston, C. W. Lin, and A. J. Bard, *J. Electrochem. Soc.*, **135**, 785 (1988). Copyright 1988 The Electrochemical Society.] (c) SEM picture of holes etched in GaAs by generation of Br_2 at tip electrode. [Reprinted with permission from D. Mandler and A. J. Bard, *J. Electrochem. Soc.*, **137**, 2468 (1990). Copyright 1990 The Electrochemical Society.]

2.6. CONCLUSIONS

In this chapter we have discussed some of the materials used for constructing integrated chemical systems and several methods that can be used for their assembly. A variety of materials are available and resolution in the Z direction at the atomic level is already possible. There are still very few examples of actual multicomponent systems with high resolution in the X and Y directions. Self-organizing systems and those constructed within naturally occurring templates seem especially promising for initial studies.

REFERENCES

1. T. A. Skotheim, ed., *Handbook of Conducting Polymers*, Marcel Dekker, New York, 1986.
2. J. E. Frommer and R. R. Chance, *Encycl. Polym. Sci. Eng.*, **5**, 462 (1986).
3. J. Heinze, *Top. Curr. Chem.*, **152**, 1 (1990).
4. F. Garnier, *Angew. Chem. Intl. Ed. Engl.*, **28**, 513 (1989).
5. J. H. Fendler, *Membrane Mimetic Chemistry*, Wiley, New York, 1982.
6. D. H. Everett, *Basic Principles of Colloid Science*, Royal Society of Chemistry, Letchworth, 1988.
7. H. van Olphen, *An Introduction to Clay Colloid Chemistry*, Wiley, New York, 1977.
8. G. W. Brindley and G. Brown, eds., *Crystal Structures of Clay Minerals and their X-ray Identification*, Monograph No. 5, Mineralogical Society, London, 1980.
9. R. M. Barrer, *Hydrothermal Synthesis of Zeolites*, Academic Press, Orlando, 1982.
10. R. Szostak, *Molecular Sieves, Principles of Synthesis and Identification*, Van Nostrand Reinhold, New York, 1989.
11. W. M. Meier and D. H. Olson, *Atlas of Zeolite Structure Types*, Butterworths, London, 1988.
12. M. S. Whittingham and A. J. Jacobson, *Intercalation Chemistry*, Academic Press, New York, 1982.
13. F. Levy, ed., *Intercalated Layered Materials*, Reidel, Dordrecht, 1979.
14. P. Laszlo, ed., *Preparative Chemistry Using Supported Reagents*, Academic Press, New York, 1987.
15. G. W. Bridger and G. C. Chinchen, in *Catalyst Handbook*, Wolfe Scientific Books, London, 1970, p. 64.
16. B. C. Gates, J. R. Katzer, and G. C. A. Schuit, *Chemistry of Catalytic Processes*, McGraw-Hill, New York, 1979.
17. P. G. Schultz, R. A. Lerner, and S. J. Benkovic, *Chem. Eng. News*, May 28, 1990, p. 26.
18. R. W. Murray, in *Electroanalytical Chemistry*, A. J. Bard, ed., Marcel Dekker, New York, 1984, Vol. 13, p. 191.

REFERENCES

19. J. Rebek, Jr., *Science*, **235**, 1478 (1987).
20. J-M. Lehn, *Science*, **227**, 849 (1985).
21. D. J. Cram, *Science*, **219**, 1177 (1983).
22. (a) M. L. Bender and M. Komiyama, *Cyclodextrin Chemistry*, Springer-Verlag, New York, 1978; (b) R. Breslow, in *Inclusion Compounds*, J. L. Atwood, J. E. D. Davies, and D. D. MacNicol, eds., Academic Press, New York, 1984, Vol. 3, p. 473.
23. T. Matsue, M. Fujihira, and T. Osa, *J. Electrochem. Soc.*, **126**, 500 (1979).
24. D. W. Urry, *Angew. Chem. Intl. Ed. Engl.*, **32**, 819 (1993).
25. (a) C. K. Dyer, *Nature*, **343**, 547 (1990); T. Mallouk, ibid., p. 515.
26. H. Fraenkel-Conrat and R. C. Williams, *Proc. Natl. Acad. Sci. USA*, **41**, 690 (1955).
27. See J. S. Lindsey, *New J. Chem.*, **15**, 153 (1991), and the many references contained therein.
28. G. L. Gaines, Jr., *Insoluble Monolayers at Liquid–Gas Interfaces*, Wiley-Interscience, New York, 1966.
29. H. Kuhn, D. Mobius, and H. Bucher, *Physical Methods of Chemistry*, A. Weissberger and W. B. Rossiter, eds., Wiley, New York, 1972, Vol. I, Part 3B, pp. 577ff.
30. A. Ulman, *Ultrathin Organic Films*, Academic Press, New York, 1991.
31. J. Sagiv, *J. Am. Chem. Soc.*, **102**, 92 (1980).
32. F. Garnier, A. Yassar, R. Hajlaoui, G. Horowitz, F. Deloffre, B. Servet, S. Ries, and P. Alnot, *J. Am. Chem. Soc.*, **115**, 8716 (1993).
33. (a) C. Liu, H. Pan, A. J. Bard, and M. A. Fox, *Science*, **247**, 897 (1993); (b) B. A. Gregg, M. A. Fox, and A. J. Bard, *J. Phys. Chem.*, **94**, 1586 (1990).
34. (a) Gao, H. G. Hong, and T. E. Mallouk, *Acc. Chem. Res.*, **25**, 420 (1992); (b) H. Lee, L. J. Kepley, H-G. Hong, and T. E. Mallouk, *J. Am. Chem. Soc.*, **110**, 618 (1988); (c) H. Lee, L. J. Kepley, H-G. Hong, S. Akhter, and T. E. Mallouk, *J. Phys. Chem.*, **92**, 2597 (1988).
35. (a) R. E. Liesegang, *Naturwiss. Wochenschr.*, **11**, 353 (1896); (b) E. S. Hedges, *Liesegang Rings and Other Periodic Structures*, Chapman & Hall, London, 1932.
36. K. F. Mueller, *Science*, **225**, 1021 (1984).
37. D. W. DeWulf and A. J. Bard, *J. Electrochem. Soc.*, **135**, 1977 (1988).
38. M. Krishnan, J. R. White, M. A. Fox, and A. J. Bard, *J. Am. Chem. Soc.*, **105**, 7002 (1983).
39. S. Mazur and S. Reich, *J. Phys. Chem.*, **90**, 1365 (1986).
40. S. Wolfram, *Sci. Am.*, **251**, 188 (1984).
41. G. Ozin, *Adv. Mater.*, **4**, 612 (1992).
42. M. S. Whittingham and A. J. Jacobson, *Intercalation Chemistry*, Academic Press, New York, 1982.
43. J. D. Hawkins, *Gene Structure and Expression*, Cambridge University Press, Cambridge, UK, 1991.
44. See, for example: Y. T. Tan and G. Challa, *Encycl. Polym. Sci. Eng.*, **16**, 554 (1989).
45. A. Weiss, *Angew. Chem. Intl. Ed. Engl.*, **20**, 850 (1981).

46. (a) Z. Cai, J. Lei, W. Liang, V. Menon, and C. R. Martin, *Chem. Mater.*, **3**, 960 (1991); (b) C. R. Martin, L. S. Van Dyke, Z. Cai, and W. Liang, *J. Am. Chem. Soc.*, **112**, 8976 (1990).
47. C. J. Brumlik and C. R. Martin, *J. Am. Chem. Soc.*, **113**, 3174 (1991).
48. Y. N. C. Chan, R. R. Schrock, and R. E. Cohen, *J. Am. Chem. Soc.*, **114**, 7295 (1992).
49. Y. N. C. Chan, R. R. Schrock, and R. E. Cohen, *Chem. Mater.*, **4**, 27 (1992).
50. See, for example: (a) D. J. Elliot, *Integrated Circuit Fabrication Technology*, McGraw-Hill, New York, 1982; (b) P. Van Zant, *Microchip Fabrication*, Semiconductor Services, San Jose, CA, 1984.
51. (a) M. Peckerar, ed., *Electron-Beam, X-Ray, and Ion-Beam Submicrometer Lithographies for Manufacturing*, SPIE Proceedings Vol. 1465, 1991; (b) Y. Yamamura, T. Fujisawa, and S. Namba, ed., *Nanometer Structure Electronics*, North-Holland, Amsterdam, 1985.
52. N. L. Abbott, J. P. Folkers, and G. M. Whitesides, *Science*, **257**, 1380 (1992).
53. A. Kumar, H. A. Biebuyck, N. L. Abbott, and G. M. Whitesides, *J. Am. Chem. Soc.*, **114**, 9188 (1992).
54. J. Orloff, *Sci. Am.*, Oct. 1991, 96.
55. See, for example: J. W. Diggle, T. C. Downie, and C. W. Goulding, *Chem. Rev.*, **69**, 365 (1989).
56. L. T. Canham, *Appl. Phys. Lett.*, **57**, 1046 (1990).
57. L. M. Goldman, B. Blanpain, and F. A. Spaepen, *J. Appl. Phys.*, **60**, 1374 (1986).
58. J. C. Angus, U. Landau, S. H. Liao, and M. C. Yang, *J. Electrochem. Soc.*, **133**, 1152 (1986).
59. D. M. Kolb, in *Advances in Electrochemistry and Electrochemical Engineering*, H. Gerischer and C. W. Tobias, ed., Wiley, New York, 1978, Vol. 11, p. 125.
60. (a) B. W. Gregory and J. L. Stickney, *J. Electroanal. Chem.*, **300**, 543 (1991); (b) B. W. Gregory, D. W. Suggs, and J. L. Stickney, *J. Electrochem. Soc.*, **138**, 1279 (1991).
61. (a) G. Binnig and H. Rohrer, *Rev. Mod. Phys.*, **59**, 615 (1987); (b) R. J. Behm, N. Garcia, and H. Rohrer, eds., *Scanning Tunneling Microscopy and Related Methods*, Kluwer Academic Publishers, Dordrecht, 1990; (c) D. A. Bonnell, ed., *Scanning Tunneling Microscopy and Spectroscopy*, VCH Publishers, New York, 1993.
62. (a) A. J. Bard, F-R. F. Fan, J. Kwak, and O. Lev, *Anal. Chem.*, **61**, 132 (1989); (b) A. J. Bard, F-R. F. Fan, D. T. Pierce, P. R. Unwin, D. O. Wipf, and F. Zhou, *Science*, **254**, 68 (1991).
63. (a) C. F. Quate, *NATO Science Forum '90, September 16–21, 1991*, NATO Series, Plenum Press, New York, 1991; (b) H. K. Wickramasinghe, ed., *Scanned Probe Microscopy*, AIP Conference Proceedings No. 241, American Institute of Physics, New York, 1992.
64. D. M. Eigler and E. K. Schweizer, *Nature*, **344**, 524 (1990).
65. T. R. Albrecht, M. M. Dovek, C. A. Lang, P. Crutter, C. F. Quate, S. W. Kuan, C. W. Frank, and R. F. Pease, *J. Appl. Phys.*, **64**, 1178 (1988).

66. Y-T. Kim and A. J. Bard, *Langmuir,* **8,** 1096 (1992).
67. (a) D. H. Craston, C. W. Lin, and A. J. Bard, *J. Electrochem. Soc.,* **135,** 785 (1988); (b) O. E. Hüsser, D. H. Craston, and A. J. Bard, *J. Vac. Sci. Technol. B,* **6,** 1873 (1988); (c) O. E. Hüsser, D. H. Craston, and A. J. Bard, *J. Electrochem. Soc.,* **136,** 3222 (1989).
68. D. Mandler and A. J. Bard, *J. Electrochem. Soc.,* **137,** 2468 (1990).

Chapter 3
CHARACTERIZATION OF INTEGRATED CHEMICAL SYSTEMS

One of the things which distinguishes ours from all earlier generations is this, that we have seen our atoms.
—Karl K. Darrow
The Renaissance of Physics
Macmillan, New York, 1936

The characterization of surfaces, interfaces and thin films is a most elaborate and often difficult task.
—M. Grasserbauer and H. W. Werner, Eds.,
Analysis of Microelectronic Materials and Devices
Wiley, New York, 1991

3.1. INTRODUCTION

The characterization of an integrated chemical system after its fabrication (i.e., the determination of its composition, structure, and properties) is critical to an understanding of the system and its mode of operation. A large number of analytical techniques are available for the study and characterization of surfaces, interfaces, and thin films, and only a very brief discussion of these will be given here. The powerful methods for the study of microelectronic materials and devices, such as integrated circuits, are clearly of use with ICSs (1). These are usually *ex situ* methods, in which the system to be investigated is removed from its usual environment, frequently into an ultrahigh vacuum (UHV) for analysis (2). A wide variety of other techniques (*in situ* methods), including spectroscopic (3) and electrochemical (4), can be used for systems in liquid or gaseous media. Characterization encompasses many areas. In addition to the obvious analytical and structural questions, such as those concerning the composition (elements and compounds) of the system, the structure of the components, and the spatial distribution of species, are questions concerning film thickness, electrical and optical properties, and the rates of mass transport and other processes within the system. Thus we begin by considering what we mean by "characterization." We continue with a brief description of methods that can be applied to ICSs and give a few examples of their use.

3.2. CHARACTERIZATION

When one speaks of "characterization" of chemical systems, including ICSs, one usually includes a number of different aspects of the system:

Elemental Analysis. It is often useful to identify what elements are present in a system or on a surface. For example, in the modification of a metal surface by covalent attachment of various species, the course of the modification can be monitored by X-ray photoelectron spectroscopy to identify the elements present in the linking agents or modifying groups. Sometimes it is also possible to find out the oxidation states of the atomic species.

Determination of Composition. Analytical techniques can be applied to identify the chemical species (elements and compounds) that make up the system initially and during its operation. These can be classified into several categories: stable stoichiometric species, stable nonstoichiometric species, and unstable intermediates. Stable stoichiometric species encompass well-defined molecular and macromolecular compounds. When these are present in monolayer and submonolayer amounts, the analytical chemical problems can be quite challenging. Spectroscopic and electrochemical techniques of high sensitivity are available, however. Systems will often contain stable materials that are not stoichiometric. For example, a film of tungsten oxide prepared on a tungsten metal surface is best represented as WO_{3-x}, where $1 < x < 0$. Even a stoichiometric film of WO_3, if reduced electrochemically by holding an electrode covered with oxide at a negative potential in an acidic solution or by treating it with hydrogen at elevated temperatures, changes to a composition $H_y WO_{3-y}$. This "doping" of the WO_3 film with hydrogen at levels from ppm (parts per million) to several percent is typical of the behavior of many stoichiometric materials when subjected to a chemical or electrochemical reduction or oxidation. Other examples of nonstoichiometric species include the oxidation of graphite, where anions can be incorporated into the surface structure by intercalation, and the oxidation of a film of polypyrrole, where anions become counter ions to positive sites within the polymer. The nature of a platinum electrode surface following anodization probably cannot be described in terms of known, stoichiometric species (e.g., PtO or PtO_2), and thus it is sometimes identified in terms of "adsorbed oxygen," Pt—OH, or "adsorbed hydroxyl radicals." Unstable intermediates are often formed during the operation of a system. These include radicals, such as $CH_3^{\cdot}$ formed during oxidation of acetate ion, or adsorbed atoms (adatoms) formed during the first stages of the electrodeposition of a metal.

Spatial Information. The way in which species are distributed across the surface (in the X-Y plane or two-dimensionally) or in the Z or depth dimension is frequently of interest. Information about the depth distribution can sometimes be obtained by sputtering away the surface layers while analyzing for surface composition, such as by Auger electron spectroscopy. Surface distributions are usually obtained by electron microscopy, sometimes combined with energy-dispersive X-ray analysis.

Orientational Information. This includes information about what crystal faces are exposed at an interface, such as the basal plane versus the edge plane of highly oriented pyrolytic graphite or the (111) versus the (100) surface of Pt. One would also often like to know about the configuration of molecules on a surface, such as whether adsorbed aromatic molecules are oriented with their rings parallel or perpendicular to the surface or whether particles of clay are oriented with the aluminosilicate layers parallel to the surface or randomly aligned. Similarly, the angle at which the molecules of a surfactant in an LB film are oriented with respect to the substrate surface can be determined.

Energetic and Dynamic Characterization. Energetic information includes the redox potentials of surface species, acid-base properties, bond energies, and the location of the band edges and the band gaps of semiconductor materials. Dynamic information pertains to the rates and activation energies of the different reactions that occur in an ICS. These include mass-transfer rates of different species and charge mobilities through films as well as charge-transfer rates at different sites or at interfaces. Note that the energetics and kinetics of surface species may be very different from those of the same species in a bulk phase.

3.3. METHODS OF CHARACTERIZATION

Phase Properties. It is often useful to have information about the bulk properties of the phases in a system, such as refractive index, electrical properties, magnetic properties, and density. For surface films, the thickness of the layer is frequently important. This can be obtained for rather hard films on the surface of a solid by profilometry, where a metal stylus is traced over the surface. Commercial instruments can measure thicknesses over a range of several hundred angstroms to a few micrometers. Thicknesses are more difficult to measure when the film is soft, as are some polymers or membranes, or when a film is immersed in a liquid. Ellipsometric techniques, described below, can often be used and in favorable cases can determine thicknesses of a few angstroms. Atomic force microscopy can also be employed to determine surface topography. The surface or interfacial area is also usually of interest. The area is clearly a fractal property that depends on the method of determination. The easiest area to determine is the projected or geometric area; this is the area measured by a macroscopic ruler, or in general, by using a measurement means whose dimensions are large compared to any surface asperities. The "true" or total surface area, measured, for example, by adsorption of small molecules (as in the BET method), is usually much larger. The ratio of total-to-projected area is called the "roughness" factor. Electrochemical methods can be employed to measure both projected and total surface areas of electrodes.

Quantitative Analysis. It is often useful to know the total amounts of given species within an ICS. This would include surface concentrations of materials (mol/cm^2 or molecules/μm^2) or concentrations within films (mol/cm^3 or molecules/μm^3).

3.3. METHODS OF CHARACTERIZATION

Many techniques are available for the characterization of ICSs. These can broadly be classified as optical or spectroscopic methods (including surface spectroscopy, ellipsometry, and magnetic resonance methods), mass spectroscopy, microscopy, and electrochemical methods. Some of these methods, especially the microscopic ones, give direct information about the structure and topography of the system. In others, like electrochemical or spectroscopic techniques, structural information requires fitting of data to a model. Some techniques, such as in the ESR or luminescence results described below, can give detailed information about the nature of the environment in the ICS or the mobility of species within the system. It is beyond the scope of this work to discuss in any detail the methods that are available to characterize ICSs and interfaces. Thus I will only list a number of techniques that have been applied to such systems, describe the basic principles of the methods, and give examples of their application. Details about many of the methods described below and extensive bibliographies are available (1-4).

3.3.1. Electrochemical Methods

Electrochemical methods are especially useful for investigating species or films on conductive substrates and studies of electron-transfer reactions at the solid–liquid interface. Their high sensitivity allows investigations of electroactive surface species in amounts down to about a tenth of a monolayer. They are particularly appropriate for obtaining information about electronic energies (e.g., redox potentials) and mass- and charge-transfer rates, within the usual time scale for electrochemical measurements (microseconds to minutes). While surface species can be identified through their potentials for oxidation or reduction, electrochemical methods are not useful for elemental analysis, nor can they provide direct information about structure or orientation, except by use of inference and models. Generally, electrochemical methods involve an electrical or other perturbation to the system of interest (e.g., a potential pulse or sweep) and observation of the electrical response (e.g., the current). Through well-developed theory and models, this response can be used to obtain useful analytical, thermodynamic, and kinetic information about the system under investigation. A number of widely used electrochemical techniques are listed in Table 3.3.1. The basic principles of these methods and discussions of the theory and instrumentation are given in several monographs and reviews (4, 5). A number of examples of their application to modified electrodes are given in Chapter 5.

3.3.2. UHV Surface Spectroscopic Methods

These methods are all based on placement of the sample in an ultrahigh vacuum (UHV) and measurement of the electron or ion fluxes that result from irradiation (or bombardment) of the sample with photons, electrons, or ions. The general principles are illustrated in Figure 3.3.1, and the various techniques and their scope are listed in Table 3.3.2. An important consideration in the use of these techniques is their sensitivity (i.e., detectable concentrations), spot size (which limits X-Y resolution), and the depth that is sampled. These are shown in Figure 3.3.2.

TABLE 3.3.1. Typical Electrochemical Methods

Name of Method	Excitation Signal	Response Signal	Information Obtained	Remarks
Conductometry Dielectrometry	Ac potential (typically 1–100 kHz)	Ac current	Resistance, conductance, capacitance, dielectric constant	Used mainly in analytical devices
Potentiometry	Concentration	Potential	Concentration, redox potentials ($E°$)	Used in mass- and charge-transfer rate measurements
Potential step chronoamperometry or chronocoulometry	Potential step	Current or coulombs vs. time	Amount, D, $E°$, kinetics	Especially useful in studies of surface species; higher sensitivity by use of superimposed square wave or potential pulses
Voltammetry or cyclic voltammetry	Potential sweep	Current vs. potential	Amount, D, $E°$, kinetics	
Chronopotentiometry	Current step	Potential vs. time	Amount, D, $E°$, kinetics	Less frequently used
Alternating-current voltammetry, impedance methods, impedance spectroscopy	Ac potential + dc potential	Ac current	Amount, D, $E°$, kinetics	Often used to model films or interfaces in terms of electrical equivalent circuit

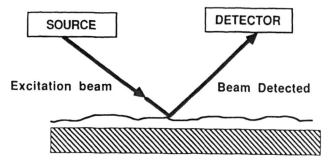

FIGURE 3.3.1. General principle of UHV surface spectroscopic techniques:

Technique	Excitation	Detect
XPS	X-ray $h\nu$	Electrons
UPS	UV $h\nu$	Electrons
Auger ES	Electrons	Electrons
SEM	Electrons	Electrons
SIMS	Ions	Ions
Laser desorption MS	$h\nu$	Ions

X-ray Photoemission Spectroscopy (XPS). XPS has been a widely used technique to study the course of surface modification, because it can provide elemental composition as well as information about the oxidation states of the elements (6). When an X-ray photon of energy, $h\nu$, strikes an atom in the sample, an electron is emitted. Although the X-ray beam penetrates quite deeply into the sample, the photoemitted electron that is detected originates in the outer 10–30 Å of the surface. This electron is collected and its energy (E_{kin}) measured. The binding energy of the electron in the atom, E_B, can then be determined (with correction for the work function of the spectrometer) as

$$E_B = h\nu - E_{kin} \quad (3.3.1)$$

The binding energy of an electron is characteristic of the particular element and is also sensitive to the oxidation state of the element and its environment. Generally, E_B increases with increasing positive charge, or higher oxidation state, of the element. An example of the use of XPS in studying the progressive modification of an SnO_2 surface is shown in Figure 3.3.3 (7). On the fresh SnO_2 surface, a large signal for Sn ($3d_{5/2}$) is observed, with a small Si ($2p$) signal from the glass substrate. When the surface is treated with a nitrogen-containing silinization linking agent, the Si signal increases and a N ($1s$) signal, characteristic of an $-$NH group appears. When the linking agent is modified further, a new N ($1s$)

3.3. METHODS OF CHARACTERIZATION

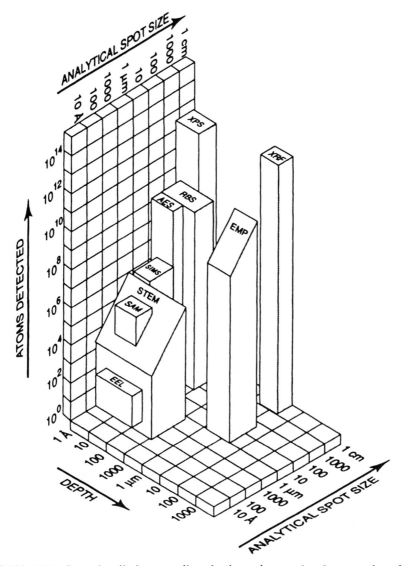

FIGURE 3.3.2. Detection limits, sampling depth, and spot size for several surface spectroscopic techniques. Rutherford backscattering (RBS), secondary-ion mass spectroscopy (SIMS), X-ray fluorescence (XRF), X-ray photoelectron spectroscopy (XPS), Auger electron spectroscopy (AES), scanning Auger microscopy (SAM), electron microprobe (EMP), electron energy loss (EEL), scanning transmission electron microscopy (STEM). [Adapted from *Texas Instruments Materials Characterization Capabilities*, Texas Instruments, Richardson, TX, 1986.]

TABLE 3.3.2. Ultra-High-Vacuum Surface Spectroscopic and Optical Methods

Name	Acronym	Irradiate with	Detect	Physical Basis	Information Obtained	Sensitivity and Resolution
X-ray and UV photoelectron spectroscopy	XPS (ESCA) UPS	Photons X rays, UV (21 eV)	Photoelectrons	Electron emission from inner shells	Surface elemental composition, oxidation states	5×10^{19} atoms/cm^3 10–30 Å depth spot ~ 5 × 5 mm
Auger electron spectroscopy	AES	Electrons (2–3 keV)	Electrons	Electron emission from surface atoms	Surface composition	5×10^{19} atoms/cm^3 10–30 Å depth spot ~ 0.1 mm
Rutherford backscattering spectroscopy	RBS	H or He ions (1–3 MeV)	H or He ions	Scattering of monoenergetic ion beam from surface	Surface composition (heavy elements)	5×10^{18} atoms/cm^3 1 μm depth spot ~ 5 mm
X-ray Diffraction	XRD	X rays	X rays	Diffraction of X rays from surface	Phase identification	1% composition μm depth large spot
X-ray fluorescence	XRF	X rays	X rays	Emission from electronic transmission (inner shell)	Elemental composition	5×10^{18} atoms/cm^3 μm depth large spot

Electron microscopy (scanning)	SEM TEM STEM	Electrons (100–300 keV)	Electrons (X rays)	Electron scattering	Surface imaging	Atomic resolution possible
Low-energy electron diffraction	LEED	Electrons (10–300 eV)	Electrons	Diffraction and scattering of electrons	Atomic surface structure	
Secondary-ion mass spectroscopy	SIMS	Ions (1–20 keV)	Sputtered atomic and molecular ions	Ion-beam-induced ejection of surface ions	Surface composition	1×10^{15} atoms/cm^3 100 Å; spatial resolution ~1 µm
Laser microprobe mass spectroscopy	—	Visible–UV light	Molecular ions	Light-induced desorption of molecules and ions	Surface composition	
Extended X-ray absorption fine structure	EXAFS SEXAFS	X rays	X rays	Absorbance and scattering of X rays from surface	Atomic structure of surface and absorbates; near-neighbor distances	

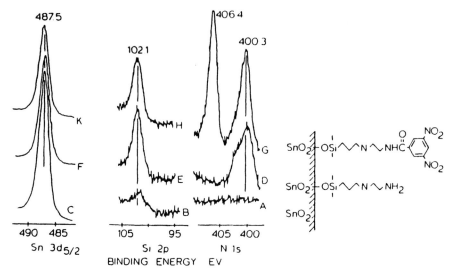

FIGURE 3.3.3. The use of XPS to follow the course of surface modification of a clean SnO_2 surface (A, B, C) following treatment with ethylenediamine silane (D, E, F) and then dinitrobenzoyl chloride (G, H, K). Note growth of Si and N(H) peaks following the first treatment and further increase of N peaks, with appearance of a new N(O) peak, following the second. [Reprinted with permission from R. W. Murray, *Acc. Chem. Res.*, **13**, 135 (1980). Copyright 1980 American Chemical Society.]

signal appears; this one identifies the presence of a $-NO_2$ group. A second example, modification of an oxide-coated Pt electrode with a monolayer of a $Ru(bpy)_3^{2+}$ species, is shown in Figure 3.3.4 (8).

It is more difficult to get quantitative information about amounts of surface species with XPS, although relative intensities are informative, as shown in the examples. XPS can also give some measure of the thickness of a film, by noting the attenuation of a signal arising from an underlying substrate after it is coated with a film, such as the changes in the Sn and Si signals in Figure 3.3.3. This requires careful calibration and empirical adjustments and is probably less accurate than alternative methods of measuring film thickness. UPS is a similar method, with excitation by UV photons.

Auger Electron Spectroscopy (AES). Auger electron peaks arise when electrons, following excitation by X-ray or electron beams, relax and lose energy by emission of secondary (or Auger) electrons. As with XPS, the energies of these electrons are diagnostic of the emitting atom and originate from within 10-30 Å of the surface. AES is often used to observe the surface of a sample as layers of the surface are removed

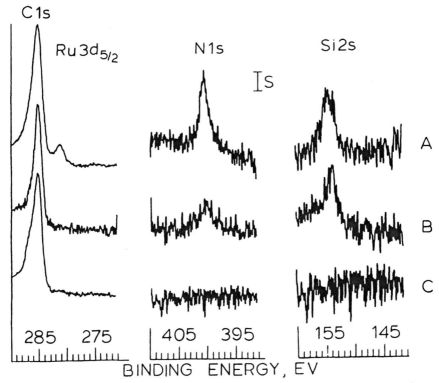

FIGURE 3.3.4. XPS of a Pt/PtO electrode in the course of modification with a layer containing the Ru(bpy)$_3^{2+}$ species. (A) Electrode with Ru species attached via an ethylenediamine silane (en-silane). (B) Electrode with only en-silane. (C) Pt electrode before treatment. The C 1s signal observed for the bare electrode represents organic contamination of the electrode surface. [Reprinted with permission from H. D. Abruña, T. J. Meyer, and R. W. Murray, *Inorg. Chem.*, **11**, 3233 (1979). Copyright 1979 American Chemical Society.]

by sputter etching with a beam of inert-gas ions. In this way, a depth profile of the sample over several thousand angstroms can be obtained. A typical example is given in Figure 3.3.5, where a sample of Si that had been coated with a film of Pt and then subjected to annealing was analyzed (9). Another example is that of a film containing the polymer PQ^{2+} and Pd on a Si substrate (Fig. 3.3.6) (10). Conversion of the sputtering time to an actual depth in the film and sample requires careful calibration with a known sample under similar conditions, since different materials will be sputtered away at different rates. Differential migration of species to the surface and redeposition of sputtered material on the surface can also cloud the interpretation of sputtering profiles. XPS and

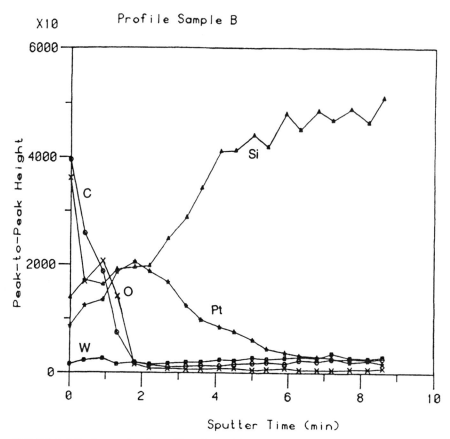

FIGURE 3.3.5. Auger depth profile of specimen of Si coated with a thin layer of Pt and annealed at 400°C for 30 min. The thickness of the Pt/Si layer, sputtered away by the Ar ion beam in about 6 min, is about 150 Å thick. Note C contamination of surface layer [Reprinted with permission from F-R. F. Fan, R. G. Keil, and A. J. Bard, *J. Am. Chem. Soc.*, **105**, 220 (1983). Copyright 1983 American Chemical Society.]

UPS can also be studied during sputtering to obtain depth profiles. Scanning Auger microscopy (SAM), in which the excitation electron beam is rastered across an area of the surface, can also be employed for surface analysis and can resolve features of ~100 Å.

3.3.3. *In Situ* Spectroscopic Methods

A number of other spectroscopic and optical methods, many involving the possibility of *in situ* examination of the sample, have been employed (Table 3.3.3). A few examples with ICSs follow.

3.3. METHODS OF CHARACTERIZATION

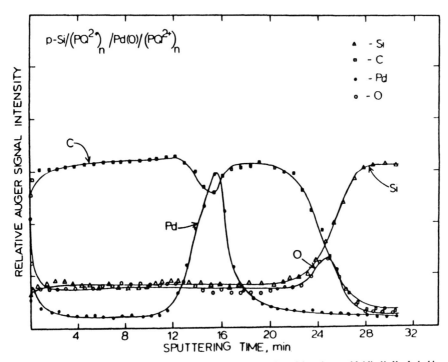

FIGURE 3.3.6. Auger depth profile for a p-Si electrode with a layer N,N'-dialkyl-4,4'-bipyridinium (PQ^{2+})-based polymer, followed by plating with Pd(0) and then functionalized again with PQ^{2+}; the structure is thus $Si/SiO_x/(PQ^{2+}\ 2Cl^-)_n/Pd/(PQ^{2+}\ 2Cl^-)_n$. [Reprinted with permission from J. A. Bruce, T. Murahashi, and M. S. Wrighton, *J. Phys. Chem.*, **86**, 1552 (1982). Copyright 1982 American Chemical Society.]

UV–Visible Spectroscopy. The absorbance spectra of films on transparent substrates have been used to study changes in the systems with time and to identify species within the films. For example, the viologen systems (e.g., MV^{2+} in a cation exchange membrane or in a clay film, or the polymer PQ^{2+}) show large color changes on reduction to the radical cation forms (11). The spectrum of the intensely colored radical cation depends on the extent of dimerization of this species. Similarly, the dimerization of tetrathiafulvalene (TTF) radical cations in polymer films can be probed spectroscopically.

Reflectance Spectroscopy. Photoacoustic Spectroscopy (PAS). Photothermal Spectroscopy (PTS). When the system under examination is not transparent, such as with films on thicker metal or semiconductor substrates, the light that is reflected from the surface of a sample can be examined (12). This is most easily accomplished by specular reflectance,

TABLE 3.3.3. *In Situ* Spectroscopic and Optical Methods

Name	Acronym	Irradiate with	Detect	Physical Basis	Information Obtained
Fourier transform infrared spectroscopy	FTIR	IR photons	IR photons	Absorption of energy characteristic of vibrations and rotations	Molecular identification of adsorbed molecules
Raman and surface-enhanced Raman spectroscopy	Raman SERS	Visible photons	Visible photons	Raman scattering of photons; resonant scattering	Vibrational spectra of adsorbed molecules
UV–visible spectroscopy	UV–vis	UV–vis photons	UV–vis photons	Absorption via electronic transitions	Electronic spectra of bulk and adsorbed molecules
Photoluminescence spectroscopy	PL	UV–vis photons	UV–vis photons	Emission on electronic relaxation	High sensitivity detection of surface species; environment
Photoacoustic and photothermal spectroscopy	PAS PTS	IR and UV–vis photons	Acoustic or thermal signal	Electronic transitions produce thermal and pressure variations	Spectra of solid polycrystalline substances

Reflectance spectroscopy	—	IR–UV–vis photons	IR–UV–vis photons	Electronic transitions; scattered photons	Spectra of solid polycrystalline substances
Second harmonic generation	SHG	UV–vis photons (laser)	UV–vis photons	Nonlinear effects at surface induced by high photon field	Spectra of species at interface
Electron-spin resonance	ESR	Microwave (magnetic field)	Microwave	Spin transitions (unpaired electrons)	Identify radicals orientation; environment
Electroluminescence	EL	Electrochemical	UV–vis photons	Electron–hole pair recombination	Energy levels in semiconductors
Inverse photoemission	IPE	Electrochemical	UV–vis photons	Electron–hole pair recombination	Energy levels in metals
Ellipsometry	—	Linearly polarized light	Elliptically polarized light	Changes in polarization on reflection	Film thickness and refractive index

when the sample surface is polished to avoid light-scattering effects. Higher sensitivities for surface films can be obtained by multiple reflections; this approach is widely used, for example, in multiple internal reflection infrared spectroscopy. For polycrystalline samples, light scattering makes measurement of the reflected radiation more difficult, although integrating sphere attachments to conventional spectrometers for collection of the scattered light are available. An alternative approach is to measure the heating, either directly or indirectly, produced by radiationless processes that occur on light absorption by the sample. In PTS (13), the temperature rise is measured directly, such as with a thermistor attached to the sample or by the deflection of a laser beam in the vicinity of the sample because of thermally produced refraction changes. In PAS (14), a chopped incident light beam is employed, and the varying pressure changes induced by the thermal changes in the sample are detected with a microphone or piezoelectric transducer. Although these latter methods are quite sensitive, they are best used for qualitative or semiquantitative studies, since the observed temperature or pressure changes are functions of both the optical and thermal properties of the samples. The photoacoustic effect can also be used for imaging in photoacoustic microscopy. A related technique, scanning acoustic microscopy, is based on the response of a sample to a focused ultrasonic beam (15).

Fourier Transform Infrared (FTIR) and Raman Spectroscopy. Infrared spectra, in favorable cases, such as for CO adsorbed on metal surfaces or for LB films of surfactants, can be used to examine adsorbates at monolayer coverages. FTIR has been used in both transmission and reflection modes to study changes in electrode surfaces during electrochemical processes. Grazing-angle and attenuated total reflectance (ATR) IR methods have been particularly useful in studying thin organic films (16). For example, alkanethiol monolayers that spontaneously assemble on gold have been studied by grazing-angle FTIR (17). By studying changes in the C—H stretching modes, one can obtain information about the conformation of the alkanethiol chains on the surface.

Raman spectroscopy, especially in the resonance Raman and surface-enhanced (SERS) modes, can also be employed to study ICSs and adsorbed layers. Differences between the Raman spectrum of a substance in solution and within or on a solid phase can provide information about the nature of the environment. For example, resonance Raman spectroscopy could be used to follow the changes that occur when a TTF-TCNQ electrode was oxidized and reduced (18) or to argue that $Ru(bpy)_3^{2+}$ adsorbed on clay (hectorite) was not greatly perturbed as compared to this ion in solution (Fig. 3.3.7) (19). SERS is a very

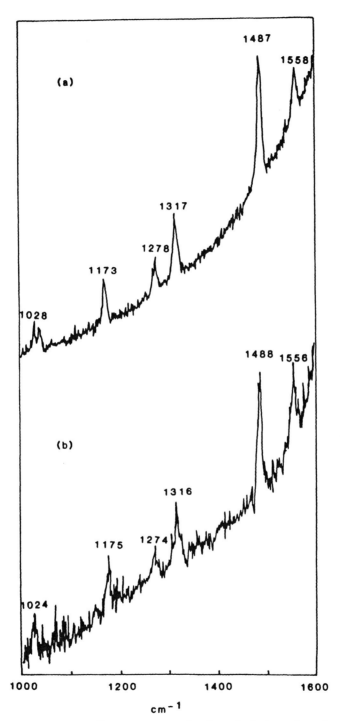

FIGURE 3.3.7. Resonance Raman spectra showing similarity of environment of $Ru(bpy)_3^{2+}$ in (a) water and (b) sodium hectorite (clay) suspension. [Reprinted with permission from P. K. Ghosh and A. J. Bard, *J. Phys. Chem.*, **88**, 5519 (1984). Copyright American Chemical Society.]

sensitive surface technique and has been useful in the characterization of monolayer films (16) for certain substrates, such as slightly roughened Ag surfaces.

Photoluminescence Spectroscopy. The luminescence spectrum of an emitting species and the lifetime of the excited state depend on the environment. For example, the cationic species (11-(3-hexyl-1-indolyl)undecyl)trimethylammonium bromide can probe the microenvironment of micelles, membranes, and anionic polyelectrolytes (20, 21). This species and others were used, for example, to study the environment and interactions in a Nafion film (22).

The luminescence properties of Ru(II) chelates, such as with bpy and related ligands, have been studied extensively as functions of the environment of the excited state. A particularly striking case is the extreme sensitivity of the quantum yield for emission of Ru(bpy)$_2$dppz^{2+}* (dppz = dipyrido[3,2:a-2'3':c]phenazine) to its environment (23). This species shows no photoluminescence in aqueous solution, but strongly luminesces when it binds to double strand helical DNA. Another example of the effect of environment on the quenching of luminescence involves Ru(bpy)$_3^{2+}$ in a suspension of the clay montmorillonite (19). Ru(bpy)$_3^{2+}$ has been widely studied in many solvents because of its interesting luminescent and photochemical properties. When this molecule is excited with light of about 460 nm, an excited state is formed that emits (in a ligand-to-metal transition) at about 580 nm. The time dependence of emission for the molecule intercalated in montmorillonite is more complex than the single exponential decay found in solution, and the emission intensity is decreased, especially at high light fluxes. This was ascribed to enhanced self-quenching (i.e., interaction of two excited states to produce ground states) because the Ru(bpy)$_3^{2+}$ was segregated within single layers of the clay particles to produce greatly enhanced effective concentrations of this species. Even stronger evidence for this segregation effect was obtained from studies of the quenching of Ru(bpy)$_3^{2+}$ luminescence by MV^{2+}. In solution, MV^{2+} is a very effective quencher ($k_q = 3 \times 10^8$ M^{-1} s^{-1}). However, when clay is introduced into a mixture of Ru(bpy)$_3^{2+}$ and MV^{2+}, the quenching rate is greatly decreased ($k_q = 1.1 \times 10^6$ M^{-1} s^{-1}). This decreased quenching rate was ascribed to segregation of Ru(bpy)$_3^{2+}$ and MV^{2+} into different layers of the clay (Fig. 3.3.8).

Electron Spin Resonance (ESR). ESR methods have been employed to identify paramagnetic species (radicals, radical ions, transition-metal ions) within ICSs. ESR can also provide information about the mobility

3.3. METHODS OF CHARACTERIZATION

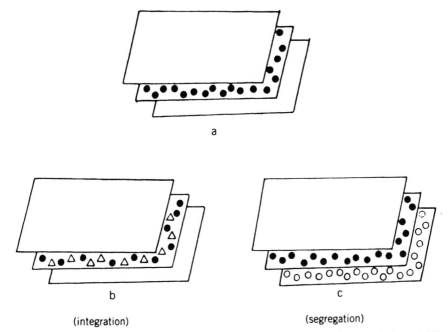

FIGURE 3.3.8. Schematic representation of the different modes of intercalation of different cations in a platelet of clay (sodium hectorite). (a) Ru(bpy)$_3^{2+}$ (●) segregated in a single sheet and (b) integrated with Zn(bpy)$_3^{2+}$ (△), an ion of similar size; (c) segregation of species of two different sizes: Ru(bpy)$_3^{2+}$ and MV^{2+} (○). Na$^+$ ions are not indicated. [Reprinted with permission from P. K. Ghosh and A. J. Bard, *J. Phys. Chem.*, **88**, 5519 (1984). Copyright American Chemical Society.]

(tumbling) of species, their electron-transfer rates, and the spatial orientation of probes within systems. In ESR measurements, the sample is held in a magnetic field and irradiated with microwave radiation. Radiation is absorbed when an unpaired (or odd) electron undergoes a spin flip in the field. In the absence of interactions with any nuclei with magnetic moments, only a single line, representing this transition, results. If the odd electron interacts with nuclei with a magnetic moment in a molecule [e.g., H$_1$($I = \frac{1}{2}$), N$_{14}$($I = 1$)], this single line is split to give a multiple-line spectrum. The magnitude of the splitting is a measure of the extent to which the unpaired electron interacts with these nuclei (the hyperfine interactions). A typical solution ESR spectrum of MV$^{·+}$ is shown in Figure 3.3.9 (curve *a*), where the odd electron interacts with two N and 14 H nuclei in the molecule. Since $I = 0$ for C^{12}, there is no magnetic interaction with this nucleus. To observe the hyperfine splitting, it is necessary that the molecule be free to tumble so that magnetic dipole–dipole interactions between molecules are av-

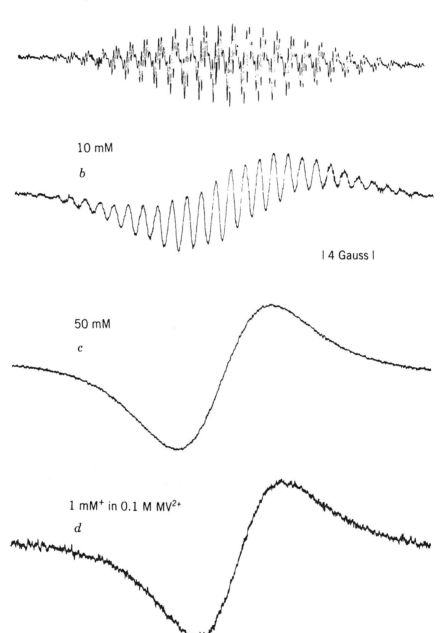

FIGURE 3.3.9. ESR spectra of MV$^{\cdot+}$ in aqueous solution: (a) 1 mM; (b) 10 mM; (c) 50 mM, showing the effect of dipole–dipole interaction; (d) 1 mM MV$^{\cdot+}$ in the presence of 0.1 M MV^{2+}, showing effects of fast electron transfer. [Reprinted with permission from J. G. Gaudiello, P. K. Ghosh, and A. J. Bard, *J. Am. Chem. Soc.*, **107**, 3027 (1985). Copyright 1985 American Chemical Society.]

3.3. METHODS OF CHARACTERIZATION

eraged to zero. If the movement of the molecule is restricted, a different spectrum, more characteristic of the molecule in a solid, results. Thus the nature of the ESR spectrum found for a paramagnetic species provides information about the local mobility of the species. For example, a comparison of the ESR spectra of $MV^{\cdot+}$ in two different types of polymer films is shown in Figure 3.3.10 (24). The spectrum of $MV^{\cdot+}$ in the Nafion film resembles that in solution, indicating free tumbling of the molecule in this environment. However, the viologen species in a $PQ^{\cdot+}$ film shows only a broad singlet, characteristic of restricted motion of the species in this polymer.

The ESR spectra also provide information about the rates of intermolecular electron transfer (et). If the radical ion is involved in a rapid et reaction, each hyperfine line will be broadened, and the hyperfine components of the spectrum will merge into a single broad line. A further increase in the rate of et will cause this single line to narrow (exchange narrowing). This effect is observed, for example, when MV^{2+} is added to a dilute solution of $MV^{\cdot+}$ (Fig. 3.3.9, curve d). Here, the rapid et between MV^{2+} and $MV^{\cdot+}$ causes the hyperfine structure to disappear. However, for $MV^{\cdot+}$ in a Nafion film, this broadening is not observed, even when the film is only partially reduced to $MV^{\cdot+}$ and the concentration of MV^{2+} is high (24). The conclusion from this experiment is that the rate of et between MV^+ and MV^{2+} in the polymer is much lower than that in solution. This point is discussed in Chapter 5.

Finally, ESR experiments produce information about the orientation of molecules. Experiments of this type employ spin probes or spin labels, such as tempamine, that contain an unpaired electron. In tempamine, the unpaired electron is largely localized within the nitroxide group and the interaction with the N nucleus produces a three-line spectrum in solution (Fig. 3.3.11A). When the molecule is held rigidly in a solid film, the spectrum is different, and the magnitude of the hyperfine splitting depends on the orientation of the spin probe to the magnetic field axis. Spectra of tempamine held in clay films of two different types are also given in Figure 3.3.11 (25). Clay consists of aluminosilicate particles composed of individual layers with negative charges that are compensated by cations that are intercalated between the layers (see Chapter 4). Tempamine was adsorbed into or on the surface of the clay particles, which were cast in the form of a film on ITO (indium tin oxide). If one assumes that the orientation of the tempamine is fixed within the clay particles, the ESR spectra provide information about the clay particle orientation with respect to the ITO surface. Note that when the clay suspension was cast from a solution containing poly(vinyl alcohol) (PVA), the ESR spectra for the film oriented parallel and per-

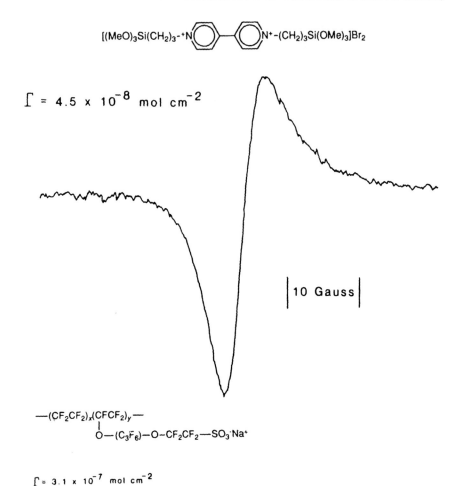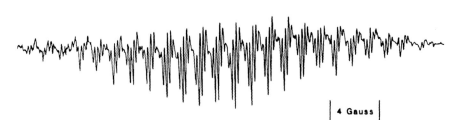

FIGURE 3.3.10. A comparison of viologen polymer films. Top, fully reduced PQ^+ film produced from monomer shown. The absence of hyperfine splittings shows restricted mobility of the viologen centers. Bottom, $MV^{\cdot +}$ in Nafion film. The similarity of this spectrum to that in solution (Fig. 3.3.9, curve a) suggests that the viologen is free to tumble and dipole–dipole interactions are averaged out. [Reprinted with permission from J. G. Gaudiello, P. K. Ghosh, and A. J. Bard, *J. Am. Chem. Soc.*, **107**, 3027 (1985). Copyright 1985 American Chemical Society.]

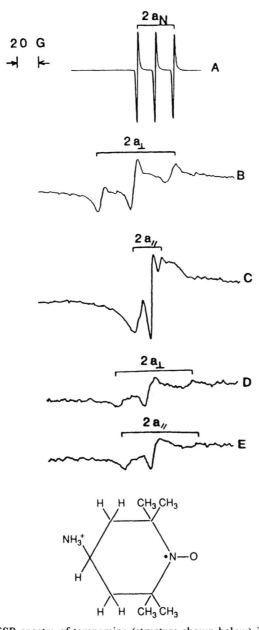

FIGURE 3.3.11. ESR spectra of tempamine (structure shown below) incorporated into clay (sodium montmorillonite) films. (A) Aqueous solution of 0.9 mM tempamine in absence of clay; (B) cast clay film oriented perpendicular to the magnetic field, H; (C) same as (B), but oriented parallel to H; (D) cast PVA/clay film oriented perpendicular to H; (E) same as (D), but oriented parallel to H. [Reprinted with permission from D. Ege, P. K. Ghosh, J. R. White, J-F. Equey, and A. J. Bard, *J. Am. Chem. Soc.*, **107**, 5644 (1985). Copyright 1985 American Chemical Society.]

pendicular to the magnetic field were the same (curves D and E). However, for films cast without addition of PVA, the spectra were different (curves B and C). The difference in splitting for the two orientations was used to estimate the extent of ordering of the film; the difference found in this case, 21.7 G, indicates a highly ordered film in the absence of PVA. The proposed structures of the films are shown in Figure 3.3.12. Note that clay films cast with PVA also show different X-ray diffraction patterns, which indicate wider spacings between the sheets of the clay layers.

Ellipsometric Methods. The reflection of a beam of linearly polarized light from a surface will generally produce elliptically polarized light, because the parallel and perpendicular components of the incident beam are reflected with different intensities and phase shifts (26). These changes can be measured and expressed in terms of two parameters, Ψ

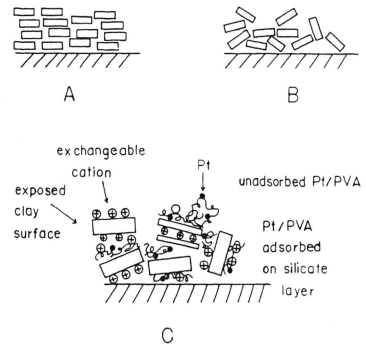

FIGURE 3.3.12. Proposed orientation of the clay films based on the ESR spectra. (*A*) Clay film; (*B*) clay/PVA film; (*C*) schematic of proposed microstructure for the clay/PVA film [also containing Pt particles (●)]. [Reprinted with permission from D. Ege, P. K. Ghosh, J. R. White, J-F. Equey, and A. J. Bard, *J. Am. Chem. Soc.*, **107**, 5644 (1985). Copyright 1985 American Chemical Society.]

3.3. METHODS OF CHARACTERIZATION

and Δ. For a thin film on a substrate, the magnitudes of these parameters depend on the refractive indices (substrate, film, and the liquid or gas contacting the film) and the film thickness. By making independent measurements of the substrate and contacting medium, and often by making measurements at several angles of incidence, wavelengths, or film thicknesses, it is possible to determine the film thickness and film refractive index. Under favorable circumstances even monolayer film thicknesses, such as of LB films, can be obtained (16).

An example of the use of ellipsometry in a study of the growth and thickness of polymer films on an electrode surface is shown in Figure 3.3.13 (27). These show the variations of Ψ and Δ as the film is deposited electrochemically; the point marked "0" represents the parameters for the electrode-solution interface before film formation occurs. Ellipsometric methods are useful because they can be employed to determine film thicknesses while the film is immersed in a liquid medium and to study the changes in thickness that occur while the film is undergoing a chemical process. For example, one can monitor the changes in the thickness of a film of poly(vinylferrocene) (PVF) when it is oxidized to the ferrocenium form and incorporates anions and solvent:

$$\text{PVF} + x\text{A}^- - xe \rightarrow \text{PVF}^{x+}(\text{A}^-)x \qquad (3.3.2)$$

3.3.4. Other Spectroscopic Methods

A number of other spectroscopic methods can be applied to the characterization of ICSs (1-3), including *nuclear magnetic resonance* (NMR) (especially NMR of solids based on magic angle spinning) and *Mössbauer spectroscopy*. Mass spectroscopic methods [including secondary-ion mass spectrometry (SIMS)] have also been applied to study molecules adsorbed on surfaces, for example, by using a laser or ion beam to cause desorption of the species that are ionized and detected.

3.3.5. Microscopy

Microscopic methods are among the most useful for imaging the structural arrangement in an ICS. The conventional optical (visible light) microscope is limited by diffraction effects of a resolution of about $\lambda/2.3$ at best and produces a two-dimensional image of the surface. Extensions of the optical microscope, however, have extended its capabilities. For example, in the confocal microscope, an intense light source, such as a laser, illuminates only a small region of the sample, and the detector is

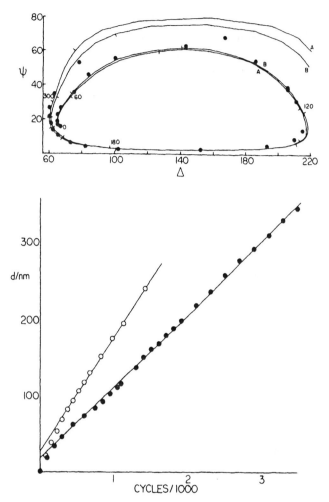

FIGURE 3.3.13. Top: variation of ellipsometric parameters during electrochemical deposition of a polybipyrazine film by oxidation of bipyrazine in aqueous sulfuric acid. The lines show the simluated curves based on the thicknesses indicated on the curves (nm) and an assumed refractive index and the points (●) show the experimental results. Bottom: film thicknesses during growth of films, obtained from ellipsometric data, for dry (○) and *in situ* (●) films. [Reprinted with permission from C. M. Carlin, L. J. Kepley, and A. J. Bard, *J. Electrochem. Soc.*, **132**, 353 (1985). Copyright 1985 The Electrochemical Society.]

arranged so that only light from a small region is collected (28). A series of digitized slice images from different depths of the sample are collected as the focal plane is moved across the sample. This results in an improvement in longitudinal resolution by a factor of about 2 and also yields a 3D (three-dimensional) image of the specimen. Even higher

3.3. METHODS OF CHARACTERIZATION

resolution (of the order of tens of nanometers) is promised by near-field scanning optical microscopy (NFSOM) where a localized optical probe, such as a very small aperture or a glass tip or fiber of small diameter, is rastered across the surface (29, 30). NFSOM is actually a form of scanned probe microscopy (like STM and AFM discussed below) and is limited by the requirement of scanning the optical probe very near (~ 10 nm) the specimen and detecting rather low light levels.

Higher resolution is possible by using shorter wavelengths, that is, X-ray or electron beams. *SEM* and *TEM* (scanning and transmission electron microscopies) involve energetic (~ 100–300-keV) beams of electrons that are focused through electromagnetic lenses and passed through the specimen (1, 31, 32). The selective absorption of these electrons form patterns that can be recorded on film. These techniques are widely used to examine many different types of surfaces. They are invaluable in obtaining information about surface features and topography, in favorable cases with angstrom resolution (see, e.g., Fig. 2.5.10). In most cases, the sample is contained in an UHV, which may lead to a structure that is different than that for the same sample immersed in a liquid or exposed to air. For high-resolution TEM or STEM studies, sample pretreatment is usually necessary to enhance contrast, and only very thin samples, supported on grids, can be used. Scanning electron microscopes often are equipped with accessories that increase their utility as analytical tools. For example, in the electron microprobe (EMP), X rays and secondary electrons of a given energy that are emitted during an SEM scan can be detected to yield spatial elemental information that can be superimposed over the SEM image. They can also be equipped to monitor other signals induced by the electron beam, such as visible light or electric current flow (EBIC, or electron-beam-induced current). While SEM is basically a UHV technique, there are microscopes equipped with efficient differential pumping arrangements that allow imaging of samples, even wet ones, at reasonable gas pressures.

Scanning probe microscopies (SPM) (33), in which very small tips are scanned across a surface and the response is employed to obtain topographic information about the structure, are capable of very high resolution. *Scanning tunneling microscopy* (STM), already mentioned in Chapter 2, is probably the most widely used SPM method. In this technique, a metal tip (usually made by etching or cutting a tungsten or platinum wire to form a very fine point) is scanned over the surface of a conducting or semiconducting substrate (Fig. 2.5.15). When a potential is applied between tip and substrate and the tip is a few angstroms from the surface, a tunneling current of several nanoamperes will flow. The distance between tip and surface can be controlled by the

z-piezoelectric element, to within fractions of an angstrom, by applying a voltage to the piezoelectric. A plot of the z-piezo voltage needed to maintain a constant tunneling current–scan distance in the X–Y plane (also controlled by piezoelectric drives) is related to the surface topography of the sample and the electron distributions or work functions at the surface. It is also possible to scan in the constant height mode, where the tunneling current is measured as a function of X–Y position. STM studies have been carried out with the sample in UHV, in air or other gases, or immersed in a liquid. While the technique of STM is limited to conductive samples, a related technique involving the measurement of the force between the tip and sample, *atomic force microscopy* (AFM), can also be used with insulating substrates.

While STM and AFM are among the most useful microscopic techniques for probing surface topography, one must be aware of their limitations. Neither provides any chemical or elemental information about the surface (although, in the spectroscopic mode, STM can yield electronic state densities and work functions). Indeed, it is frequently difficult to tell from an STM or AFM image what species is being seen. Moreover, the scanning tip can interact strongly with the surface, especially when attempting high resolution (i.e., close tip–substrate spacing), and thus can perturb the surface under study. Finally, the usual tip shapes do not allow probing of deep narrow pores or sharp features. Other scanning probe microscopies have been developed, such as the *scanning electrochemical microscope* (SECM) discussed briefly in Section 2.5.7, which are more sensitive to surface chemistry and reactivity.

Another, more specialized, tunneling technique that can be applied to solid films with adsorbates that can be subjected to low temperatures is inelastic tunneling spectroscopy. In this method, the tunneling current between two conductive films across a thin insulating solid, such as a film of Al_2O_3, is measured. Variations of the tunneling current with applied potential can provide information about bond vibrations of species adsorbed or bonded to the insulating film.

3.3.6. Other Methods

Sometimes it may be useful to characterize systems and surfaces by the more "classical" methods (16, 34). For example, the total surface area can be obtained from measurements of the extent of adsorption of gases, dyes, or radioactive tracers. The surface properties can also be probed by surface tension (e.g., through contact angle measurements) or calorimetric measurements. Often, the electrical or magnetic properties of the system will be of importance. In general, the full repertoire of physical and chemical techniques that have been widely used in the study of homogeneous phases and interfaces can also be applied to the characterization of ICSs.

REFERENCES

1. M. Grasserbauer and H. W. Werner, eds., *Analysis of Microelectronic Materials and Devices*, Wiley, New York, 1991.
2. J. H. Block, A. M. Bradshaw, P. C. Gravelle, J. Haber, R. S. Hansen, M. W. Roberts, N. Shepperd, and K. Tamaru, *Pure Appl. Chem.*, **62**, 2297 (1990).
3. W. N. Delgass, G. L. Haller, R. Kellerman, and J. H. Lunsford, *Spectroscopy in Heterogeneous Catalysis*, Academic Press, New York, 1979.
4. A. J. Bard and L. R. Faulkner, *Electrochemical Methods*, Wiley, New York, 1980.
5. P. T. Kissinger and W. R. Heineman, eds., *Laboratory Techniques in Electroanalytical Chemistry*, Marcel Dekker, New York, 1984.
6. (a) T. A. Carlson, *Photoelectron and Auger Spectroscopy*, Plenum, New York, 1975; (b) C. R. Brundle and A. D. Baker, eds., *Electron Spectroscopy; Theory, Techniques and Applications*, Academic Press, New York, 1977; (c) D. Briggs and M. P. Seah, eds., *Practical Surface Analysis by Auger and X-Ray Photoelectron Spectroscopy*, Wiley, New York, 1990.
7. R. W. Murray, *Acc. Chem. Res.*, **13**, 135 (1980).
8. H. D. Abruña, T. J. Meyer, and R. W. Murray, *Inorg. Chem.*, **11**, 3233 (1979).
9. F-R. F. Fan, R. G. Keil, and A. J. Bard, *J. Am. Chem. Soc.*, **105**, 220 (1983).
10. J. A. Bruce, T. Murahashi, and M. S. Wrighton, *J. Phys. Chem.*, **86**, 1552 (1982).
11. (a) D. C. Bookbinder and M. S. Wrighton, *J. Electrochem. Soc.*, **130**, 1080 (1983); (b) J. R. White and A. J. Bard, *J. Electroanal. Chem.*, **197**, 233 (1986).
12. (a) J. D. E. McIntyre, in *Advances in Electrochemistry and Electrochemical Engineering*, R. H. Muller, ed., Wiley-Interscience, New York, 1973, pp. 61–166; (b) R. E. Hummel, in *Analysis of Microelectronic Materials and Devices*, M. Grasserbauer and H. W. Werner, eds., Wiley, New York, 1991, pp. 719–732.
13. G. H. Brilmyer and A. J. Bard, *Anal. Chem.*, **52**, 685 (1980).
14. A. Rosencwaig, *Photoacoustics and Photoacoustic Spectroscopy*, Wiley, New York, 1980.
15. C. F. Quate, A. Atalar, and H. K. Wickramasinghe, *Proc. IEEE*, **67**, 1092 (1979).
16. A. Ulman, *Ultrathin Organic Films*, Academic Press, New York, 1991, pp. 6–17.
17. M. D. Porter, T. B. Bright, D. Allara, and C. E. D. Chidsey, *J. Am. Chem. Soc.*, **110**, 6136 (1988).
18. W. L. Wallace, C. D. Jaeger, and A. J. Bard, *J. Am. Chem. Soc.*, **101**, 4840 (1979).
19. P. K. Ghosh and A. J. Bard, *J. Phys. Chem.*, **88**, 5519 (1984).
20. G. M. Edelman and W. O. McClure, *Acc. Chem. Res.*, **1**, 65 (1968).

21. N. J. Turro and T. Okubo, *J. Am. Chem. Soc.*, **104,** 2985 (1982).
22. N. E. Preito and C. R. Martin, *J. Electrochem. Soc.*, **131,** 751 (1984).
23. (a) J-C. Chambron and J-P. Sauvage, *Chem. Phys. Lett.*, **182,** 603 (1991); (b) A. E. Friedman, J-C. Chambron, J-P. Sauvage, N. J. Turro, and J. K. Barton, *J. Am. Chem. Soc.*, **112,** 4960 (1990).
24. J. G. Gaudiello, P. K. Ghosh, and A. J. Bard, *J. Am. Chem. Soc.*, **107,** 3027 (1985).
25. D. Ege, P. K. Ghosh, J. R. White, J-F. Equey, and A. J. Bard, *J. Am. Chem. Soc.*, **107,** 5644 (1985).
26. R. M. A. Azzam and N. M. Bashara, *Ellipsometry and Polarized Light*, North-Holland, Amsterdam, 1977.
27. C. M. Carlin, L. J. Kepley, and A. J. Bard, *J. Electrochem. Soc.*, **132,** 353 (1985).
28. T. Wilson and C. J. R. Sheppard, *Theory and Practice of Scanning Optical Microscopy*, Academic Press, London, 1984.
29. E. Betzig, J. K. Trautman, T. D. Harris, J. S. Weiner, and R. L. Kostelak, *Science*, **251,** 1468 (1991).
30. A. Lewis and K. Lieberman, *Nature*, **354,** 214 (1991).
31. J. I. Goldstein, D. E. Newbury, P. Echlin, D. C. Joy, C. E. Fiori, and E. Lifshin, *Scanning Electron Microscopy and X-Ray Microanalysis*, Plenum, New York, 1981.
32. D. C. Joy, A. D. Romig, and J. I. Goldstein, eds., *Principles of Analytical Electron Microscopy*, Plenum, New York, 1986.
33. (a) H. K. Wickramasinghe, ed., *Scanned Probe Microscopy*, AIP Conference Proceedings No. 241, American Institute of Physics, New York, 1992; (b) D. A. Bonnell, ed., *Scanning Tunneling Microscopy and Spectroscopy*, VCH Publishers, New York, 1993.
34. A. W. Adamson, *Physical Chemistry of Surfaces*, Wiley-Interscience, New York, 1990.

Chapter 4
CHEMICALLY MODIFIED ELECTRODES

> This lovely assembly is not a technological object. But it is an artistic one for a scientific clientele. And it shows beautifully the level of control of both structure and function that can be imposed on an electrode.
> —Larry Faulkner
> *Chem. Eng. News*,
> Feb. 27, 1984, p. 28

4.1. INTRODUCTION

We have already been introduced to chemically modified electrodes (CMEs) by several examples in Chapter 1. Such CMEs represent some of the more highly developed and investigated ICSs, and their construction and characterization probably typify other systems. A number of reviews have appeared in the area of CMEs (1), and this chapter does not pretend to cover this field exhaustively. Rather, the basic principles involved in the construction and application of CMEs will be discussed along with a number of examples. Novel structures that can be considered ICSs will be stressed.

CMEs result from the purposeful modification of a conductive substrate to produce an electrode suited to a particular function and whose properties are different from those of the unmodified substrate. The strong, and sometimes irreversible, adsorption of a species to an electrode surface has long been known to modify the electrochemical behavior of the electrode. For example, the adsorption of CN^- on Pt increases the potential at which hydrogen is evolved (i.e., increases the hydrogen overpotential) and thus extends the range of the electrode to more negative potentials in electroanalytical applications. Conversely, adsorption of alkaloids and proteins on Hg decreases the hydrogen overpotential; polarographic waves produced in this way with solutions containing cobalt ions and small amounts of proteins or other sulfhydryl-containing species (Brdicka waves) were investigated 50 years ago (2). Purposeful covering of electrode surfaces with adsorbed layers or films have also been used to change the electron-transfer rates at the electrode surface. A Pt electrode immersed in an acidic solution containing Sn(IV), when subjected to potentials where hydrogen is evolved, becomes coated with a layer of hydrous tin oxide because of the pH increase at the electrode surface. This layer increases the hydrogen overpotential and allows a Pt electrode to be used for the coulometric generation of Sn(II) from Sn(IV); this process is not possible at a bare Pt electrode because of concomitant hydrogen evolution (3).

In the 1970s, interest arose in the modification of electrode surfaces by covalent attachment of monolayers of different species to electrode surfaces and the characterization of such electrodes by electrochemical and other means. In 1978, electrodes modified with thicker polymeric films were introduced and this has become a very active area of research. Paralleling this work was activity in the field of electronically conductive polymers and organic metals, many of which could be produced electrochemically. Electrode surfaces modified with inorganic materials with well-defined structures (e.g., clays, zeolites, Prussian blue) have also been studied. Recent work with many of these materials has stressed the development of more complex structures (bilayers, arrays, biconductive films) that can be classified as ICSs.

Interest in CME is based mainly on possible applications. Electrocatalysis has been of prime interest. For example, an electrode of an inexpensive and rugged material that could reduce oxygen to water at a reasonable rate at a potential near the thermodynamic one would find wide use in fuel cells, batteries, and other electrochemical systems. Modified electrodes could also be employed in displays. Materials that change color on oxidation and reduction form the basis of electrochromic displays. There are also examples of surface films that emit light on electrochemical excitation (electrogenerated chemiluminescence). These have the potential for use as active displays. Modifying layers can also protect the underlying substrate from corrosion or attack during use. Layers of different materials on semiconductor electrodes have been suggested for this application and are discussed in Chapter 6. CMEs can also serve as analytical sensors and reference electrodes; several examples have already been discussed in Chapter 1. Finally, there is growing interest in molecular electronic devices: electrochemical systems that can mimic the behavior of diodes, transistors, and electrical networks. In addition to these applications, however, CMEs have proved useful in the characterization of electron- and mass-transfer processes in polymers and other materials (as discussed in Chapter 5) and in gaining insight into how surface structures can be designed to carry out specified reactions or processes.

4.2. SUBSTRATES

All CMEs start with a particular electronically conductive substrate material that is then modified: carbon, a metal, a semiconductor, a conductive polymer, or an organic metal. There are several desirable characteristics for a good substrate material. The material should show high electrical conductivity. This is seldom a problem with bulk metals or carbon, but is a consideration with semiconductors or with thin films of metal on a nonconductive substrate. The nature of the contact to the substrate may also be important, especially with semiconductors, since the junction may show significant contact resistance and rectifying properties. The substrate should show good resistance to corrosion and chemical attack by the contacting solution. The high stability of Pt, Au, and C accounts for their widespread use as electrode materials. Stability becomes an important consideration for electrodes used as anodes at very positive potentials and under high temperature and high acidity conditions. Electrodes should also be mechanically stable. Mechanical and chemical stability also implies that the surface remains unchanged with time during use as an electrode. Surfaces can change because of slow chemical processes (e.g., surface oxidation) or by adsorption of small amounts of impurities from the contacting solution. Since amounts of only $\sim 10^{-10}$ mol/cm^2 of electrode surface are needed to produce a monolayer, even low concentrations of a strongly adsorbing impurity can eventually form a monolayer on an electrode surface and change its properties. Surfaces can also reconstruct with time, even under UHV conditions. This may be especially important with single crystals. For practical applications, it is usually desirable to have a high real surface area to maximize the zone of interfacial contact with the solution. For fundamental studies, however, such as investigations by STM, atomic smoothness over large areas is desired (but usually hard to attain). Finally, if attachment of components to the substrate surface is to be carried out via covalent linkages, suitable surface chemistry of the substrate should be available for the desired linking reactions.

4.2.1. Metals

Pt and Au are frequently used for modification because they are inert and chemically stable. Before modification, the metals are frequently polished, such as with fine diamond or alumina powder using conventional metallographic polishing methods, followed by cleaning with nitric acid and washing with water, often with sonication to remove adhering particles. The actual efficacy of such procedures has not been studied in a systematic manner, but removal of surface impurities by some method is usually needed. Polishing does leave scratches and grooves in the metal surface of about the size of the polishing material (e.g., for fine particles, ~ 0.05 μm). An electrochemical treatment that is frequently used with Pt involves repeated cycling (e.g., in 1 M H_2SO_4) between potentials where H_2 and O_2 are evolved. This presumably causes the oxidation of organic impurities on the surface and also tends to "break up" the electrode surface and produce a rougher one. A number of other types of pretreatments, using, for example, selective adsorption, have been proposed to yield smooth and clean surfaces (4). The cyclic voltammogram of the substrate metal is often a good indicator of its surface condition and its suitability for use as an electrode or for modification. An example of the cyclic voltammetric signature of a pretreated polycrystalline Pt electrode is given in Figure 4.2.1. There is growing interest in the use of single-crystal metal electrodes, although these have not yet been employed extensively for surface modification. Smooth metal layers can also be prepared by vacuum evaporation or sputtering on substrates like glass or mica.

Less expensive metals are often used in practical, large scale applications. In some cases, these are treated with a suitable catalytic material to improve their properties for a particular process. For example, the dimensionally stable anode (DSA) is made by coating a Ti electrode with a RuO_2 and TiO_2 mixture along with other proprietary materials and is used for chlorine evolution in the chloralkali process. Stainless-steel cathodes are sometimes coated with Raney nickel as a catalyst to decrease the overpotential for hydrogen evolution. Sometimes a modified system is produced by coating the metal on a polymer, such as by chemical reduction (electroless plating) or by vacuum evaporation. Such structures are discussed below.

4.2.2. Carbon

A number of different forms of carbon have been used as electrode materials: single-crystal graphite, pyrolytic graphite, powdered graphite, carbon black, and glassy or vitreous carbon (5). Graphite has a layered structure (Fig. 4.2.2) consisting of hexagonal carbon units arranged in sheets. Its properties are very anisotropic; for example, the electrical conductivity is much higher along the sheets (parallel to the basal plane)

4.2. SUBSTRATES

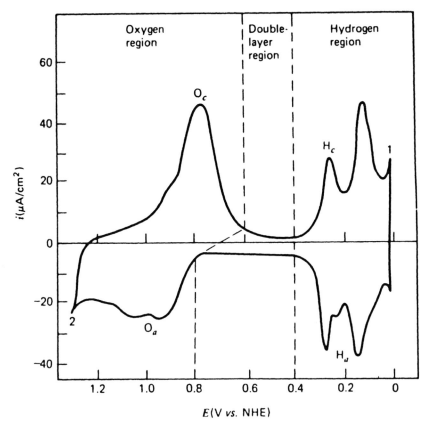

FIGURE 4.2.1. Cyclic voltammogram for a smooth polycrystalline Pt electrode in 0.5 M H_2SO_4, demonstrating a clean Pt surface. H_c and H_a show formation and removal of adsorbed hydrogen; O_a and O_c the formation and reduction of an oxide layer. (1) start of bulk hydrogen evolution; (2) start of bulk oxygen evolution. The exact shape of the curve depends on the crystal faces of Pt exposed, pretreatment of the electrode, and supporting electrolyte. The voltammogram for a single-crystal Pt electrode is different (see Ref. 4). [Reprinted with permission from A. J. Bard and L. R. Faulkner, *Electrochemical Methods*, Wiley, New York, 1980. Copyright © 1980 John Wiley & Sons.]

than perpendicular to them. Either the basal plane or the edge plane can be used as the electrode surface. Graphitic materials are also capable of intercalating various species between the carbon sheets. Intercalation is a reversible topotactic reaction where the intercalated species is introduced in various stages, as suggested by the diagram in Figure 4.2.2. During the intercalation process, the distance between the carbon sheets increases, and hence, one intersheet layer tends to be filled before the other layers. Highly ordered (or stress-annealed) pyrolytic graphite

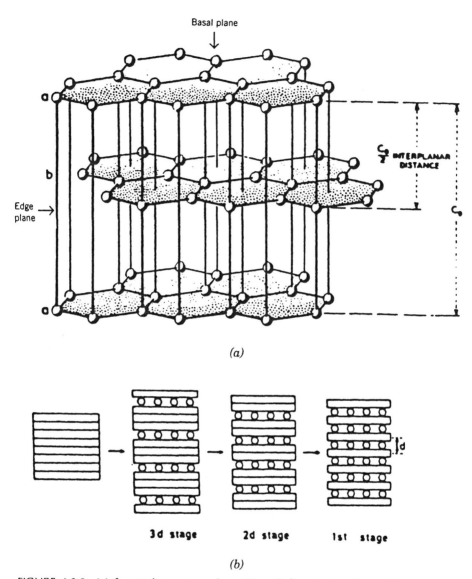

FIGURE 4.2.2. (a) Layered structure of graphite; $C_0/2$, the interplanar distance between a and b layers is 3.35 Å. (b) Intercalation of a species (○) into graphite, showing different stages (3rd, 2nd, 1st).

(HOPG) is a polycrystalline material, but it shows a high degree of alignment of the sheets of the crystallites (and hence is used to construct X-ray monochromators). Fresh surfaces of this material can be obtained by peeling off the top layers with adhesive tape; such surfaces can be very smooth, as shown by topographic scans with the scanning tunneling

4.2. SUBSTRATES

microscope (Fig. 4.2.3). However, the surface also contains steps, defects, and other types of features (6). Ordinary pyrolytic graphite is less ordered and is also available in the form of thin (~7-μm) fibers that can be used to make ultramicroelectrodes. Powdered graphite of various types can be used as porous beds for flow-through electrolysis or can be mixed with mineral oil in the construction of carbon paste electrodes. Carbon black is a finely divided, high-surface-area form of carbon that is sometimes mixed with insulating polymers to form a conductive matrix. Carbon black, mixed with Teflon and Pt, is used as a gas-diffusion electrode in some types of fuel cell. Glassy or vitreous carbon is an impermeable, amorphous material that is used in analytical applications, but is less defined than the more oriented graphitic materials for surface modification.

Before modification or use as electrodes, carbon is usually subjected to a pretreatment (5). This is most frequently an oxidative one involving

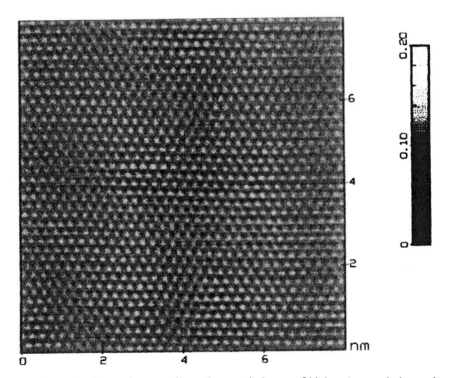

FIGURE 4.2.3. A scanning tunneling microscopic image of high-order pyrolytic graphite at atomic resolution. Note that the image shows the surface electronic distribution and that only every other C atom on the surface is imaged because of interactions with the underlying plane of C atoms (compare with Fig. 4.2.2a).

electrochemical oxidation or cycling, heating in an oxygen atmosphere, or exposure to an oxygen-containing plasma or laser beam. Such treatments probably clean the electrode surface and introduce oxygen functionalities for use in later modification steps. The activity of an electrode for electron transfer after such a treatment is frequently greatly enhanced compared to the electrode before such treatment. For example, the fresh surface of HOPG shows poor activity as an electrode for the oxidation of ferrocyanide or catechol. However, following oxidative treatment of the surface, much more facile electron-transfer kinetics for these reactions are observed. The treatment clearly produces a surface film on the graphite (as monitored, e.g., by ellipsometry) and often shows different colors during the oxidation stages because of interference effects.

4.2.3. Semiconductors

Single-crystal semiconductor materials are mainly used in connection with photoelectrochemical studies and are discussed in Chapter 6. The semiconductors most widely used in CMEs (in a highly doped and rather conductive form) are SnO_2 and In_2O_3 [or mixtures of these, called *indium tin oxide* (ITO)] as transparent thin films on glass substrates. These are usually produced by spray deposition or by sputtering. Semiconductor films can also be formed anodically on a metal substrate (e.g., WO_3 and IrO_x) or by chemical vapor deposition.

4.2.4. Conductive Polymers and Organic Metals

Electronically conductive polymers, such as polypyrrole and polyaniline (see Fig. 2.2.4), can also serve as supports, at least over a potential range where their conductivity is maintained. Organic metals, which are usually organic compounds composed of donor and acceptor species with a segregated stack structure, can also be used. A typical material of this type is TTF-TCNQ; others are listed in Table 4.2.1. These supports have the advantage of having some characteristics of a mediator couple, in addition to being a simple inert conductor, and have found use as substrates for enzyme electrodes.

4.3. TYPES OF MODIFIERS—MONOLAYERS

Modifying monolayers in CMEs can be composed of different substances, such as organic surfactants, adsorbed species, or biological materials. Representative examples of these and methods of surface

4.3. TYPES OF MODIFIERS—MONOLAYERS

TABLE 4.2.1. Some Conductive Organic Charge-Transfer Salts (Organic Metals)

Compound	Conductivity at Room Temperature $(\Omega\text{-cm})^{-1}$	Maximum Conductivity $(\Omega\text{-cm})^{-1}$
TTF-TCNQ	500	2×10^4
TMTTF-TCNQ	350	5×10^3
TMTSF-TCNQ	1200	7×10^3
TMTSF-DMTCNQ	500	5×10^3
$(TTT)_2I_3$	1000	3×10^3
$(BEDT-TTF)_2I_3$	30	1.5×10^4

modification follow. Monolayers can be formed on electrode surfaces by irreversible adsorption, covalent attachment, or by Langmuir–Blodgett (LB) and self-assembly techniques (to produce organized assemblies).

4.3.1. Adsorption

Many substances spontaneously adsorb onto a substrate surface from solution, generally because the substrate environment is energetically more favorable for that species than that in solution (Fig. 4.3.1). For example, sulfur-containing species are strongly held on mercury, gold, and other metal surfaces, because of strong metal–sulfur interactions. Thus when a mercury electrode is placed in contact with a solution containing only small (less than micromolar) amounts of cystine or a sulfur-containing protein, such as bovine serum albumin, a monolayer forms on the Hg surface. Electrochemical oxidation and reduction of the surface species can then be observed. Strong adsorption from aqueous solutions of some ions (e.g., halides, SCN^-, CN^-), and many organic compounds, especially those containing aromatic rings, double bonds, and long hydrocarbon chains, on metal or carbon surfaces also takes place. A typical example (Fig. 4.3.1B) is that of 9,10-phenanthraquinone (PAQ), which forms a monolayer on the basal plane of pyrolytic graphite simply on immersion into a solution of the PAQ in 1 M $HClO_4$ (7). The electrochemical response shows that 37 μC of charge/cm^2 of electrode surface is passed during the reduction and reoxidation steps; this is equivalent to 1.9×10^{-10} mol PAQ/cm^2 or 1.1×10^{14} molecules/cm^2 (for a two-electron redox process). Adsorption of metal ions that normally would not adsorb can take place by anion-induced adsorption (Fig. 4.3.1C). In this case, the strong adsorption of an anion

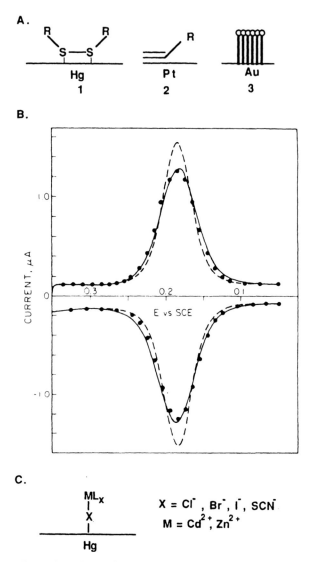

FIGURE 4.3.1. Examples of specific adsorption on electrodes. (A) Adsorption of (1) a disulfide or protein on Hg; (2) an olefin on Pt; (3) an organized LB film on Au. (B) Experimental (solid line) and theoretical (●) cyclic voltammograms for adsorbed 9,10-phenanthrenequinone on the basal plane of pyrolytic graphite at 50 mV/s in 1 M HClO$_4$ (dashed line) theoretical ideal (nernstian thin layer) behavior. (C) Adsorption of metal ion or complex through an anionic ligand bridge. [Reprinted with permission from A. P. Brown and F. C. Anson, *Anal. Chem.*, **49**, 1589 (1977). Copyright 1977 American Chemical Society.]

4.3. TYPES OF MODIFIERS—MONOLAYERS

(e.g., SCN^-) that can act as a ligand for a particular metal, M, will induce its adsorption on the substrate (e.g., Hg).

4.3.2. Covalent Attachment

Stronger attachment to the substrate surface can be accomplished by covalent linking of the desired component to surface groups present on, or formed on, the substrate. These covalent linking procedures frequently employ organosilanes and other linking agents (Fig. 2.2.7) and are discussed in detail in several references (1). The substrate surface is usually pretreated, such as by an oxidative reaction, to form surface groups (Fig. 4.3.2). The surface is then treated with the linking agent and the desired component. Favorite components linked to electrode surfaces in this way, because they show easily detected electrochemical reactions, are the ferrocenes, viologens, and $M(bpy)_x^{n+}$ species (M = Ru, Os, Fe). Typical covalent attachments and the electrochemical responses of the layers are shown in Figure 4.3.3. A table listing many examples of such electrodes is given in Ref. 1a.

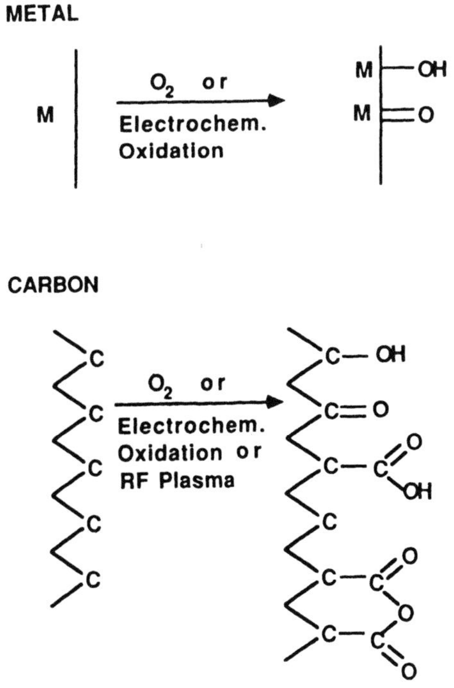

FIGURE 4.3.2. Schematic representation of formation of functional groups on a metal or carbon surface via oxidation before treatment with linking agents.

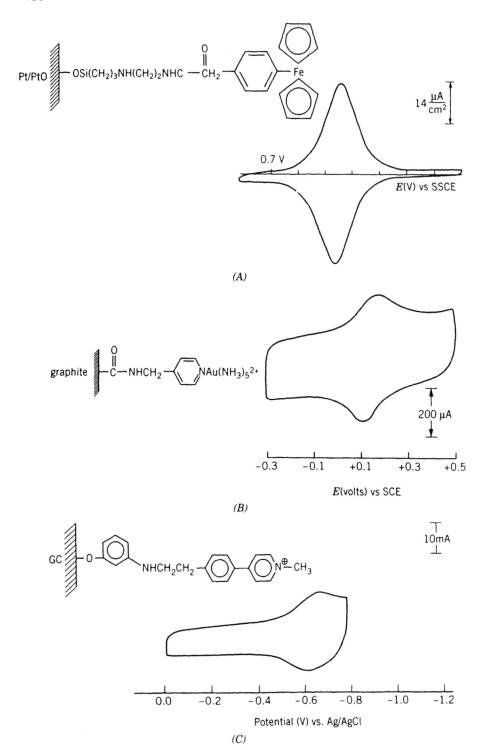

4.3.3. Organized Assemblies

Monolayers of surface-active compounds (LB films) can be transferred from the liquid-air interface to a substrate surface, as described in Section 2.4. For example, a monolayer of Os(bpy)2(4,4'-dinonadecyl-2,2'-bpy), 2.0×10^{-10} mol/cm^2, can be formed on a tin oxide electrode surface (8). The cyclic voltammetric response, representing electrochemistry of the Os(III/II) couple, is shown in Figure 4.3.4. The Os(III) form readily oxidizes Fe^{2+} to Fe^{3+} at potentials where Fe^{2+} oxidation does not occur at an unmodified SnO_2 electrode. Under these conditions, the monolayer can be considered an electrocatalyst for iron(II) oxidation. Another example of an organized monolayer, a viologen head group with a long-chain hydrocarbon tail, is shown in Figure 4.3.5 (9). As before, the monolayer was transferred to a SnO_2 surface with a film balance; the pressure-area curve is shown. Cyclic voltammetric scans were used to estimate the coverage. Note here that the area/molecule found from the film balance surface pressure curve corresponds quite closely to the number of coulombs found from the integrated area of the electrochemical response. An alternative to LB films are self-assembled monolayer films that can be formed without a film balance. These frequently involve layers of organosulfur (e.g., thiol) compounds with long-chain alkyl groups on Au or alkyl siloxane monolayers formed by treating surfaces containing hydroxyl groups with long-chain alkyl trichlorosilanes (10).

4.4. TYPES OF MODIFIERS—POLYMERS

4.4.1. Types

Several different types of polymers have been used to modify electrode surfaces (Table 4.4.1). *Electroactive polymers* contain oxidizable or reducible groups within the polymer backbone. Typical examples are

---

FIGURE 4.3.3. Cyclic voltammograms of electrodes modified by covalent attachment of monolayers of different types. (*A*) Pt electrode with attached ferrocene, 200 mV/s. [Reprinted with permission from J. R. Lenhard and R. W. Murray, *J. Am. Chem. Soc.*, **100**, 7870 (1978). Copyright 1978 American Chemical Society.]. (*B*) Graphite with attached py-Ru(NH$_3$)$_5$, 5 V/s. [Reprinted with permission from C. A. Koval and F. C. Anson, *Anal. Chem.*, **50**, 223 (1978). Copyright 1978 American Chemical Society.] (*C*) Glassy carbon with attached viologen, 100 mV/s. [Reprinted with permission from D. C. S. Tse, T. Kuwana, and G. P. Royer, *J. Electroanal. Chem.*, **98**, 345 (1979). Copyright 1979 Elsevier.]

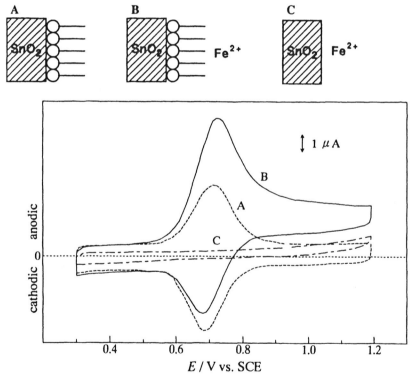

FIGURE 4.3.4. Cyclic voltammograms of tin oxide electrode modified by an LB monolayer of Os(bpy)$_2$(4,4'-dinonadecyl-2,2'-bpy) and its effect on the oxidation of Fe^{2+}: (A) 2.0 × 10^{-10} mol/cm^2 monolayer; (B) as (A), in the presence of 0.1 mM Fe^{2+}; (C) bare tin oxide electrode in same solution as (B). All solutions were 0.18 M H$_2$SO$_4$ at 111 mV/s. Results show that oxidation of Fe^{2+} is very slow at an unmodified tin oxide electrode (C), but the oxidation is catalyzed by the monolayer [i.e., via reaction of Os(III) with Fe(II)]. [Reprinted with permission from H. Daifuku, I. Yoshimura, I. Hirata, K. Aoki, K. Tokuda, and H. Matsuda, *J. Electroanal. Chem.*, **199,** 47 (1986). Copyright 1986 Elsevier.]

4.4. TYPES OF MODIFIERS—POLYMERS

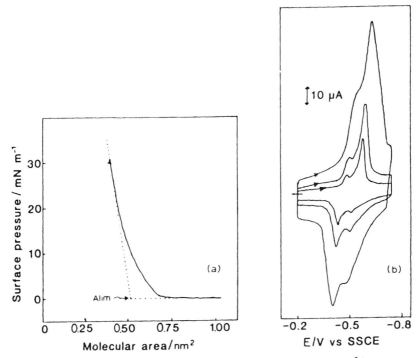

FIGURE 4.3.5. (a) Surface pressure-molecular area isotherm for $C_{16}MV^{2+}$ at air-water interface. The limiting area (A_{lim}) of 50 Å^2/molecule corresponds to 4.2×10^{-10} mol/cm^2 of $C_{16}MV^{2+}$. The film was transferred to an indium tin oxide electrode (0.52 cm^2) at a pressure of 30 mN/m. (b) Cyclic voltammogram of LB film in 0.2 M NaClO$_4$ solution at 100, 200, and 500 mV/s. The integrated areas correspond to 4.0×10^{-10} mol/cm^2, in agreement with the area found from the isotherm in (a). [Reprinted with permission from C-W. Lee and A. J. Bard, *J. Electroanal. Chem.*, **239**, 441 (1988). Copyright 1988 Elsevier.]

poly(vinylferrocene) (PVF), PQ^{2+}, and polymerized Ru(vbpy)$_3^{2+}$. *Coordinating* (ligand-bearing) *polymers*, such as poly(vinylpyridine), contain groups that can coordinate to species like metal ions and bring them into the polymer matrix. *Ion-exchange polymers* (polyelectrolytes) contain charged sites that can bring in solution species via an ion-exchange process; typical examples are Nafion, polystyrene sulfonate, and protonated poly(vinylpyridine). *Electronically conductive polymers*, such as polypyrrole and polyaniline, can also be considered as ion-exchange materials, since the polymer redox processes are usually accompanied by incorporation of ions into the polymer network. *Biological polymers*, such as enzymes and other proteins, are often useful in sensor applications. *Blocking polymers* are formed from the monomers, as by oxidation

TABLE 4.4.1. Typical Polymers Used for Electrode Modification

Name	Structure	Abbreviation
Electroactive		
Poly(vinylferrocene)	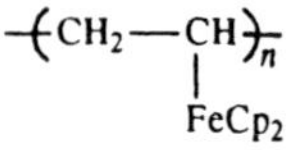 $-(CH_2-CH)_n-$ with FeCp$_2$ substituent	PVF
Poly[Ru(vbpy)$_3^{2+}$]	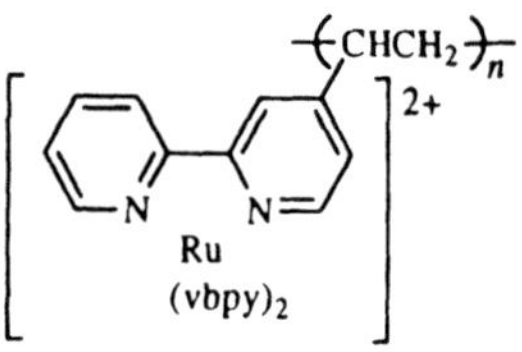 [Ru(vbpy)$_2$(4-vinyl-4'-methyl-2,2'-bipyridine)]$^{2+}$, polymerized through $-(CHCH_2)_n-$	
Poly(xylyl viologen)	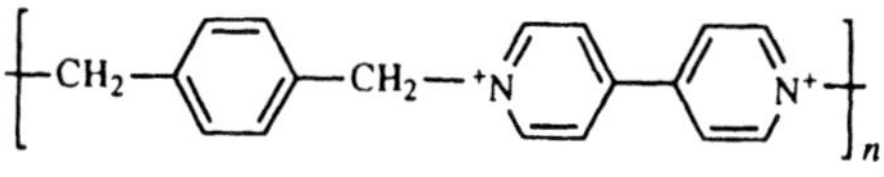 $-[CH_2-C_6H_4-CH_2-{}^+N\text{-py-py-}N^+]_n-$	
Polymerized viologen organosilane	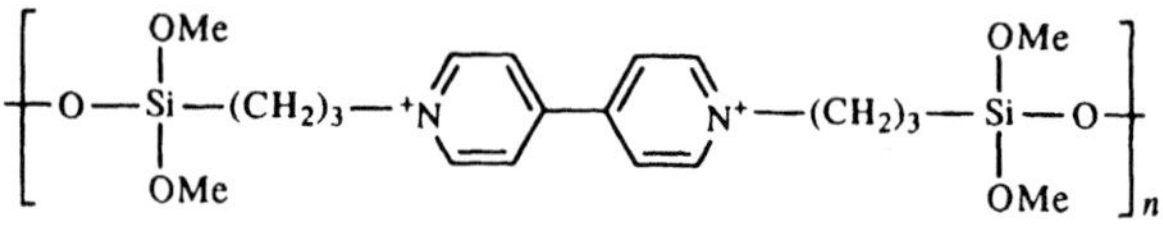 $-[O-Si(OMe)_2-(CH_2)_3-{}^+N\text{-py-py-}N^+-(CH_2)_3-Si(OMe)_2-O]_n-$	PQ^{2+}
Ion Exchange (Polyelectrolyte)		
Nafion	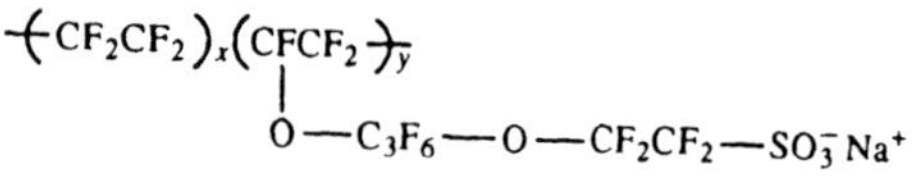 $-(CF_2CF_2)_x(CFCF_2)_y-$; side chain $O-C_3F_6-O-CF_2CF_2-SO_3^-Na^+$	NAF
Poly(styrenesulfonate)	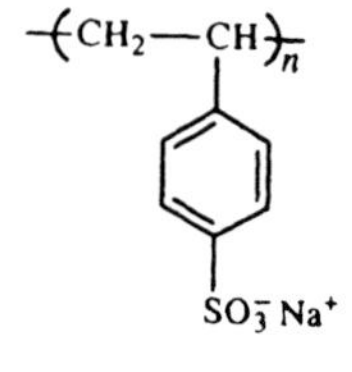 $-(CH_2-CH)_n-$ with C$_6$H$_4$-SO$_3^-$Na$^+$ substituent	PVS

4.4. TYPES OF MODIFIERS—POLYMERS

TABLE 4.4.1. (*Continued*)

Name	Structure	Abbreviation
Ion Exchange (Polyelectrolyte) (Continued)		
Poly(vinylpyridinium)	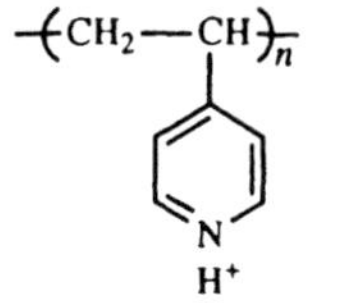 $-(CH_2-CH)_n-$ (4-pyridinium, H^+)	$PVPyH^+$
Coordinating		
Poly(vinylpyridine)	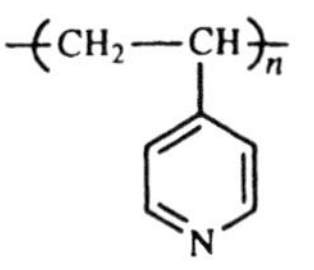 $-(CH_2-CH)_n-$ (4-pyridine)	PVPy
Electronically Conducting[a]		
Polypyrrole	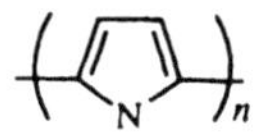 $-(pyrrole)_n-$	PP
Polythiophene	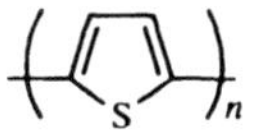 $-(thiophene)_n-$	PT
Polyaniline	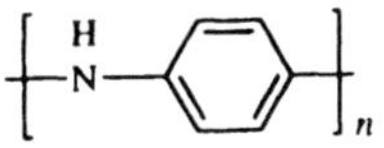 $-[NH-C_6H_4]_n-$	PANI

[a] See Figure 2.2.4.

of phenols, to produce impermeable layers and blocked or passivated surfaces.

4.4.2. Preparation

Polymer films can be formed on an electrode surface from solutions of either the polymer or monomer. Methods that start with dissolved polymer include the following:

Cast or Dip Coating. The polymer is dissolved in a suitable solvent and a few drops of the solution are placed on the substrate surface and allowed to evaporate. This very simple method seldom produces films of uniform thicknesses, since the rate of evaporation of solvent is not the same at different locations on the liquid drop surface.

Spin Coating. More uniform films can be produced by spin coating with the polymer dissolved in a volatile solvent (see Section 2.5.1).

Electrodeposition. A polymer film can be formed by oxidation or reduction of the dissolved polymer from a solvent in which the product is insoluble. For example, when PVF, dissolved in a methylene chloride solution containing $Bu_4N^+ClO_4^-$, is oxidized at an electrode, a film of $PVF^+ClO_4^-$ forms on the electrode surface (Fig. 4.4.1). If the electrode is then transferred to a CH_3CN solution, where neither the PVF nor the $PVF^+ClO_4^-$ form is soluble, both forms will remain on the surface during oxidation and reduction and produce typical surface cyclic voltammetric waves. Electrodeposition can produce uniform films with the amount of polymer

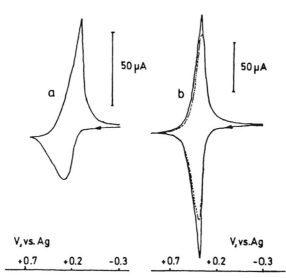

FIGURE 4.4.1. Electrodeposition of a poly(vinylferrocene) (PVF) film on a platinum substrate. (*a*) Oxidation of PVF in a 0.01 mM solution in $CH_2Cl_2/0.1$ M tetra-*n*-butylammonium perchlorate (TBAP) causes precipitation of $PVF^+ClO_4^-$. (*b*) Transfer of the film to $CH_3CN/0.1$ M TBAP and reduction produces a film of PVF. ———, first cycle; ·········, 250th cycle. [Reprinted with permission from A. Merz and A. J. Bard, *J. Am. Chem. Soc.*, **100**, 3222 (1978). Copyright 1978 American Chemical Society.]

4.4. TYPES OF MODIFIERS—POLYMERS

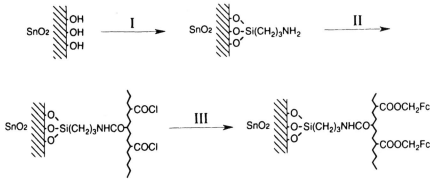

FIGURE 4.4.2. Scheme for bonding polymer to electrode with poly(methacryl chloride) anchors. (I) SnO_2 electrode (or other oxidized surface) is treated with γ-aminopropyltriethoxysilane in dry benzene at 70°C for >24 h. (II) This silanized electrode is treated with a 5% solution of poly(methacryl chloride) in dry THF at 70°C for 12 h. (III) This modified surface is reacted with a 5% solution of hydroxymethylferrocene (Fc) in dry THF at 70°C for 12 h. [Reprinted with permission from K. Itaya and A. J. Bard, *Anal. Chem.*, **50**, 1487 (1978). Copyright 1978 American Chemical Society.]

deposited closely controlled by the amount of charge passed in the deposition step.

Covalent Attachment via Functional Groups. Polymers can be linked covalently to an electrode surface as described above. An example is shown in Figure 4.4.2.

Films can also be produced by methods that start with the monomer:

Thermal Polymerization. A typical example of an organosilane monomer is given in Figure 4.4.3. This approach is simple, but the extent of coverage and nature of the film is difficult to control.

Electrochemical Polymerization. Polymerization of monomers can be induced by reduction or oxidation. For example, compounds containing vinyl groups, such as vinylferrocene and complexes containing the vinylpyridine or vinylbipyridine ligand, will polymerize on reduction. Compounds with phenolic and aromatic amine groups will often form polymers on oxidation. Many of the electronically conducting polymers are prepared from solutions of the monomer (e.g., pyrrole, thiophene, aniline) by electrochemical oxidation or by cycling the substrate potential.

Plasma Polymerization. Polymers will form in an rf plasma containing monomer (see Section 2.5.3). Films of PVF and

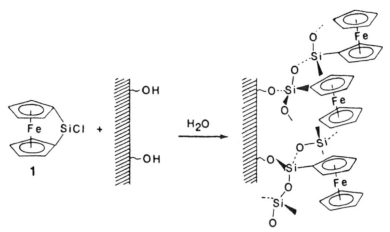

FIGURE 4.4.3. Preparation of an attached surface polymer from a polymerizable ferrocene-based reagent (1) on a Si surface. [Reprinted with permission from M. S. Wrighton, *Science*, **231**, 32 (1986). Copyright 1986 by the AAAS.]

poly(vinylpyridine), from plasmas containing vinylferrocene and vinylpyridine, respectively, have been produced. Plasma polymerization involves more energetic conditions than most of the other polymerization methods described, and this can produce more highly crosslinked films. For example, the PVF produced by plasma polymerization is less permeable to solution ions than that produced by spin coating or electrodeposition starting with PVF.

Photochemical Polymerization. There are a few examples of polymer films produced by photochemical generation of an initiator. It is also possible to induce crosslinking in films by exposure of polymer layers to high-energy radiation. Note that crosslinking can also be induced chemically. For example, a film of electrodeposited PVF can be made more chemically stable and less soluble by treatment with benzoyl peroxide and triallyltrimellitate and heating at 95°C for 2.5 h, which induces crosslinking of the PVF (11).

4.5. TYPES OF MODIFIERS—INORGANIC FILMS

Different types of inorganic films, such as metal oxide, clay, zeolite, and metal ferrocyanide, can also be formed on electrode surfaces. These films are of interest because they frequently show well-defined structures (e.g., they have unique pore or interlayer sizes), are thermally and chemically very stable, and are usually inexpensive and readily available.

4.5.1. Metal Oxides

A film of Al_2O_3 can be produced by anodic growth on an aluminum substrate immersed in a solution of H_3PO_4 (12). The thickness of the film can be controlled by the applied potential and the time of anodization (Fig. 4.5.1). This Al_2O_3 film has a porous structure, as shown, and a thin insulating barrier layer of oxide at the Al/substrate surface. The aluminum oxide film can be removed from the substrate and handled by means of nylon mesh supports. The barrier layer can be dissolved away with dilute acid and a metallic contact/substrate (e.g., Au) deposited on the porous film by vacuum evaporation. The film can then be used to serve as a support for other materials, such as poly(vinylpyridine) (PVP). The cyclic voltammogram shown is for $Fe(CN)_6^{4-}$ introduced into the protonated PVP film by ion exchange. Oxide films of other metals, such as Ti, W, and Ta, can be produced in a similar way. Oxide films can also be produced by CVD, vacuum evaporation and sputtering (See section 2.5), and anodic oxidation and deposition from colloidal solution.

Related inorganic films are those of polyoxometallates (iso- and heteropolyacids) (13). For example, the heteropolyanion $P_2W_{17}MoO_{62}K_6$ shows a number of reduction waves at a GC (glassy carbon) electrode. A film of $P_2W_{17}MoO_{62}K_6$ formed on the GC electrode surface during reduction of this compound shows electrocatalytic activity for hydrogen evolution. Since a wide variety of metallic polyanionic species (e.g., of W, Mo, V) with a rich chemistry exists, such films are promising modifiers.

4.5.2. Clays and Zeolites

Clays and zeolites, both naturally occurring and synthetic, are aluminosilicates with well-defined structures that usually show ion-exchange properties (14) (see also Section 2.2.1). In addition to their high stability and low cost, they often show catalytic properties and have been widely used as heterogeneous catalysts. They also can be used as templates for synthesis (see Section 2.4.2). The structure of a typical clay, montmorillonite, is shown in Figure 2.2.6. The basic building blocks of the individual clay sheets are Si tetrahedra and Al octahedra that are interconnected as shown, through shared O atoms. When atoms of a lower charge (e.g., Mg and Li) are substituted for the Al atoms, the individual

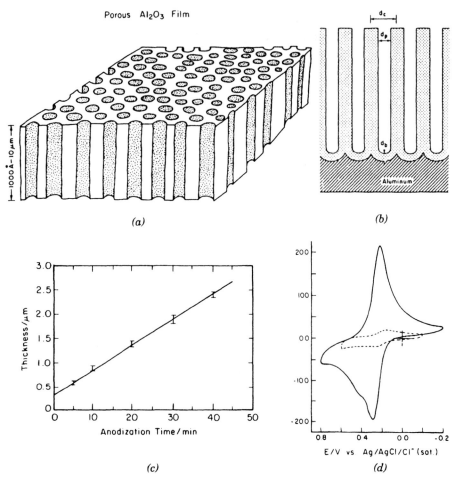

FIGURE 4.5.1. Structure of porous aluminum oxide film prepared by anodic oxidation of Al in phosphoric acid: (a) Pictorial representation; (b) schematic cross section showing pore diameter, d_p, and barrier layer thickness, d_b; (c) dependence of film thickness on anodization time in 4% H_3PO_4 at 23°C and 65 V; (d) cyclic voltammogram of $Fe(CN)_6^{3-/4-}$ at Au modified by Al_2O_3 film with poly(vinylpyridinium) layer (solid line), compared to behavior for a solution of $Fe(CN)_6^{4-}$ (dashed line). [Reprinted with permission from C. Miller and M. Majda, *J. Electroanal. Chem.*, **207**, 49 (1986). Copyright 1986 by Elsevier.]

layer sites take on a net negative charge that is compensated by cations that are held outside the sheets. Montmorillonite is an example of a 2:1 layered smectite, since each sheet contains 2 Si—O tetrahedral layers and 1 Al—O octahedral layer, with about 15% of the Al(III) substituted by Mg(II). Each particle of clay, of a size of the order of micrometers, contains several sheets. The structure of the clay particles can be deter-

4.5. TYPES OF MODIFIERS—INORGANIC FILMS

mined by X-ray diffraction. Dry montmorillonite with exchanged Na^+ has a basal spacing (the distance between the same point on repeating 2:1 units) of 13.6 Å. On heating to 400°C for several hours, interlamellar water is removed and the films collapse to the thickness of the silicate sheet, 9.8 Å. Thus, the interlayer separation for the dry, but unbaked, clay is about 3.8 Å. On soaking in water, the basal plane spacing increases to 14.7 Å, and treatment with a polymer that adsorbs on the clay, such as poly(vinylalcohol) (PVA), increases the spacing further, to ~20.5 Å. Clay layers can be "pillared" by treatment with appropriate inorganic or organic agents, such as polyoxyanions of Fe, Al, or Zr, to form structures between the silicate layers that maintain the interlayer spacing at a given value (e.g., ~17 Å). Films of clay can be cast on substrate surfaces and will remain intact when they are used as electrodes. The conditions of casting and the presence of additives, such as PVA in the film forming mixture, affects the structure of the film that is produced (see Fig. 3.3.12).

Electroactive cations [e.g., $Ru(bpy)_3^{2+}$ or MV^{2+}] can be exchanged into the clay film and show typical cyclic voltammetric responses of surface-confined species (Fig. 4.5.2) (15, 16). As discussed in connection with Figure 3.3.8, clay particles show interesting structural effects that can produce segregation of ions. Stereoselective effects have also been reported for clay-modified electrodes (17). The species $Ru(phen)_3^{2+}$ exists as two different optical isomers, designated Λ and Δ (Fig. 4.5.3a), both of which can serve as exchangeable ions in a montmorillonite film. When a film is immersed in a racemic mixture of the ions, the total amount that is incorporated into the film is much greater than that incorporated from a solution of an equal concentration containing only one isomer. Presumably the packing of the different isomers into the interlamellar space is better than that for one isomeric species, as suggested in Figure 4.5.3b. Moreover, the clay layer containing one intercalated isomer shows stereoselectivity in electrochemical reactions carried out at the clay-modified electrode:

$$rac\text{-}Co(phen)_3^{2+} - e \xrightarrow[Ru(phen)_3^{2+}]{\text{montmorillonite}} \Lambda\text{-}Co(phen)_3^{3+} \quad (4.5.1)$$
(7% optical purity)

$$C_6H_{11}\text{-}S\text{-}Ph - 2e \xrightarrow[Ru(phen)_3^{2+}]{\text{montmorillonite}} (+)\Delta\text{-}C_6H_{11}\text{-}SO\text{-}Ph \quad (4.5.2)$$
(15% optical yield)

Zeolites are aluminosilicates with structures consisting of well-defined cages and pores (Fig. 4.5.4) (18). These also show ion exchange prop-

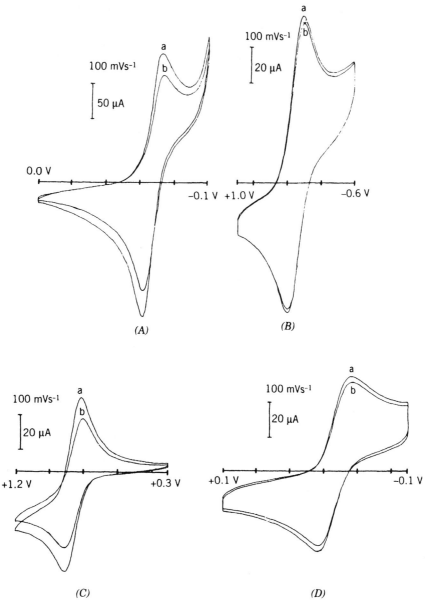

FIGURE 4.5.2. Cyclic voltammograms for several electroactive compounds incorporated into montmorillonite/polyvinylalcohol/Pt films on SnO_2 electrode. (A) Methyl viologen: (a) 1st, (b) 100th scan. (B) $Ru(NH_3)_6^{3+}$: (a) 1st, (b) 105th scan; (C) $Fe(bpy)_3^{2+}$: (a) 1st, (b) 40th scan. (D) Trimethylammonium ferrocene: (a) 1st, (b) 40th scan. Recorded for electrode in 0.1 M Na_2SO_4, pH 7 at scan rate of 100 mV/s. [Reprinted with permission from P. K. Ghosh and A. J. Bard, *J. Am. Chem. Soc.*, **105**, 5691 (1983). Copyright 1983 American Chemical Society.]

4.5. TYPES OF MODIFIERS—INORGANIC FILMS

Λ - $M(1,10\text{-phen})_3^{2+}$ (M=Fe, Ru)

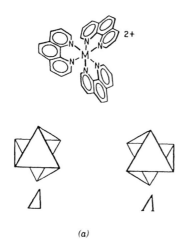

(a)

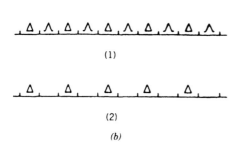

(1)

(2)

(b)

FIGURE 4.5.3. (a) Structure and schematic absolute configurations of Δ and Λ forms of tris-1,10-phenanthroline complex. (b) Suggested difference in packing of racemic (1) and enantiomeric (2) chelates on the surface of a clay particle. [Reprinted with permission from A. Yamagishi and A. Aramata, *J. Chem. Soc. Chem. Commun.*, **1984**, 452 (1984). Copyright 1984 The Royal Society of Chemistry.]

erties and can be employed as modifying layers on electrode surfaces (15, 19). For example, a suspension of zeolite Y particles, ~1 μm in diameter, containing a small amount of polystyrene as a binder in THF can be used to cast a film on a SnO_2 electrode surface. The film, of about 60 μm total thickness, shows most of the polystyrene forming a porous layer at the outer (solution) surface of the film, with only a first few micrometers of zeolite on the inner (electrode) side active for electrochemical reactions. Ions, such as $Ru(bpy)_3^{2+}$ and $Co(CpCH_3)_2^+$, can be incorporated into the film by soaking the formed electrode in the appropriate solution or by presoaking the zeolite particles before the film is cast. Typical results for a zeolite film with these species are shown in Figure 4.5.5. The $Ru(bpy)_3^{2+}$ is bound on the surface of the zeolite particles and shows only a small wave attributable to that adsorbed on zeolite particles directly contacting the SnO_2 surface [(b) lower curve].

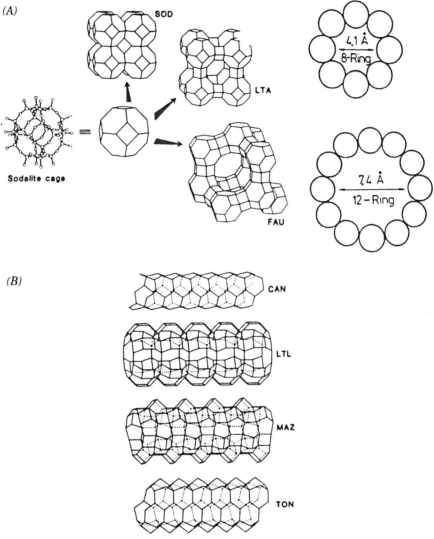

FIGURE 4.5.4. Zeolite structures. (A) Buildup of three different zeolite structures from a sodalite cage (complete SiO_4^{4-} and AlO_4^{5-} tetrahedra based framework shown on left and usual reduced representation in middle). SOD, sodalite; LTA, Linde A; FAU, Faujasite (zeolite Y). Pore sizes of LTA and FAU shown on right. (B) Some other zeolites: CAN, Cancrinite; LTL, Linde L; MAZ, Mazzite (ZSM-4); TON, ZSM-22. [Adapted from J. M. Newsam, *Science*, **231,** 1093 (1986), copyright 1986 by the AAAS and from W. Hölderich, M. Hesse, and N. Näumann, *Angew. Chem. Intl. Ed. Engl.*, **27,** 226 (1988). Copyright 1988 VCH Verlagsgesellschaft.]

4.5. TYPES OF MODIFIERS—INORGANIC FILMS

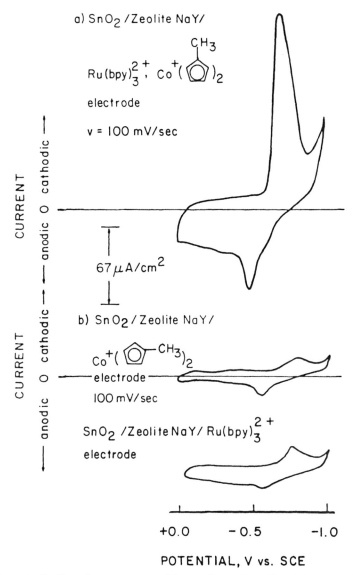

FIGURE 4.5.5. Cyclic voltammograms of ion-exchanged zeolite Y-filmed SnO_2 electrodes in 1 mM KH_2PO_4 solution. Film containing (*a*) both $Ru(bpy)_3^{2+}$ and $Co(CpCH_3)_2^+$; (*b*) $Co(CpCH_3)_2^+$ alone (top) and $Ru(bpy)_3^{2+}$ alone (bottom). [Reprinted with permission from Z. Li and T. E. Mallouk, *J. Phys. Chem.*, **91**, 643 (1987). Copyright 1987 American Chemical Society.]

The $Co(CpCH_3)_2^+$ can penetrate the pores of the zeolite particles, but only a small fraction of the material adsorbed in the zeolite film is electroactive [(b) upper curve]. A film that contains both $Ru(bpy)_3^{2+}$ and $Co(CpCH_3)_2^+$ shows a larger wave (a), since now surface-bound $Ru(bpy)_3^{2+}$ can transfer charge to a larger quantity of bulk $Co(CpCH_3)_2^+$ in the zeolite pores.

4.5.3. Transition-Metal Hexacyanides

Thin films of materials such as Prussian blue (PB) (a lattice of ferric ferrocyanide) and related materials can be formed on electrode surfaces and show interesting properties (20). PB can be deposited on a suitable substrate by immersion in a solution of $FeCl_3$ and $K_3Fe(CN)_6$ to produce $FeFe(CN)_6$ by electrochemical reduction. The blue film, which is given the formula $Fe_4[Fe(CN)_6]_3$ or $KFeFe(CN)_6$, can be oxidized in a KCl solution to form Berlin green and reduced to form Everitt's salt by the following electrode reactions:

$$Fe_4^{III}[Fe^{II}(CN)_6]_3 + 4e + 4K^+ \rightarrow K_4Fe_4^{II}[Fe^{II}(CN)_6]_3$$

or

$$KFe^{III}Fe^{II}(CN)_6 + e + K^+ \rightarrow K_2Fe^{II}Fe^{II}(CN)_6$$
(Everitt's salt)

$$Fe_4^{III}[Fe^{II}(CN)_6]_3 - 3e + 3Cl^- \rightarrow Fe^{III}[Fe^{III}(CN)_6Cl]_3$$

or

$$KFe^{III}Fe^{II}(CN)_6 - e - K^+ \rightarrow Fe^{III}Fe^{III}(CN)_6$$
(Berlin green)

A typical cyclic voltammogram is shown in Figure 4.5.6a. The composition of the film at different stages of oxidation and reduction was determined by surface spectroscopy (XPS and Auger) as well as in situ Mössbauer studies on iron 57-enriched films. The film structures are zeolitic in nature, and this affects the movement of cations through them. For example, while the film can be cycled repeatedly in supporting electrolytes containing K^+ or Cs^+, the cyclic voltammetric behavior is distorted and repeated cycling is not possible in Na^+ and Li^+ electrolytes (Fig. 4.5.6b). PB electrodes show electrocatalytic properties (e.g., for the reduction of oxygen) and the color changes that occur suggest possible electrochromic applications.

4.5. TYPES OF MODIFIERS—INORGANIC FILMS

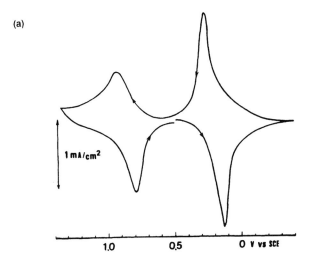

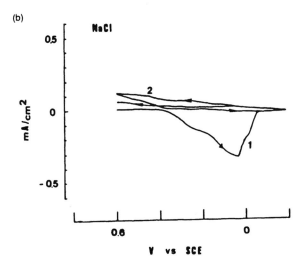

FIGURE 4.5.6. Cyclic voltammograms of a Prussian blue modified electrode (SnO_2 substrate). Solution pH 4 and contains (a) 1 M KCl, (b) 0.1 M NaCl. Radii of hydrated ions: K^+, 1.25 Å; Na^+, 1.83 Å. Prussian blue has a zeolitic structure with a channel diameter of about 3.2 Å, so that transport of ions with radii above about 1.6 Å is impeded. [(a) Reprinted with permission from K. Itaya, H. Akahashi, and S. Toshima, *J. Electrochem. Soc.*, **129**, 1498 (1982). Copyright 1982 The Electrochemical Society. (b) Reprinted with permission from K. Itaya, T. Ataka, and S. Toshima, *J. Am. Chem. Soc.*, **104**, 4767 (1982). Copyright 1982 American Chemical Society.]

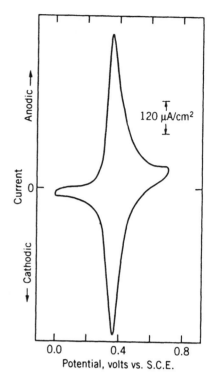

FIGURE 4.5.7. Cyclic voltammogram of a nickel surface chemically modified by oxidation in the presence of ferricyanide to form a layer of KNiFe(CN)$_6$. [Reprinted with permission from B. D. Humphrey, S. Sinha, and A. Bocarsly, *J. Phys. Chem.*, **88**, 736 (1984). Copyright 1984 American Chemical Society.]

Other metal ferricyanide films have also been studied. A film of nickel ferricyanide can be deposited by oxidizing a nickel electrode in the presence of ferricyanide:

$$\mathrm{Ni} + \mathrm{Fe^{III}(CN)_6^{3-}} + \mathrm{K^+} - 2e \rightarrow \mathrm{K[Ni^{II}Fe^{III}(CN)_6]}$$

This film shows reversible cyclic voltammetric behavior involving reduction and reoxidation of the Fe centers with concomitant movement of cations and solvent into and out of the film (Fig. 4.5.7) (21). Related films of ferric ruthenocyanide and osmocyanide have also been deposited and studied electrochemically.

4.6. TYPES OF MODIFIERS—BIOLOGICALLY RELATED MATERIALS

Many electrodes modified with biologically derived materials have been described, usually in connection with the preparation of electrochemical sensors (22). The basic approach in such biosensors involves the immobilization of a biologically sensitive coating (e.g., an enzyme, antibody, DNA) which can interact with ("recognize") a target analyte and in the process produce an electrochemically detectable signal. Probably the most highly developed are electrodes containing surface-confined enzymes, such as that described in Chapter 1 (23). Related types of electrodes involve suspensions of bacteria and slices of tissue. In many cases, the enzyme or suspension is simply held in the vicinity of the electrode by a permeable polymer membrane, such as a dialysis membrane. Alternative methods of immobilization include entrapment in a gel, encapsulation, adsorption, and covalent linkage. Because the extensive literature in this area has been the subject of a number of reviews, such electrodes will not be discussed further here.

4.7. MODIFIED ELECTRODES WITH CHEMICALLY SENSITIVE CENTERS

Electrode surfaces modified with selective centers are of interest in both analytical and synthetic applications. As indicated in Section 2.2.6, such centers include crown ethers, cryptands, cyclodextrins, many other macrocylic molecules, as well as specific chelating ligands. Several of these have been used to modify electrode surfaces. For example, there are several reports of surface modification with cyclodextrins (24). These are naturally occurring polysaccharides (more specifically, cyclic 1,4-linked D-glucopyranose oligomers) that can be isolated with high purity and occur in several different forms with unique cavity sizes (Fig. 2.2.9). These can be adsorbed or bonded to the surface of a graphite electrode [e.g., Eq. (2.2.1)]. In the application of these to electrochemical synthesis, it was shown that the chlorination of anisole at an electrode with a layer of α-CD leads to a larger ratio of p- to o-chloroanisole than at an unmodified electrode because of the specific orientation of the anisole at the electrode surface induced by the CD (Fig. 2.2.9). The selective partitioning of CD into a Nafion-modified electrode has also been reported and used to distinguish between o- and p-nitrophenol in aqueous solutions in voltammetric reduction (24b). Whereas the selectivity ratio, defined as the ratio of the peak currents for reduction of the isomers, in a solution containing CD at a bare electrode is 1.3; at a Nafion-covered electrode under the same conditions it is 33. In this case, the preferred binding of the p isomer by the CD and the decreased partitioning of the CD-bound isomer into the Nafion layer are responsible for the drastic decrease in the height of the p-nitrophenol reduction wave.

Electrodes can also be modified with specific chelating agents. For example, a film of a copolymer of quaternized vinylpyridine and vinylferrocene can be deposited by electrooxidation (25). This cationic film can take up anionic chelating agents, such as sulfonated bathocuprine, by ion exchange (Fig. 4.7.1). The immobilized chelate will then interact with certain metal ions [e.g., Fe(II), Cu(I)] that can then be detected by cyclic voltammetry. Metal-ion concentrations in the ppm range were determined by this approach. Metal ions and complexes, e.g., Ru(edta)$^-$ and Ru(bpy)$_3^{2+}$, can be incorporated into films of polymers such as poly(vinylpyridine) and Nafion by complexation or ion-exchange effects, but with somewhat lower selectivity.

An alternative approach involves incorporation of suitable ligands into LB or self-assembled monolayers. For example, the chelate 2,2′-thiobis(ethylacetoacetate) (TBEA) and a monolayer former, n-ocatadecyltrichorosilane (OTS), can be cast on a gold electrode to produce a layer that will specifically complex certain metal ions (Cu^{2+}) and not others (Fe^{2+}) (Fig. 4.7.2) (26). Monolayers can also be formed from mixtures in which the film-forming molecule is diluted with a second species that forms defects or gate sites in the layer. Electron transfer to solution species can occur at the gate sites, and the electrochemical behavior depends on the ratio of defect-forming species to blocking molecule in the mixture used to prepare the monolayer. Examples of these kinds of modified electrodes are LB layers containing ubiquinone as a gate molecule (27) and self-assembled monolayers of 1-hexadecanethiol containing 4-hydroxythiophenol as the defect-inducing molecule (28). Such modified electrodes have been suggested as ion-selective sensors.

Chemically sensitive groups within a surface layer are also useful in the controlled release of species from an electrode surface, as illustrated in Fig. 4.7.3 (29). In this application, a surface film of a composition that can be switched electrochemically into a state where a component of the film is released into the solution is used. As shown, this can involve reduction of a bond in the polymer, as in the release of dopamine, or reduction of a site to cause expulsion of an anion by electrostatic effects. Specificity into the controlled release process is gained by the use of selective centers such as those described above. Electrochemically switched cation binding, for example, has been described for solution species. A representative case, shown in Figure 4.7.4, is that of pentaoxa[13]-ferrocenophane which can bind Na^+ in its reduced state, but releases it in its oxidized (ferrocenium) state (30). Similar controlled release of alkali metal ions has been observed for electrochemically reduced nitrobenzene-substituted lariat ethers (31).

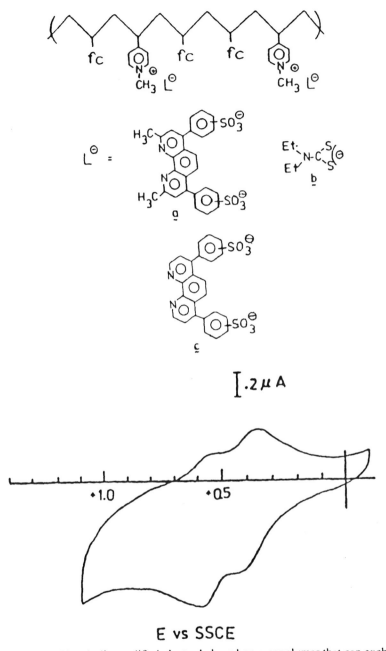

FIGURE 4.7.1. Chemically modified electrode based on a copolymer that can exchange chelating agents specific for given metals. Copolymer, containing ferrocene groups (fc) shown with different ligands (L^-): (a) sulfonated bathocuproine; (b) diethyldithiocarbamate; (c) sulfonated bathophenanthroline. Cyclic voltammogram shows such a modified electrode with sulfonated bathocuproine in a MeCN/TBAP solution containing 5 × 10^{-5} M solution of Cu(I) that is preconcentrated into the polymer film. [Reprinted with permission from A. R. Guadalupe and H. Abruña, *Anal. Chem.*, **57**, 142 (1985). Copyright 1985 American Chemical Society.]

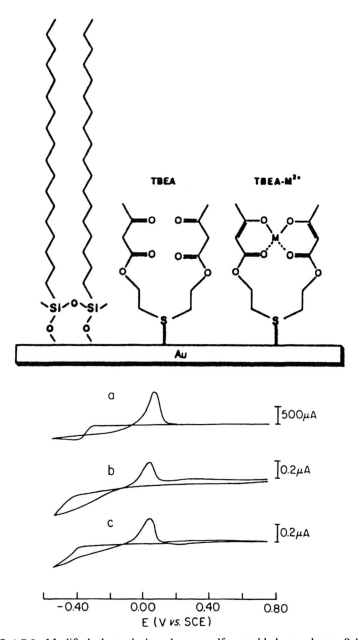

FIGURE 4.7.2. Modified electrode based on a self-assembled monolayer. Schematic representation of the monolayer components on a gold surface: OTS, n-ocatadecyltrichorosilane; TBEA, 2,2'-thiobis(ethylacetoacetate); TBEA-M^{2+}, TBEA with bound metal ion. The cyclic voltammograms show this modified electrode in aqueous sulfuric acid solutions containing (a) 1 mM Cu^{2+}; (b) 1 μM Cu^{2+}; (c) 1 μM Cu^{2+} + 3 mM Fe^{2+}. [Reprinted with permission from S. Steinberg and I. Rubinstein, *Langmuir*, **8**, 1183 (1992). Copyright 1992 American Chemical Society.]

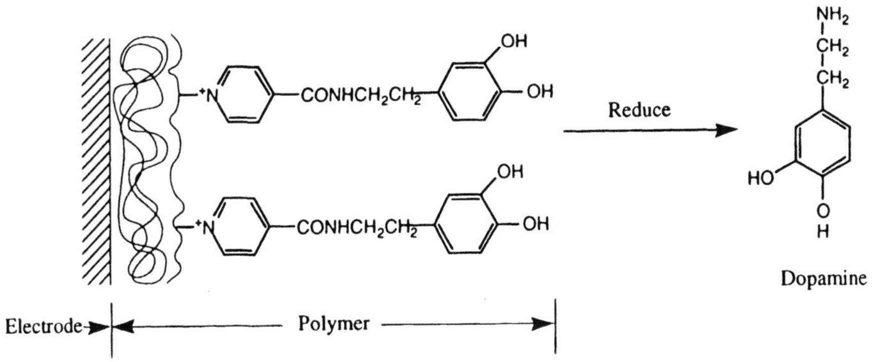

FIGURE 4.7.3. A redox polymer that releases dopamine on electrochemical reduction based on a modified polystyrene backbone holding N-(2-(3,4-dihydroxyphenyl)ethyl)isonicotinaminde units (see Ref. 29).

4.8. MORE COMPLEX MODIFIED ELECTRODE STRUCTURES

In addition to the modified electrodes described in the previous sections, which usually involve a conductive substrate and a single film of modifying material, more complicated structures have been described. These are more closely related to integrated chemical systems. Typical examples (Fig. 4.8.1) include multiple films of different polymers (bilayer structures), metal films formed on the polymer layer (sandwich structures), multiple conductive substrates under the polymer film (electrode arrays), intermixed films of ionic and electronic conductor (biconductive layers), and polymer layers with porous metal or minigrid supports (solid polymer electrolyte or ion-gate structures) (1b, 1f). These often show different electrochemical properties than the simpler modified electrodes and may be useful in applications such as switches, amplifiers, and sensors.

4.8.1. Porous Metal Films on Polymers

Porous metal films, such as Pt or Au, can be deposited on free-standing polymer membranes or on polymer films on an electrode surface by chemical reduction or vacuum evaporation (32). For example, a porous film of Pt can be deposited on a Nafion membrane by clamping it between a solution of $PtCl_6^{2-}$ and a solution of a reducing agent, such as hydrazine. The reducing agent diffuses through the membrane and causes precipitation of metallic Pt on the membrane surface and partially within the membrane structure as well. A similar electrochemical method for deposition of Ag on a polyimide film on an electrode surface was pre-

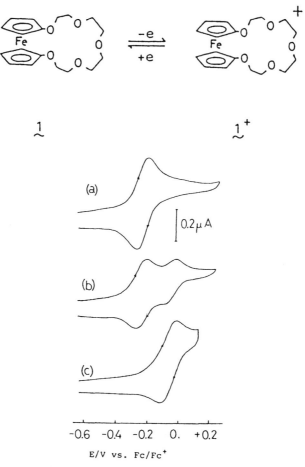

FIGURE 4.7.4. Principle of controlled release of a cation by electrochemical switching. **1**, pentaoxa[13] ferrocenophane will coordinate with Na^+, but releases it on oxidation to the $\mathbf{1}^+$ form. The cyclic voltammograms show solutions of **1** (0.2 mM) in CH_2Cl_2/0.1 M TBAPF$_6$ in the absence of Na^+ (*a*); with 1 mM NaClO$_4$ after stirring for (*b*) 5 min and (*c*) 1 h. The shift in the wave shows coordination with the Na^+, with the uncharged form coordinating 740 times more strongly than the $\mathbf{1}^+$ form. [Reprinted with permission from T. Saji, *Chem. Lett.*, **1986**, 275. Copyright 1986 The Chemical Society of Japan.]

viously described (see Fig. 2.4.9) (33). Such structures are of interest as electrodes in *solid polymer electrolyte* (SPE) cells, employed, for example, in fuel cells and for water electrolysis. In the former application, two porous Pt films on either side of a Nafion membrane form the anode (for hydrogen oxidation) and the cathode (for oxygen reduction) in the cell. Note that in this arrangement a solution of liquid elec-

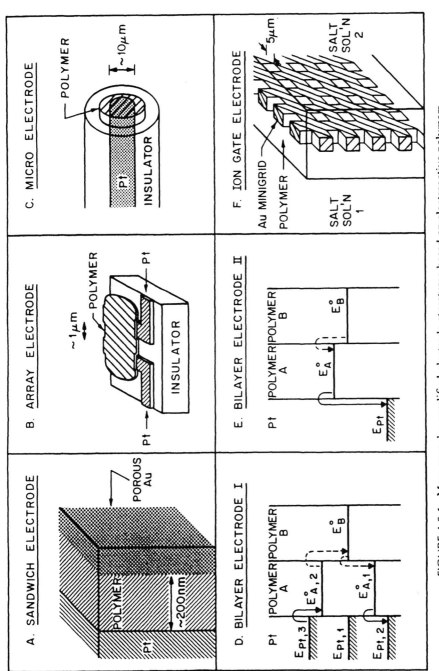

FIGURE 4.8.1. More complex modified electrode structures based on electroactive polymers. (A) Sandwich electrode; (B) array electrode; (C) microelectrode; (D, E) bilayer electrodes; (F) ion-gate electrode. [Reprinted with permission from C. E. D. Chidsey and Royce W. Murray, *Science*, **231**, 25 (1986). Copyright 1986 by the AAAS.]

trolyte is not required and the electrodes are contacted only by the bathing gases, H_2 and O_2, and the product, water, is removed from the cell. Such an arrangement has also been suggested for electrochemical synthesis in highly resistive solutions, such as those with very low concentrations of supporting electrolyte (Fig. 4.8.2) (34, 35). In this arrange-

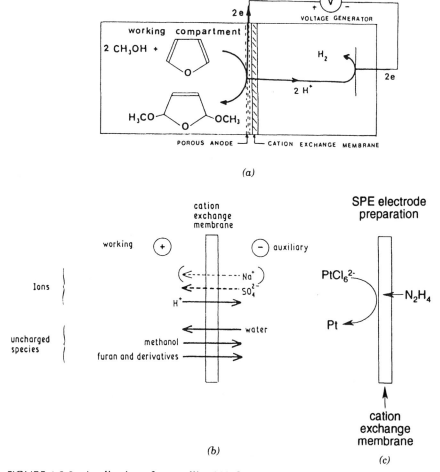

FIGURE 4.8.2. Application of a metallized Nafion membrane for electrolysis of organic species (methoxide addition to furan via oxidation) in the absence of a supporting electrolyte. (a) Schematic diagram of cell arrangement. The Pt/Nafion membrane separates the aqueous solution containing the cathode from the nonaqueous solution. (b) Detail of the fluxes of the different species through the membrane during electrolysis. [Reprinted with permission from E. Raoult, J. Sarrazin, and A. Tallec, *J. Appl. Electrochem.*, **15**, 85 (1985). Copyright 1985 Chapman & Hall.] (c) Schematic diagram of the method of preparation of the platinized Nafion membrane.

ment the porous metal film is the working electrode with the counter and reference electrodes contained in a separate solution in the auxiliary chamber. The working electrode solution with the substrate of interest is in another chamber, separated from the auxiliary chamber by the SPE membrane. During electrolysis, charge is compensated by the movement of cations through the membrane. This arrangement can be used in a voltammetric mode to study electrode reactions in highly resistive solvents, such as toluene or THF (32a), and should also be useful for electrode reaction studies in the gas phase.

4.8.2. Sandwich Structures

Deposition of a porous metal on top of a polymer film formed on a conductive substrate leads to a structure in which a thin polymer film is sandwiched between two conductive films, at least one of which is porous to solution or gas-phase species. While this structure is formally similar to that used in SPE electrolysis, the polymer films in these are much thinner and are usually deposited on a nonporous substrate before formation of the porous metal layer. A typical construction is shown in Figure 4.8.3 (36). A film (~100–1000 nm thick) of polymer is deposited on a Pt electrode by electrochemical reduction of an MeCN solution of $Os(bpy)_2(p\text{-cinn})_2^{2+}$ (p-cinn = 4-py-NHCOCH=CHPh) or $Os(bpy)_2(vpy)_2^{2+}$. A film of Au 20–100 nm thick is formed on the polymer surface by vacuum evaporation. The Pt substrate acts as one electrode and the porous Au as a second electrode, with reference and auxiliary electrodes placed in the bulk solution in which the sandwich is immersed. When a sufficiently large potential is applied between the Pt and Au, a current flows, representing oxidation of Os(II) at the Pt and reduction of Os(III) at the Au, as shown in Figure 4.8.3. Charge is carried through the polymer layer by electron hopping between Os(II) and Os(III) sites in a process that formally is equivalent to diffusion with an effective diffusion coefficient (related to the electron-transfer rate) of D_{ct} (see Section 5.4). The magnitude of the steady-state current that flows is a function of D_{ct} and can be used to measure this quantity, if the polymer film thickness is known. Relating D_{ct} to the electron-transfer rate constant is much less straightforward.

Such sandwich electrode structures can also pass current between substrate and porous film through the polymer film when no solvent or supporting electrolyte is present (37). For example, if the film initially contains a 1:1 mixture of the Os(III) and Os(II) species, when a small potential (e.g., 200 mV) is applied across the film, a current will flow (Fig. 4.8.4A). Os(III) is reduced to the II state at the cathode, Os(II) is

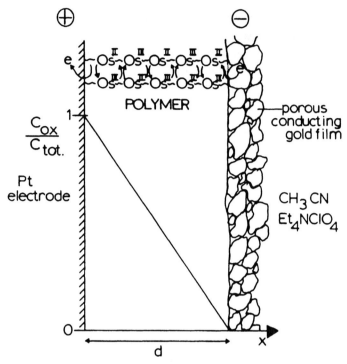

FIGURE 4.8.3. Schematic representation of a sandwich electrode produced by deposition of a film of poly-$[Os(bpy)_2(vpy)_2]^{2+}$ on a Pt substrate and then deposition of a porous Au film. Conduction through the film occurs by electron hopping between Os(III) and Os(II) centers. The concentration profile shown is for the Pt positive and the Au negative of $E^{\circ\prime}$ for the couple. [Reprinted with permission from P. G. Pickup and R. W. Murray, *J. Electrochem. Soc.*, **131**, 833 (1984). Copyright 1984 The Electrochemical Society.]

oxidized to the III state at the anode, and the anion, A^- (e.g., ClO_4^-) moves from cathode toward anode to maintain electroneutrality in the film. A steady-state current and stationary concentration profiles are eventually attained, as shown. The situation is different if the film is initially all in the Os(II) form (Fig. 4.8.4*B*). Since no anion is available in the gas phase bathing the electrode, a small potential applied across the substrate and porous film will not cause a current to flow. However, when the potential across the film attains a value (~ 2.5 V) where Os(II) can be reduced to the I state at one electrode and oxidized to the III state at the other, with movement of A^- to compensate charge, current will flow. The resulting concentration profiles are shown in Figure 4.8.4*B*. The effective diffusion coefficient, again, can be determined

4.8. MORE COMPLEX MODIFIED ELECTRODE STRUCTURES

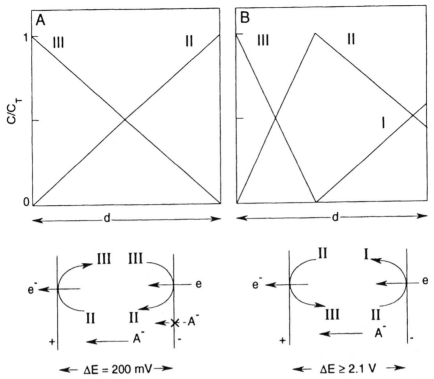

FIGURE 4.8.4. Concentration profiles of different oxidation states of Os in sandwich cell of poly-$[Os(bpy)_2(vpy)_2]^{2+}$ with ClO_4^- (A^-) as anion contacting a gas or solvent with no electrolyte, so that only A^- in the film is available to compensate charge (no A^- penetrates the Au film. (A) Film originally containing 1:1 Os(III):Os(II); (B) film originally all in the Os(II) form (37). [Adapted from J. C. Jernigan, C. E. D. Chidsey, and R. W. Murray, J. Am. Chem. Soc., **107**, 2824 (1985). Copyright 1985 American Chemical Society.]

from measurements of the steady-state current and was found to be a function of the gas contacting the electrode.

Fabrication of sandwich electrode structures of the type described above requires great care to avoid deposition of the evaporated metal in pinholes in the polymer film, with subsequent shorting (short-circuiting) of the current through the polymer layer. This frequently means that relatively thick films of polymer are needed. Note, however, that earlier sandwich structures involving layers of organized assemblies deposited by Langmuir–Blodgett techniques with deposited metal films, in which tunneling measurements were made, have been claimed to be essentially pinhole-free (38).

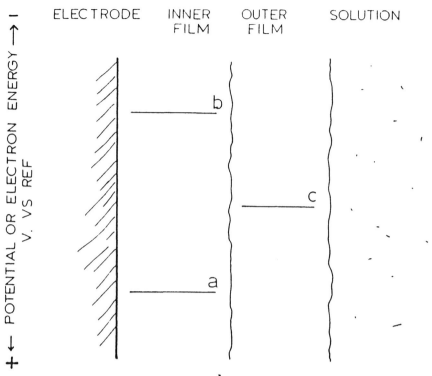

FIGURE 4.8.5. Schematic diagram of bilayer electrode, for example, Pt/poly-[Ru(vbyp)$_3$]$^{2+}$/poly-[Os(bpy)$_2$(vpy)$_2$]$^{2+}$, showing redox potentials in the polymer layers. [Reprinted with permission from P. G. Pickup, C. R. Leidner, P. Denisevich, and R. W. Murray, *J. Electroanal. Chem.*, **164**, 39 (1984).]

4.8.3. Bilayer Structures

A bilayer structure consists of two different films deposited on a substrate (Fig. 4.8.5). A typical system is one consisting of a Pt substrate with an electrodeposited film of poly-[Ru(vbpy)$_3^{2+}$] on which a film of poly-[Os(bpy)$_2$(vpy)$_2^{2+}$] is electrodeposited (39). The electrochemical response of such a system, when immersed in a MeCN solution, is shown in Figure 4.8.6. The behavior is governed by the $E°$ values of the different half-reactions of the polymer couples, as indicated in the figure. The inner layer Ru(III) can oxidize the outer layer Os(II) to Os(III) on a positive-going scan, but the Os(III) thus formed cannot be reduced back to Os(II) until a fairly negative potential is attained, that is, where reduction of Ru(II) to Ru(I) in the inner film occurs. Thus on oxidation, charge is "trapped" in the outer film. This is shown on the cyclic voltammogram, where the anodic peaks at potentials around +1 V,

4.8. MORE COMPLEX MODIFIED ELECTRODE STRUCTURES

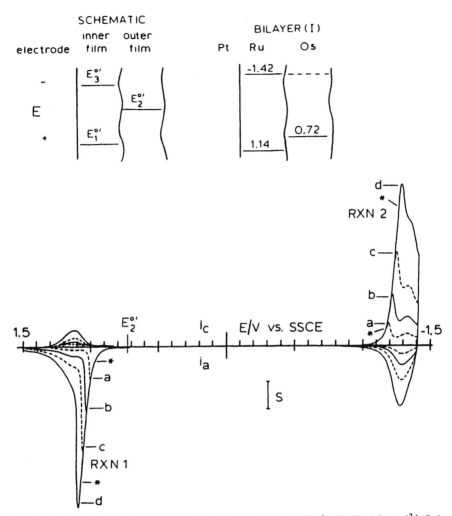

FIGURE 4.8.6. Cyclic voltammetry of the bilayer electrode Pt/poly-[Ru(vbyp)$_3$]$^{2+}$ (5.6 × 10^{-9} mol/cm^2)/poly-[Os(bpy)$_2$(vpy)$_2$]$^{2+}$ (30 × 10^{-9} mol/cm^2)/0.1 M Et$_4$NClO$_4$, MeCN at scan rates (a–d) of 20, 50, 100, and 200 mV/s. Potentials of the Ru(I/II), Ru(II/III) and Os(II/III) couples are shown above. [Reprinted with permission from C. R. Leidner, P. Denisevich, K. W. Willman, and R. W. Murray, *J. Electroanal. Chem.*, **164**, 63 (1984). Copyright 1984 Elsevier.]

obtained on an initial anodic scan, represent Os(II) → Os(III) and Ru(II) → Ru(III), while the cathodic peak at this potential represents only Ru(III) → Ru(II). The cathodic peaks at about −1.4 V, obtained following the anodic scan, represent release of the trapped charge: Ru(II) → Ru(I) and Os(III) → Os(II). Finally, the anodic peak at this potential on scan reversal represents the reaction Ru(I) → Ru(II), which resets the system back to the initial condition. The detailed electrochemical response of the bilayer structure, such as that seen on the initial anodic scan, depends on the rate of electron transfer at the polymer–polymer interface, so that such structures can be used to probe such reactions. Although their response is rather slow compared to solid-state devices, these bilayer structures may find applications as memory and electrochromic devices, and as rectifying junctions.

4.8.4. Electrode Arrays

Arrays of conductive electrodes (e.g., Pt or Au) on an insulating substrate (e.g., SiO_2 or Si_3N_4), with widths and spacings of the order of micrometers (Fig. 4.8.7), can be formed by photolithographic techniques (as described in Section 2.5.1) (Fig. 4.8.8) and are commercially available (40). An example of such an array, covered with a film of poly(3-methyl thiophene) and employed as a microelectronic device, was discussed in Section 1.2.6. Such arrays can take on a number of configurations. For example, the pair of electrodes bridged by a polymer film is the "open face" version of the sandwich electrode and is probably somewhat easier to fabricate in a pinhole-free form. Alternatively, a different polymer can be deposited on each electrode of an array pair to form a bilayer-like arrangement. Moreover, three-electrode devices to produce a structure equivalent to a field-effect transistor (FET) can be fabricated.

An example of the FET-type device is shown in Figure 4.8.9. The closely spaced parallel fingers of Au are covered with a layer of the polymer polypyrrole (see Fig. 2.2.4), which can be produced by electrochemical oxidation of a solution of pyrrole in MeCN either at constant potential or by cycling. The conductivity of the polypyrrole depends on its state of oxidation, which can be controlled by the potential applied to the gate electrode with respect to an external reference electrode, V_G. In its oxidized state, polypyrrole is "doped" with anions and is electronically conductive, while in the reduced state it is not. Thus the current flow between the two outer electrodes, labeled (in FET terms) as *source* and *drain*, is a function of the state of the polypyrrole and is thus controllable by the potential V_G. The response time of such a device depends on the amount of polymer that must be switched from the insulating to conductive state and the spacing between the electrodes. While direct photolithography produces interelectrode spacings of ~1 μm, the gap can be closed by electrodeposition of more metal onto

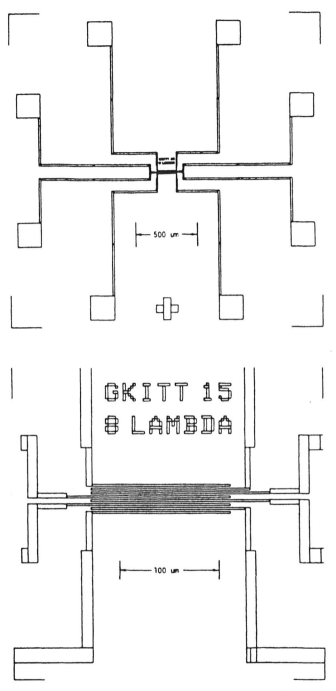

FIGURE 4.8.7. Layout of typical microelectrode array at two different size scales, showing (above) the contact pads and (below) the details of the electrode array structure. Typical arrays comprise electrodes of Pt or Au formed on oxidized Si wafers. [Reprinted with permission from G. P. Kittlesen, H. S. White and M. S. Wrighton, *J. Am. Chem. Soc.*, **106**, 7389 (1984). Copyright 1984 American Chemical Society.]

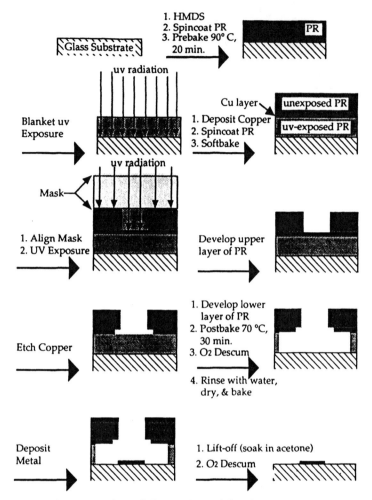

FIGURE 4.8.8. Representation of the method of forming microelectrode array on a glass substrate by photolithographic procedure. [Reprinted with permission from I. Fritsch-Faules and L. R. Faulkner, *Anal. Chem.*, **64**, 1118 (1992). Copyright 1992 American Chemical Society.]

the electrodes or by "shadow evaporation" of additional metal on the array (41).

An example of the bilayer arrangement is shown in Figure 4.8.10, where the viologen polymer, BPQ^{2+}, is cathodically deposited on one electrode and poly(vinylferrocene) (PVF) is anodically deposited on the other (41a). The structure shown in the figure also demonstrates the use of electrodeposition of Pt on the original Au array electrodes to decrease the interelectrode gap. The electrochemical behavior of this structure is

4.8. MORE COMPLEX MODIFIED ELECTRODE STRUCTURES

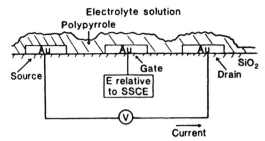

FIGURE 4.8.9. Electrochemical transistor based on a polypyrrole polymer coating over a three-terminal gold array. The potential on the gate electrode with respect to a reference electrode in the solution (SSCE; saturated sodium chloride calomel electrode) controls the state of oxidation of the polypyrrole and hence the flow of current between the source and drain electrodes. Note that the same effect can also be accomplished in a two-electrode device (see Fig. 1.2.19). [Reprinted with permission from C. E. D. Chidsey and R. W. Murray, *Science*, **231**, 25 (1986). Copyright 1986 by the AAAS.]

shown in Figure 4.8.11. Ideally, current flow should be unidirectional (i.e., the device should show diode-like behavior), because BPQ^{2+} can only be reduced (to the +1, or at more negative potentials, to the 0 state) and PVF can only be oxidized (to the +1 state) within the available potential window. Thus current flow begins when the potential applied between the two array electrodes is about the difference of the

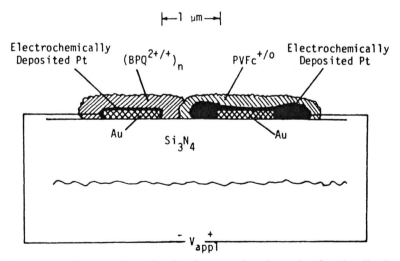

FIGURE 4.8.10. Cross section of pair of array microelectrodes functionalized with different polymers (open-face sandwich arrangement) that shows diode-like behavior. [Reprinted with permission from G. P. Kittlesen, H. S. White, and M. S. Wrighton, *J. Am. Chem. Soc.*, **107**, 7373 (1985). Copyright 1985 American Chemical Society.]

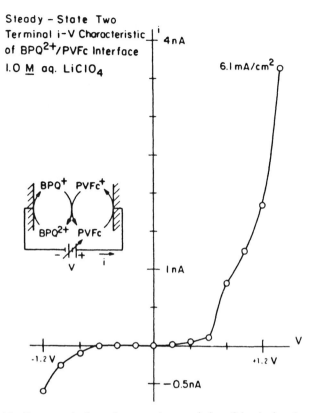

FIGURE 4.8.11. Two-terminal steady-state characteristics of the device shown in Figure 4.8.10. The right-hand side corresponds to the connections and bias as shown in the insert; and the left-hand side, to the opposite bias (PVF side negative). [Reprinted with permission from G. P. Kittlesen, H. S. White, and M. S. Wrighton, *J. Am. Chem. Soc.*, **107**, 7373 (1985). Copyright 1985 American Chemical Society.]

standard potentials of the two redox couples, with the occurrence of the following reactions:

At anode: $\quad\quad\quad PVF - e \to PVF^+ \quad\quad\quad$ (4.8.1)

At cathode: $\quad\quad\quad BPQ^{2+} + e \to BPQ^+ \quad\quad\quad$ (4.8.2)

At interface: $\quad BPQ^+ + PVF^+ \to PVF + BPQ^{2+} \quad$ (4.8.3)

When the potential is applied in the opposite direction, only a small current flow is found for the same magnitude of applied potential (~ 0.9 V); the finite reverse current is ascribed to impurities in the water or the onset of water decomposition. This reverse current increases at larger applied potentials, so the diode behavior in this device is nonideal.

4.8.5. Membranes and Ion Gates

An important capability in certain ICSs is the selectivity and control of the flows of species across an interface or through a membrane. The permeability of a membrane to charged and neutral species can be controlled by electric fields across the membrane, either generated externally or chemically (i.e., by pH or complexation effects) (42). For example, the permeability of a poly(methacrylic acid) membrane to insulin is affected by the ionization state of the membrane and the pH gradient across the membrane. Many other examples of similar systems have been reported.

One form of an ion gate (Fig. 4.8.12) consists of a porous conductive support on which is deposited a polymer film whose oxidation state can be changed electrochemically and whose permeability to ions is a function of its oxidation state (43), for example, a Au minigrid (with 5-μm pores) onto which has been electrodeposited a film of poly(pyrrole) ($\sim$ 10 μm thick). When the film is in the reduced (nonconductive) state, the flux of Cl$^-$ across the film from a solution of 1 M KCl to one of 1 M KNO$_3$ is 100 times smaller than that across the film when it is in the oxidized state, where anions are incorporated in the film as counterions.

Electrochemically controlled changes in polymer permeability could

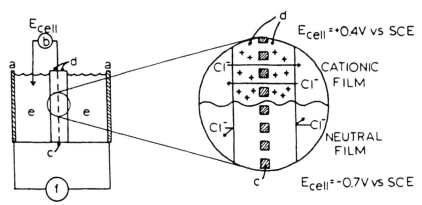

FIGURE 4.8.12. Ion-gate membrane arrangement. The cell on the left shows the membrane, consisting of a polypyrrole film (d) on the gold minigrid (c). This separates two 1 M KCl solutions (e). The state of the membrane (oxidized or reduced) is controlled by power supply (b). The membrane impedance is determined by measuring the ac current flow between Pt electrodes (a) with apparatus (f). A schematic view of the membrane is shown on the right, indicating the change from conducting (oxidized) state on top to the neutral (reduced) state at the bottom. [Reprinted with permission from P. Burgmayer and R. W. Murray, *J. Am. Chem. Soc.*, **104**, 6139 (1982). Copyright 1982 American Chemical Society.]

probably also be accomplished with other types of polymers, either through changes in the ionic content of the polymer film (e.g., PVF in oxidized and reduced states) or through changes in the structure of the polymer film on oxidation and reduction of electroactive centers within the film. For example, the permeability of a Nafion film to cations and neutral substances is decreased by the addition of the species $Ru(bpy)_3^{2+}$, presumably because its introduction changes the film structure through "crosslinking" (44). The extent of structural change depends on the state of oxidation of the complex ion, and this introduces the possibility of electrochemical control. Solid polymer electrolyte structures, as described in Section 4.8.1, should also be useful in the fabrication of ion gates.

The control of an ionic flux across these ion gates parallels similar processes in biological membranes. The response time of the ion gate depends on the time required to convert the polymer film from one state to the other. With the thick films discussed above, this can be of the order of minutes. Faster response would result from thinner (e.g., bilayer) membranes, analogous to biological ones, assuming that they can be made sufficiently leak-free. For example, an ion-gate monolayer membrane, based on the liposome-like amphiphile bis(ω-mercaptoundecyl) phosphate self-assembled on a gold electrode, has been described (Fig. 4.8.13) (45). This film was shown to block the penetration of $Fe(CN)_6^{3-}$ to the gold substrate to an extent that depended on the mode of film formation and presumably the film structure. For some films,

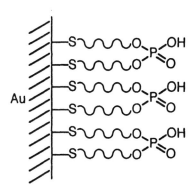

FIGURE 4.8.13. Schematic representations of monolayer of bis(ω-mercaptoundecyl) phosphate self-assembled on a gold electrode. [Reprinted with permission from N. Nakashima, T. Tagujchi, Y. Takada, K. Fujio, M. Kunitake, and O. Manabe, *J. Chem. Soc. Chem. Commun.*, **1991**, 232 (1991). Copyright 1991 The Royal Society of Chemistry.]

$Fe(CN)_6^{3-}$ penetration to the Au was blocked in neutral or alkaline solutions, but was allowed under acidic conditions, demonstrating a pH-controlled ion gate. However, this layer was on an electrode surface, rather than separating two solutions, and was really closer to the electrodes with chemically sensitive centers discussed in Section 4.7.

4.8.6. Biconductive Films

Multicomponent film structures that contain both electronic and ionic conductors, called *biconductive films* (or mixed-conductivity composites), can be fabricated electrochemically. Interest in these stems from the possibility of improving the rates of charge transport through the films and of incorporating catalysts or semiconductor particles within an ionically conducting polymer layer. An early example of this kind of structure was the deposition of the electronically conducting solid tetrathiafulvalenium bromide (TTF^+Br^-) within a Nafion layer on an electrode (46). When TTF^+ is first exchanged into a Nafion layer, as the counter cation to the $\sim SO_3^-$ groups, it shows the usual cyclic voltammetric response for reduction to TTF in the presence of KBr as supporting electrolyte:

$$\sim SO_3^-TTF^+ + K^+ + e \rightarrow \sim SO_3^-K^+ + TTF \qquad (4.8.4)$$

However, oxidation of the TTF occurs with formation of the electronically conductive bromide:

$$TTF + 0.7Br^- \rightarrow TTF^+Br_{0.7}^- + 0.7e \qquad (4.8.5)$$

This species forms with the production of colored zones and micrometer-sized needles (Fig. 4.8.14), and the film shows unique cyclic voltammetric behavior (Fig. 4.8.15) that appears to be connected with the structural changes occurring on oxidation and reduction of the electroactive zones within the film.

Electronically conductive polymers can also be deposited within ionically conductive supports. For example, polypyrrole can be formed inside Nafion or clay layers by electrochemical oxidation of a pyrrole solution (47). These films show electrochemical and mechanical properties that are different from those of the single component (Nafion or polypyrrole). Oxidation of pyrrole by Fe(III) chloride in an aqueous solution of methycellulose produces a colloidal dispersion that can be used to prepare conductive thin films (48).

An alternative approach to such films involves polymerization of mol-

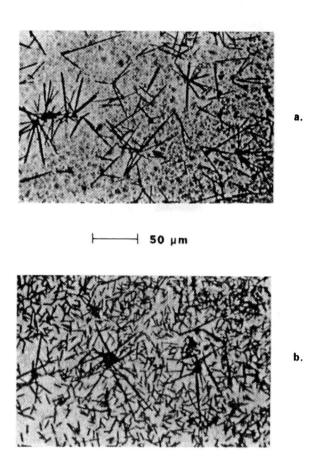

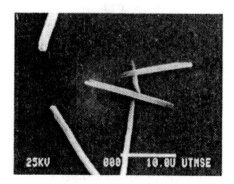

FIGURE 4.8.14. Top (*a*, *b*): optical micrographs of SnO_2/Nafion, TTF^+ electrodes that were cycled 10 times in a 1 M KBr solution and then removed and observed in the (*a*) oxidized and (*b*) reduced form. Bottom: scanning electron micrograph of needles formed in Nafion, TTF^+ cycled as above. [Reprinted with permission from T. P. Henning, H. S. White, and A. J. Bard, *J. Am. Chem. Soc.*, **104,** 5862 (1982). Copyright 1982 American Chemical Society.]

4.8. MORE COMPLEX MODIFIED ELECTRODE STRUCTURES

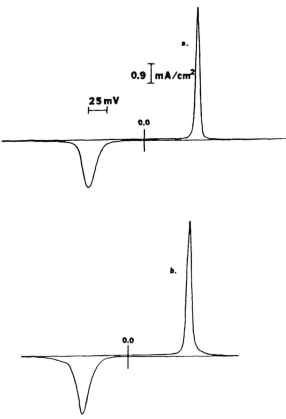

FIGURE 4.8.15. Cyclic voltammograms of (a) Pt/Nafion, TTF$^+$ electrode in 1 M KBr; (b) Pt/TTF film (~800 Å thick) in 1 M KBr. Scan rate, 10 mV/s. [Reprinted with permission from T. P. Henning, H. S. White, and A. J. Bard, *J. Am. Chem. Soc.*, **103**, 3937 (1981). Copyright 1981 American Chemical Society.]

ecules that contain the basic components of both ionic and electronic polymers within the same species. For example, the species Ru(bpy)L$_2^{2+}$, where L is a bipyridine substituted with a pyrrole, is shown below:

On oxidation of $Ru(bpy)L_2^{2+}$ from a MeCN solution at a Pt electrode, a polymer film forms on the surface that can be considered a polypyrrole backbone containing substituted and charged Ru(III) groups (49). The film shows typical surface waves in cyclic voltammetry characteristic of both the $Ru(bpy)^{2+}$ centers and pyrrole centers, although the electronic conductivity of the dry film is orders of magnitude smaller than that of polypyrrole. This suggests that the electronic conductivity in this film is disrupted by the pendant Ru centers and that it is not a true biconductive film. The strategy employed, however, could be of use in the fabrication of such films. Another approach would involve starting with a polypyrrole film in the oxidized, conductive form and bringing in as counter anions to the positive centers of the polypyrrole species that could generate an ionically conductive structure. For example, species like the anionic tetrasulfonated iron phthalocyanine can be incorporated into a polypyrrole film; such a film shows improved catalytic behavior for O_2 reduction (50). An anionic species that could be induced to polymerize or bind other cations electrostatically, could form a biconductive film. Biconductive films are also formed when metal centers are formed inside conductive polymers, such as the deposition of Cu and Ag onto poly-$[Ru(bpy)_2(vpy)_2]^{2+}$ (51). Another example of metal deposition within a polymer was discussed in Section 2.4 (see Fig. 2.4.9).

REFERENCES

1. General references—chemically modified electrodes: (a) R. W. Murray, "Chemically Modified Electrodes," in *Electroanalytical Chemistry*, A. J. Bard, ed., Marcel Dekker, New York, Vol. 13, 1984, p. 191; (b) C. E. D. Chidsey and R. W. Murray, "Electroactive Polymers and Macromolecular Electronics," *Science*, **231**, 25 (1986); (c) R. W. Murray, "Chemically Modified Electrodes," *Acc. Chem. Res.*, **13**, 135 (1980); (d) R. W. Murray, Andrew G. Ewing, and Richard A. Durst, "Chemically Modified Electrodes. Molecular Design for Electroanalysis," *Anal. Chem.*, **59**, 379A (1987); (e) M. S. Wrighton, "Prospects for a New Kind of Synthesis: Assembly of Molecular Components to Achieve Functions," *Inorg. Chem.*, **4**, 269 (1985); (f) M. S. Wrighton, "Surface Functionalization of Electrodes with Molecular Reagents," *Science*, **231**, 32 (1986); (g) L. R. Faulkner, "Chemical Microstructures on Electrodes," *Chem. Eng. News*, **62**(9), 28 (1984); (h) A. J. Bard, "Chemical Modification of Electrodes," *J. Chem. Ed.*, **60**, 302 (1983); (i) A. J. Bard and W. E. Rudzinski, "Electrochemistry at Modified Electrode Surfaces," in *Preparative Chemistry Using Supported Reagents*, P. Laszlo, ed., Academic Press, San Diego, 1987, pp. 77-97; (j) M. Fujihira, "Modified Electrodes," in *Topics in Organic Electrochemistry*, A. J. Fry and W. E. Britton, eds., Plenum, New York, 1986; (k) I. Rubinstein, "Electrochemical Processes at Polymer-Coated Electrodes," in *Applied Polymer Analysis and Characterization*, J. Mitchell, Jr., ed., Hanser: Munich, 1992, Vol. II, Part III, Chapter 1; (l) R.

M. Murray, ed., *Molecular Design of Electrode Surfaces*, Wiley, New York, 1992; (m) G. Inzelt, "Mechanism of Charge Transport in Polymer Modified Electrodes," in *Electroanalytical Chemistry*, A. J. Bard, ed., Marcel Dekker, New York, 1994, Vol 18, pp. 89-241.

2. I. M. Kolthoff and J. J. Lingane, *Polarography*, Interscience, New York, 1952, p. 291.
3. A. J. Bard, *J. Electroanal. Chem.*, **2**, 117 (1962).
4. See, for example: (a) D. Zurawski, L. Rice, M. Hourani, and A. Wieckowski, *J. Electroanal. Chem.*, **230**, 221 (1986); (b) J. Clavilier, *J. Electroanal. Chem.*, **107**, 211 (1980) and references cited therein.
5. R. L. McCreery, in *Electroanalytical Chemistry*, A. J. Bard, ed., Marcel Dekker, New York, 1991, Vol. 17, p. 221.
6. See, for example: H. Chang and A. J. Bard, *Langmuir*, **7**, 1143 (1991).
7. A. P. Brown and F. C. Anson, *Anal. Chem.*, **49**, 1589 (1977).
8. H. Daifuku, I. Yoshimura, I. Hirata, K. Aoki, K. Tokuda, and H. Matsuda, *J. Electroanal. Chem.*, **199**, 47 (1986).
9. C-W. Lee and A. J. Bard, *J. Electroanal. Chem.*, **239**, 441 (1988).
10. See, for example: (a) W. A. Zisman, *Advances in Chemistry Series*, No. 43, American Chemical Society, Washington, DC, 1964, p. 1; (b) R. G. Nuzzo, B. R. Zegarski, and L. H. Dubois, *J. Am. Chem. Soc.*, **109**, 733 (1987); (c) M. D. Porter, T. B. Bright, D. L. Allara, and C. F. D. Chidsey, ibid., 3559; (d) R. Maoz and J. Sagiv, *J. Coll. Interfac. Sci.*, **100**, 465 (1984); (e) H. O. Finklea, S. Avery, M. Lynch, and T. G. Furtsch, *Langmuir*, **3**, 409 (1987); (f) G. M. Whitesides and G. S. Ferguson, *Chemtracts—Org. Chem.*, **1**, 171 (1988).
11. R. M. Kannuck, J. M. Bellama, E. A. Blubaugh, and R. A. Durst, *Anal. Chem.*, **59**, 1473 (1987).
12. C. Miller and M. Majda, *J. Electroanal. Chem.*, **207**, 49 (1986).
13. (a) B. Keita and L. Nadjo, *J. Electroanal. Chem.*, **243**, 87 (1988); (b) P. J. Kulesza, G. Roslonek, and L. R. Faulkner, *J. Electroanal. Chem.*, **280**, 233 (1990) and references cited therein.
14. P. Laszlo, ed., *Preparative Chemistry Using Supported Reagents*, Academic Press, San Diego, 1987.
15. A. J. Bard and T. Mallouk, in *Molecular Design of Electrode Surfaces*, R. M. Murray, ed., Wiley, New York, 1992, Chapter 6.
16. P. K. Ghosh and A. J. Bard, *J. Am. Chem. Soc.*, **105**, 5691 (1983).
17. (a) A. Yamagishi and A. Aramata, *J. Chem. Soc. Chem. Commun.*, **1984**, 452 (1984); (b) *J. Electroanal. Chem.*, **191**, 449 (1985); (c) A. Yamagishi, *J. Coord. Chem.*, **16**, 131 (1987).
18. (a) J. M. Newsam, *Science*, **231**, 1093 (1986); (b) W. Hölderich, M. Hesse, and F. Näumann, *Angew. Chem. Intl. Ed. Engl.*, **27**, 226 (1988).
19. (a) C. G. Murray, R. J. Nowak, and D. R. Rolison, *J. Electroanal. Chem.*, **164**, 205 (1984); (b) H. A. Gemborys and B. R. Shaw, *J. Electroanal. Chem.*, **208**, 95 (1986); (c) Z. Li and T. E. Mallouk, *J. Phys. Chem.*, **91**, 643 (1987).
20. (a) K. Itaya, I. Uchida, and V. D. Neff, *Acc. Chem. Res.*, **19**, 162 (1986); (b) K. Itaya, H. Akahashi, and S. Toshima, *J. Electrochem. Soc.*, **129**, 1498 (1982); (c) K. Itaya, T. Ataka, and S. Toshima, *J. Am. Chem. Soc.*, **104**, 4767 (1982).

21. B. D. Humphrey, S. Sinha, and A. Bocarsly, *J. Phys. Chem.*, **88,** 736 (1984).
22. (a) E. A. H. Hall, *Biosensors*, Prentice-Hall, Englewood Cliffs, NJ, 1991; (b) G. A. Rechniz, *Anal. Chem.*, **54,** 1194A, 1982; *Science*, **214,** 287 (1981); (c) M. E. Meyerhoff and Y. M. Fraticelli, *Anal. Chem.*, **54,** 27 (1982).
23. (a) L. C. Clark, Jr. and C. Lyons, *Ann. NY Acad. Sci.*, **102,** 29 (1962); (b) P. W. Carr and L. D. Bowers, *Immobilized Enzymes in Analytical and Clinical Chemistry*, Wiley-Interscience, New York, 1980; (c) N. Lakshminarayanaiah, *Membrane Electrodes*, Academic Press, New York, 1976; (d) P. L. Bailey, *Analysis with Ion Selective Electrodes*, Heyden, London, 1976.
24. (a) T. Matsue, M. Fujihira, and T. Osa, *J. Electrochem. Soc.*, **126,** 500 (1979); (b) T. Matsue, U. Akiba, and T. Osa, *Anal. Chem.*, **58,** 2096 (1986).
25. A. R. Guadalupe and H. Abruña, *Anal. Chem.*, **57,** 142 (1985).
26. (a) I. Rubinstein, S. Steinberg, Y. Tor, A. Shanzer, and J. Sagiv, *Nature*, **332,** 426 (1988); (b) S. Steinberg and I. Rubinstein, *Langmuir*, **8,** 1183 (1992).
27. R. Bilewicz and M. Majda, *J. Am. Chem. Soc.*, **113,** 5464 (1991).
28. O. Chailapakul and R. M. Crooks, *Langmuir*, **9,** 884 (1993).
29. (a) A. N. K. Lau and L. L. Miller, *J. Am. Chem. Soc.*, **105,** 5271 (1983); (b) B. Zinger and L. L. Miller, ibid., **106,** 6861 (1984).
30. T. Saji, *Chem. Lett.*, **1986,** 275 (1986).
31. A. Kaifer, L. Echegoyen, D. A. Gustowski, D. M. Goli, and G. W. Gokel, *J. Am. Chem. Soc.*, **105,** 7168 (1983).
32. (a) D. W. DeWulf and A. J. Bard, *J. Electrochem. Soc.*, **135,** 1977 (1988); (b) H. Takenaka and H. Torikai, Kakai Tokkyo Koho (Jpn. Patent) 55, 38934 (1980).
33. S. Mazur and S. Reich, *J. Phys. Chem.*, **90,** 1365 (1986).
34. E. Raoult, J. Sarrazin, and A. Tallec, *J. Appl. Electrochem.*, **15,** 85 (1985).
35. Z. Ogumi, S. Ohashi, and Z. Takehara, *Electrochim. Acta*, **30,** 121 (1985).
36. P. G. Pickup and R. W. Murray (a) *J. Am. Chem. Soc.*, **105,** 4510 (1983); (b) *J. Electrochem. Soc.*, **131,** 833 (1984).
37. J. C. Jernigan, C. E. D. Chidsey, and R. W. Murray, *J. Am. Chem. Soc.*, **107,** 2824 (1985).
38. H. Kuhn, *J. Photochem.*, **10,** 111 (1979).
39. (a) P. G. Pickup, C. R. Leidner, P. Denisevich, and R. W. Murray, *J. Electroanal. Chem.*, **164,** 39 (1984); (b) C. R. Leidner, P. Denisevich, K. W. Willman, and R. W. Murray, ibid., **164,** 63 (1984).
40. (a) G. P. Kittlesen, H. S. White, and M. S. Wrighton, *J. Am. Chem. Soc.*, **106,** 7389 (1984); (b) H. S. White, G. P. Kittleson, and M. S. Wrighton, ibid., **106,** 5375 (1984); (c) I. Fritsch-Faules and L. R. Faulkner, *Anal. Chem.*, **64,** 1118 (1992) and references cited therein.
41. (a) G. P. Kittlesen, H. S. White, and M. S. Wrighton, *J. Am. Chem. Soc.*, **107,** 7373 (1985); (b) E. T. Turner-Jones, O. M. Chyan, and M. S. Wrighton, ibid., **109,** 5526 (1987).
42. A. J. Grodzinsky and A. M. Weiss, *Sep. Purif. Methods*, **14,** 1 (1985).
43. (a) P. Burgmayer and R. W. Murray, *J. Am. Chem. Soc.*, **104,** 6139 (1982); *J. Electroanal. Chem.*, **147,** 339 (1983).
44. M. Krishnan, X. Zhang, and A. J. Bard, *J. Am. Chem. Soc.*, **106,** 7371 (1984).

45. N. Nakashima, T. Tagujchi, Y. Takada, K. Fujio, M. Kunitake, and O. Manabe, *J. Chem. Soc. Chem. Commun.*, **1991**, 232 (1991).
46. (a) T. P. Henning, H. S. White, and A. J. Bard, *J. Am. Chem. Soc.*, **103**, 3937 (1981); (b) **104**, 5862 (1982).
47. F-R. F. Fan and A. J. Bard, *J. Electrochem. Soc.*, **133**, 301 (1986).
48. R. B. Bjorklund and B. Liedberg, *J. Chem. Soc. Chem. Commun.*, **1986**, 1293 (1986).
49. S. Cosnier, A. Deronzier, and J. C. Moutet, *J. Electroanal. Chem.*, **193**, 193 (1985).
50. R. A. Bull, F-R. Fan, and A. J. Bard, *J. Electrochem. Soc.*, **131**, 687 (1984).
51. P. G. Pickup, K. N. Kuo, and R. W. Murray, *J. Electrochem. Soc.*, **130**, 2205 (1983).

Chapter 5
ELECTROCHEMICAL CHARACTERIZATION OF MODIFIED ELECTRODES

5.1. INTRODUCTION

Electrochemical methods are very useful in examining modified electrodes immersed in solution. These methods are sufficiently sensitive that fractions of a monolayer can be detected. They can also provide information about the thermodynamic properties of the layer, kinetics of electron transfer (et) at the layer–electrode and layer–solution interfaces, and for thicker layers, the kinetics of mass and charge transfer within the layer. They thus provide a useful complement to the powerful spectroscopic techniques discussed in Chapter 3. In this chapter, the basic concepts of several electrochemical approaches and the final equations that are used in the treatment of data will be presented; details of derivations and the finer points of the methodology will not be discussed. For those unfamiliar with electroanalytical techniques, detailed treatments and monographs on electrochemical methods and modified electrodes are available (1–6).

First, the characterization of monolayers and thin films by voltammetric techniques will be described. The permeation of solution species through films and the application of electrochemical methods to the study of film porosity and membrane behavior will then be treated. Then measurements of mass and charge transfer within the film, representing physical movement of species or hopping of electrons between sites, will be discussed. Finally, et at the electrode–film interface (heterogeneous et), which is a function of the potential drop across the interface, and et and mass transfer at the film–solution interface will be treated. These processes are shown schematically in Figure 5.1.1.

5.2. ELECTROCHEMISTRY OF MONOLAYERS AND THIN FILMS

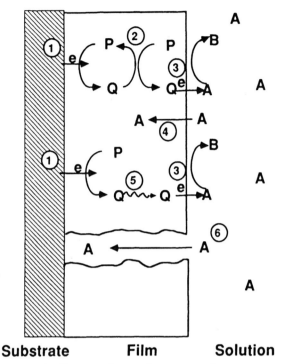

Substrate Film Solution

FIGURE 5.1.1. Schematic diagram of several processes that can occur at a modified electrode. P represents a reducible substance in a film on the substrate surface and A, a species in solution. Processes shown are (1) heterogeneous et to P to produce reduced form, Q; (2) et from Q to another P in film (electron hopping in film); (3) et from Q to A at film/solution interface; (4) penetration of A into film (where it can also react with Q or at the substrate–film interface); (5) movement (mass transfer) of Q within the film; (6) movement of A through pinhole or channel in the film to the substrate, where it can be reduced. The contributions and rates of these processes can be probed by electrochemical techniques.

5.2. ELECTROCHEMISTRY OF MONOLAYERS AND THIN FILMS

Consider a film of an electroactive substance on the surface of an inert conductive substrate (the electrode). The film can be a fraction of a monolayer of material or a layer up to several micrometers thick. We assume here that there are no species dissolved in solution that are electroactive within the potential range in which the film species is oxidized or reduced. One approach to the study of the electroactivity involves the imposition of a potential step or sweep across a region of potentials where electron transfer between electrode and film can occur.

This et process is monitored by the current that results from the potential perturbation.

5.2.1. Potential Step Experiment

Consider the case where the film initially contains only the electroactive component O, present at a concentration C_O^*, which can be reduced at a potential well negative of the standard potential, $E°$, for the half-reaction

$$O + ne = R \qquad (5.2.1)$$

When the potential is stepped to a value where the reduction of O to R occurs and where the concentration of O at the electrode–film interface is essentially zero, a cathodic current (i) will flow. The observed current–time (i–t) behavior can be classified into two regions.

> *Region A.* At very short times, before an appreciable fraction of the film has been electrolyzed, the i–t behavior follows that for semi-infinite diffusion [the Cottrell equation (7)]:
>
> $$\frac{i_d(t)}{nFAD^{1/2}C_A} = (\pi t)^{-1/2} \qquad (5.2.2)$$
>
> This behavior obtains as long as $t \ll \phi^2/2D$, where ϕ is the film thickness, and D is the diffusion coefficient of mobile electroactive species or charge in the film.[1]
>
> *Region B.* At longer times, the diffusion layer reaches the film–solution interface, and the rate of depletion of electroactive species in the film decreases.

Schematic concentration profiles for a potential step at a modified electrode and the i–t behavior, illustrating these two regions, are shown in Figure 5.2.1. The amount of charge, Q, passed for the total reduction

---

[1] We neglect here an additional component of the current, the charging or capacitive current, i_c, which represents the nonfaradaic current caused by charging of the electrical double layer at the interface. Note, however, that this current can make an appreciable contribution at very short times or when only monolayer amounts of O are present.

5.2. ELECTROCHEMISTRY OF MONOLAYERS AND THIN FILMS

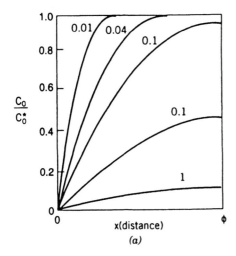

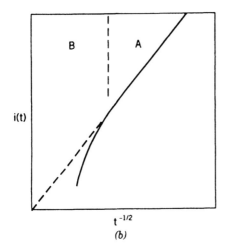

FIGURE 5.2.1. (a) Concentration profiles within a thin film of thickness ϕ at different times, indicated on the curve in terms of the dimensionless parameter Dt/ϕ^2. (b) Schematic current–time curve (in form i vs. $t^{-1/2}$) for the current transient during a potential step at a thin film electrode, showing zones A and B mentioned in text.

of all O in the film is[2]

$$Q = nFA\phi C_O^* = nFA\Gamma_O^* \tag{5.2.3}$$

where Γ_O^* is the amount of O in the film (mol/cm^2).

Typical results of an experiment of this type are shown in Figure 5.2.2 (8). In this experiment, (trimethylammonio)methyl ferrocene

---
[2] When the double-layer charge (Q_{dl}) is important $Q = nFA\Gamma_O^* + Q_{dl}$.

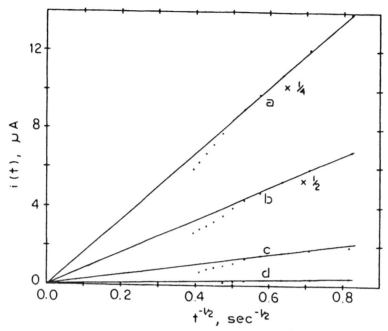

FIGURE 5.2.2. Actual current–time curves (in form i vs. $t^{-1/2}$) for potential step oxidation of Cp_2FeTMA^+ in a Nafion film on a glassy carbon substrate, with a film thickness, ϕ, of 1 μm. The different curves are for different concentrations of Cp_2FeTMA^+ in the film: (a) 0.76 M; (b) 0.19 M; (c) 35 mM; (d) 7 mM. The deviation of the curve from a straight line ("Cottrell behavior") occurs on zone B when thin layer effects become significant. [Reprinted with permission from H. S. White, J. Leddy, and A. J. Bard, J. Am. Chem. Soc., **104**, 4811 (1982). Copyright 1982 American Chemical Society.]

(Cp_2FeTMA^+) was incorporated into a 1-μm-thick film of Nafion on a glassy carbon (GC) electrode surface (by extraction from an aqueous solution of the PF_6^- salt). Plots of i versus $t^{-1/2}$ were linear and showed zero intercepts, as predicted from Eq. (5.2.2) for $t < 3$ s. The diffusion coefficient, D_S, of Cp_2FeTMA^+ in the film could be obtained from the slope of the curve and was found to be 1.7×10^{-10} cm^2/s, essentially independent of the concentration of Cp_2FeTMA^+, over a concentration range of 7 mM–1.2 M. At longer times, the current fell below the line (i was smaller than that expected from pure diffusional behavior), because the electroactive substance was depleted as the film approached thin-film behavior. Chronopotentiometric (constant current) experiments showed analogous behavior and yielded the same values for the diffusion coefficient (8).

5.2.2. Potential Sweep Experiments–Nernstian Reactions

Similar behavior is found with potential sweeps. When the sweep rate, v, is large, the i-E curve is that governed by semiinfinite diffusion (9). At small v, the system can be treated by considering that the species is depleted uniformly throughout the film (the "thin layer" approximation) (10). This approximation holds to within 1% as long as

$$|v| < (3 \times 10^{-2}) \frac{RT}{nF} \frac{D}{\phi^2} \qquad (5.2.4)$$

The predicted behavior under these conditions can be derived in a rather straightforward manner, without the need to include mass-transfer effects. Under the assumption that there are no interactions between the components of the film, O and R, and that the electron transfer at the electrode–film interface is rapid, the concentrations of O and R throughout the film will follow the Nernst equation

$$E = E^{\circ\prime} + \frac{RT}{nF} \log\left[\frac{C_O(t)}{C_R(t)}\right] = E^{\circ\prime} + \frac{RT}{nF} \log\left[\frac{\Gamma_O}{\Gamma_R}\right] \qquad (5.2.5a)$$

or

$$\frac{\Gamma_O}{\Gamma_R} = \exp\left[\frac{nF}{RT}(E - E^{\circ\prime})\right] = \eta \qquad (5.2.5b)$$

where $E^{\circ\prime}$ is the formal potential for the O/R couple on the electrode. If the film is a monolayer or less, $E^{\circ\prime}$ can be related to that of the same redox couple in solution by taking account of the energy of interaction of O and R with the electrode. For example, if the adsorption of O and R is governed by Langmuir isotherms, $E^{\circ\prime} = E^{\circ\prime}(\text{soln}) + (RT/nF)\ln(b_R/b_O)$, where b_R and b_O are adsorption parameters related to the free energy of adsorption of R and O. The rate of the electrode reaction, i/nF, can then be equated to the rate of electrolysis of O to yield

$$\frac{i}{nF} = -A \frac{d\Gamma_O(t)}{dt} = -Av \frac{d\Gamma_O}{dE} \qquad (5.2.6)$$

By combining these equations, with the knowledge that $\Gamma_O^* = \Gamma_O + \Gamma_R$, the following equation of the i–E curve is obtained:

$$i = \frac{n^2 F^2}{RT} A v \Gamma_O^* \frac{\eta}{(1+\eta)^2} \qquad (5.2.7)$$

A typical i–E response for cyclic voltammetry of a nernstian or reversible electron-transfer reaction of a surface-confined species under "thin layer" conditions is shown in Figure 5.2.3. The forward scan follows Eq. (5.2.7), and the reverse scan (where R is oxidized to O) is just the mirror image. The peak potentials of the cathodic and anodic waves, $E_{p,c}$ and $E_{p,a}$, respectively, both occur at $E°$. The peak current, i_p, is given by

$$i_p = \frac{n^2 F^2}{4RT} v A \Gamma_O^* \qquad (5.2.8)$$

and as shown, i_p is directly proportional to scan rate. This is what one would expect intuitively, since the total area under the curve, Q, is a constant and independent of v. If the scan rate is doubled (i.e., the time to traverse the wave is halved), the current all along the wave must double to pass the same amount of charge, Q. A convenient parameter for testing whether the wave is an uncomplicated, nernstian one is the peak width at one-half the peak current, $\Delta E_{p,1/2}$, given by

$$\Delta E_{p,1/2} = 3.53 \frac{RT}{nF} = \frac{90.6}{n} \text{ mV at } 25°C \qquad (5.2.9)$$

One can compare these surface waves with those characteristic of a species freely diffusing under semiinfinite conditions (Fig. 5.2.3B). The surface waves decay to zero current, are more symmetric (in both the X and Y planes), and show no peak splitting (i.e., $\Delta E_p = E_{p,a} - E_{p,c} = 0$). One should also see semiinfinite voltammetric behavior in films on an electrode, when the scan rate is sufficiently fast; see Eq. (5.2.3). Such fast scan rates are usually not possible with very thin films or monolayers, however.

The ideal, nernstian, behavior of Eq. (5.2.7) is actually not seen very often for real films on electrode surfaces. Deviations in the i–E curve can arise from kinetic limitations in the rate of electron transfer (et), interactions between the film components, complexities in the reaction mechanism, changes in film structure, and limitations caused by ionic motion.

5.2. ELECTROCHEMISTRY OF MONOLAYERS AND THIN FILMS

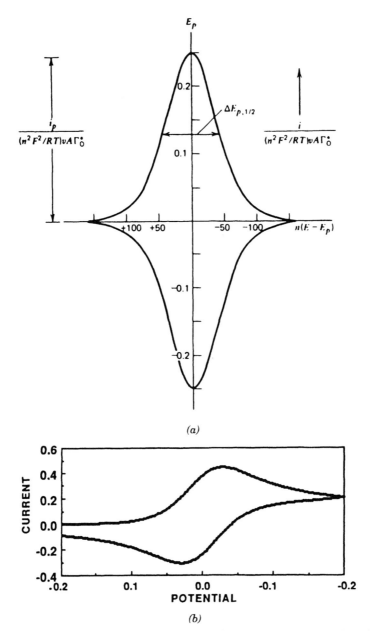

FIGURE 5.2.3 (a) Ideal current–potential curve for potential sweep (cyclic voltammetry) experiment with a thin film or adsorbed layer of a species that undergoes rapid electron transfer at the film–substrate interface (i.e., a nernstian reaction) with the thin-layer approximation valid. Current is shown in normalized form and potential axis is given for 25°C. (b) Cyclic voltammetry of a nernstian of a species dissolved in solution. [Reprinted with permission from A. J. Bard and J. R. Faulkner, *Electrochemical Methods*, Wiley, New York, 1980. Copyright 1980 John Wiley & Sons.]

5.2.3. Potential Sweep Experiments—Slow Heterogeneous Kinetics

If the rate of et at the electrode–film interface is slow, Eq. (5.2.5) no longer holds and must be replaced by a condition that relates how the rate of et varies with potential. The current–potential curves for slow heterogeneous et, with the assumption that mass transfer through the film is rapid, are asymmetric, as shown in Figure 5.2.4 (5, 11). The peak current is given by

$$i_p = n\alpha n_a \frac{F^2}{2.718RT} A v \Gamma_O^* \tag{5.2.10}$$

and is proportional to v. $E_{p,c}$ and ΔE_p are

$$E_{p,c} = E^{\circ\prime} + \frac{RT}{\alpha n_a F} \ln\left(\frac{RT}{\alpha n a F} \frac{k^\circ}{v}\right) \tag{5.2.11}$$

$$\Delta E_{p,1/2} = 2.44 \frac{RT}{\alpha n_a F} = \frac{62.5}{\alpha n_a} \text{ mV at } 25^\circ C \tag{5.2.12}$$

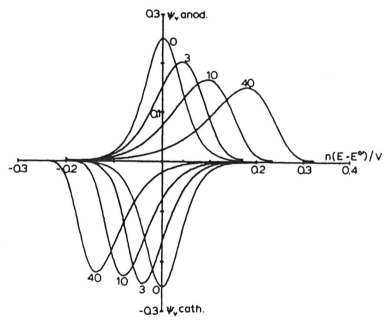

FIGURE 5.2.4. Effect of slow heterogeneous electron transfer kinetics on the shape of the cyclic voltammetric wave for a thin layer on an electrode surface. Results are for 25°C. and $\alpha = 0.6$ with values of nvF/RTk° shown on each curve. [Reprinted with permission from E. Laviron, *J. Electroanal. Chem.*, **100**, 263 (1979). Copyright 1979 Elsevier.]

5.2.4. Potential Sweep Experiments—Lateral Interactions of Film Components

When lateral interactions exist between O and R in the film, the shape of the i-E curve depends on the energies of the interactions of O with O, R with R, and O with R. The exact shape of the curve depends on the choice of the adsorption isotherm or how the interactions are taken into account. If the heterogeneous et is rapid, the problem can be addressed by using the desired isotherm to derive the appropriate form of the Nernst expression, Eq. (5.2.5). For example, if a Frumkin-type isotherm (1, 12, 13) is assumed, the relevant expression is

$$\frac{\Gamma_O}{\Gamma_R} = \frac{\theta_O}{\theta_R} \exp\left[2v\theta_O(a_{OR} - a_O) + 2v\theta_R(a_R - a_{OR})\right] \quad (5.2.13)$$

where a_{OR}, a_O, and a_R are the O—R, O—O, and R—R interaction parameters ($a_i > 0$ for an attractive interaction and $a_i < 0$ for a repulsive one), v is the number of water molecules displaced from the surface by adsorption of one O or R, and θ_O and θ_R are the fractional coverages of O and R, respectively. The expression for the i-E curve is then (14)

$$i = \frac{n^2 F^2 A v \Gamma_O^*}{RT} \frac{\theta_R(1 - \theta_R)}{1 - 2vG\theta_T\theta_R(1 - \theta_R)} \quad (5.2.14)$$

where $\theta_T = (\theta_O + \theta_R)$, $G = a_O + a_R - 2a_{OR}$, $\Gamma_O^* = \Gamma_O + \Gamma_R$, $\theta_i = \Gamma_i/\Gamma_O^*$. The potential variation in Eq. (5.2.14) arises through the variation of θ_R with E, via Eq. (5.2.5b). Typical i-E curves based on Eq. (5.2.14) are given in Figure 5.2.5. The curve shape is governed by the interaction parameter, $2vG\theta_T$. When this is 0, the behavior is that of Figure 5.2.3 and $\Delta E_{p,1/2} = 90.6/n$ mV. When $2vG\theta_T > 0$, $\Delta E_{p,1/2} < 90.6/n$; when $2vG\theta_T < 0$, $\Delta E_{p,1/2} > 90.6/n$.

The equations given above, based on the Frumkin isotherm, assume a random distribution of O and R sites in the film. If the film is structured, as in an organized monolayer deposited by the LB technique, there will be an ordered distribution of the sites. Under these conditions, a statistical mechanical approach is needed to account for the interactions and find the i-E curve (15). For example, voltammograms for a surface film consisting of a two-dimensional quasicrystalline lattice with an hexagonal close-packed arrangement of sites are shown in Figure 5.2.6. Note that for negative values of the interaction parameter at a structured film a double wave results, even for a single-electrode reaction, while the random distribution produces only a single broadened wave.

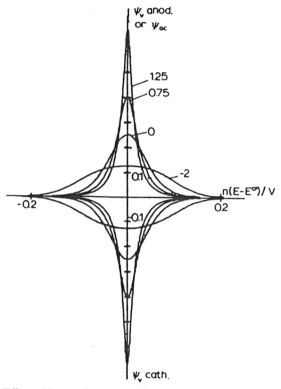

FIGURE 5.2.5. Effect of interactions on cyclic voltammetric wave shape for a thin layer on an electrode. A Frumkin isotherm is assumed and values of $vG\theta_T$ are shown on each curve. The curve with $vG\theta_T = 0$ corresponds to that in Figure 5.2.3. [Reprinted with permission from E. Laviron, *J. Electroanal. Chem.*, **100**, 263 (1979). Copyright 1979 Elsevier.]

Note that the preceding approaches to treatments of nonideal surface films rely on empirical adjustable parameters to produce waves with the shapes of experimental ones. Smith and White (16) recently pointed out that by taking account of the interfacial potential distribution (related to such factors as the dielectric constants of film and solution, the concentrations of electroactive adsorbate and supporting electrolyte, and the film thickness) the shape of the cyclic voltammetric curves could be modeled without the need for such parameters. Moreover, in actual studies of modified surfaces, the voltammetric waves frequently deviate from the above, still rather idealized, behavior, for example, by showing severe asymmetry or strongly nonparabolic shapes (17). Rarely can actual experimental voltammograms be described simply in terms of the preceding types of parameters. The overall situation is usually much

5.2. ELECTROCHEMISTRY OF MONOLAYERS AND THIN FILMS

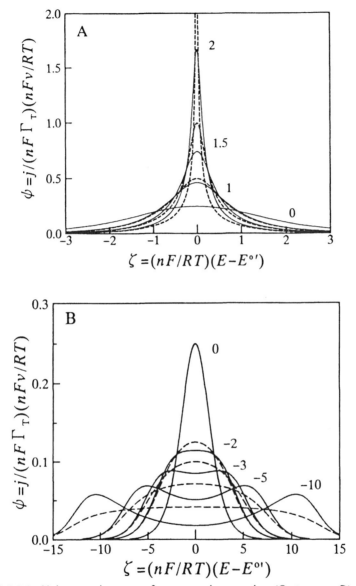

FIGURE 5.2.6. Voltammetric curves for a nernstian reaction (O + ne = R) derived for an organized film (a two-dimensional quasicrystalline lattice in an hexagonal close-packed arrangement) (solid) and a random distribution of sites (dashed) for different values of the interaction parameter, W/RT. [Reprinted with permission from H. Matsuda, K. Aoki, and K. Tokuda, *J. Electroanal. Chem.*, **217**, 15 (1987). Copyright 1987 Elsevier.]

more complicated, with factors such as inhomogeneity of the film, finite mass and charge transport through the film, and structural and resistive changes in the film on oxidation and reduction coming into play. Some of these factors will be dealt with in later sections of this chapter.

5.3. PERMEATION OF SOLUTION SPECIES THROUGH FILMS

Let us consider electrochemical approaches to the determination of permeability of a surface film. Here we assume that the film itself is not electroactive or electronically conductive, so that before a species, A, in solution can be reduced at the conductive electrode, it must penetrate the film. Two possible modes are suggested in Figure 5.1.1: permeation of species A into the film, as, for instance, by a membrane-type extraction process (process 4), or movement of A through pinholes or channels (process 6). Electrochemical measurements can provide information about these processes, and are useful in assessing the extent to which films formed on an electrode surface (e.g., thin polymer layers or organized assemblies) are free from pinholes.

5.3.1. Pinhole Model

Consider an electrode covered with a film that has continuous pores or channels from the solution to the electrode. We can ask how the electrolysis of a species in solution at such an electrode differs from that at the bare (unfilmed) electrode. The answer depends on the extent of coverage of the electrode by the film, the size and distribution of the pores, and the time scale of the experiment. The situation is shown schematically in Figure 5.3.1. Take the case where the fraction of the surface uncovered, $(1 - \theta)$, is small, the pore radius, a, is small, and the pores are spaced far apart (compared to a). When the time scale of the experiment (e.g., the time for a potential step in chronoamperometry, t) is small, so that $(Dt)^{1/2} \ll a$, the electrode will show the usual electrochemical response, except that the area will be $(1 - \theta)$ times that of the bare electrode; that is, electrolysis of solution species occurs only directly at the substrate within the pores of the film. When the time is such that the diffusion layers from the individual pores grow to the point where they overlap and merge, the electrode behavior approaches that of the unfilmed electrode with a total area of that of the bare electrode. Thus a study of the electrochemical response as the effective time scale of the experiment is varied, can provide information about θ, a, and the

5.3. PERMEATION OF SOLUTION SPECIES THROUGH FILMS

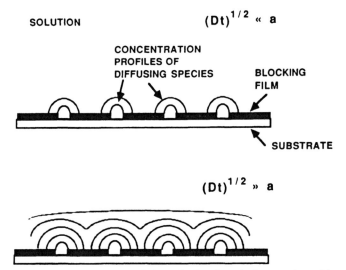

FIGURE 5.3.1. Schematic diagram of electrolysis of a solution species with a diffusion coefficient, D, at a conductive substrate covered with a blocking film containing pores of radius, a, at short times (top) and long times (bottom).

pore distribution. The actual situation can be complicated, for example, if the film thickness is large and the channels to the electrode are tortuous ones. Most of the models employed to treat these experiments assume a very thin blocking film with channels of negligible depth and a uniform distribution of pores in a specific geometry. We consider two types of experiments, chronoamperometric and rotating-disk electrode (RDE) ones.

Chronoamperometry. Consider the current, i, during a potential step to the mass-transfer limiting region for the reduction of solution species, A, with diffusion coefficient D at concentration C_A, at the film-covered electrode. The i–t curve at the bare electrode of total area, A, follows the familiar Cottrell equation (7):

$$\frac{i_d(t)}{nFAD^{1/2}C_A} = (\pi t)^{-1/2} \tag{5.3.1}$$

Different models for partially covered electrodes have been proposed. For example, Gueshi et al. (18) considered an electrode surface with uniformly distributed, circular, active regions of radius a inside hexagonal, inactive regions of total radius R. The current at such an electrode,

$i(t)$, normalized to the current at the same time at the bare electrode, is given by (19)

$$i(\tau)/i_d(\tau) = \frac{1}{\sigma^2 - 1} [\sigma \exp(-\tau) - 1 \\ + \sigma^2(\pi T)^{1/2} \exp(T)\{\text{erf}(\sigma T^{1/2}) - \text{erf}(T^{1/2})\}] \quad (5.3.2)$$

where $T = \tau/(\sigma^2 - 1)$, $\sigma = \theta/(1 - \theta)$, $\tau = lt$, and l is a function of D, the hole size and distribution, and θ, as defined in Ref. 19. A plot of $i(\tau)/i_d(\tau)$ is given in Figure 5.3.2(a) for different values of θ. Note that at short times (small τ) the current ratio attains the limiting value of $1 - \theta$. At long times, when the diffusion layer grows to a thickness that is large compared to R, the ratio approaches unity. The location of the intermediate region depends on θ and a, and thus a plot of $i(\tau)/i_d(\tau)$ versus t can be used to estimate these parameters.

Rotating-Disk Electrode. Analogous experiments can be carried out with the rotating-disk electrode (RDE). The limiting current, i_A, for the electroreduction of solution species A at the bare RDE (20) follows the Levich equation:

$$i_A = 0.62nFAC_A D^{2/3}\nu^{-1/6}\omega^{1/2} = \frac{FAC_A D}{\delta} \quad (5.3.3)$$

where δ is the thickness of the diffusion layer in solution; $\delta = 1.61 D^{1/3}\nu^{1/6}\omega^{-1/2}$. For the same electroreduction at the film-covered electrode, the limiting current, $i_{\lim}$, depends on the rotation rate and the size and distribution of the pores (or active sites) (21):

$$\frac{1}{i_{\lim}} = \frac{1}{i_A} + \sum_n A_n \frac{\tanh\left[\dfrac{x_n \delta}{r_2}\right]}{nFADC_A} \quad (5.3.4)$$

where A_n is a function of r_1 (the radius of the active site), r_2 ($\frac{1}{2}$ the distance between centers of adjacent sites), and x_n (the zero points of first-order Bessel functions). Experimental data can be analyzed by plotting $1/i_{\lim}$ against $\omega^{-1/2}$. This equation has been applied to film-covered electrodes with a discussion of the limiting conditions and limitations (22).

5.3. PERMEATION OF SOLUTION SPECIES THROUGH FILMS

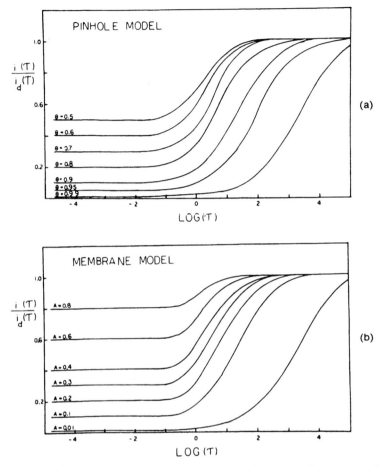

FIGURE 5.3.2. Working curves for chronoamperometric (potential step) experiments at electrodes covered with a blocking film assuming (a) pinhole and (b) membrane models. The curves are given in terms of dimensionless parameters, the current ratio, $i(\tau)/i_d(\tau)$, and τ (see text), for different coverages, θ or values of $A = K/\gamma$. [Reprinted with permission from J. Leddy and A. J. Bard, *J. Electroanal. Chem.*, **153**, 223 (1983). Copyright 1983 Elsevier.]

Other Methods. The effect of partial blockage of the electrode surface in cyclic voltammetry (CV) has also been considered (23, 24). While the basic approach is analogous to those above, the effects of heterogeneous et kinetics comes into play in CV (whereas only mass-transfer limiting currents need be considered for chronoamperometry and RDE methods carried out at sufficiently extreme potentials). The system treated assumed circular active sites of radius R_a surrounded by inactive zones of radius R_0, with R_0 small compared to the thickness of the diffusion layer. When the film coverage, θ, is not too close to unity, the CV behavior is the same as that at the bare electrode (e.g., i_p is nearly the same), but the apparent rate constant for heterogeneous et, k_{ap}°, is smaller than the actual rate constant by the factor $(1 - \theta)$. This is equivalent to saying that the diffusional process is not greatly perturbed under these conditions, but the effective current density is increased over that at the bare electrode because of blockage by the film. When $\theta \to 1$, the CV behavior becomes analogous to that found for a CEC process (where a chemical reaction precedes and follows the et process) and depends on two parameters: $k_0(1 - \theta)(RT/DFv)^{1/2}$ and $[(1 - \theta)(DRT/Fv)]^{1/2}/0.6R_0$. When the second parameter is small (i.e., very high coverage), the voltammogram will have the S shape of a polarogram with a plateau rather than a peak. The limiting (plateau) current under these conditions is

$$i_{\lim} = \frac{FAC_A D(1 - \theta)^{1/2}}{0.6R_0} \quad (5.3.5)$$

and is independent of v. Because the effect of heterogeneous et kinetics enters into consideration of the CV behavior, obtaining estimates of coverage and pore size is probably more difficult with this technique than with potential step and RDE methods (24). The effect of surface blockage on the ac (faradaic impedance) response has also been considered (25).

5.3.2. Membrane Model

An alternative way in which A in solution can traverse the film, assumed now to be pinhole-free, is via extraction into the film and diffusion through it to the electrode. The film can be treated as an immiscible liquid layer on the electrode surface into which substance A is extracted. If the extraction rate at the film–solution interface is rapid, so that equilibrium is maintained, the concentration of A at the film side of the interface $(x = \phi^-)$ is related to that on the solution side of the interface $(x = \phi^+)$ by the expression

$$C_A(\phi^-, t) = \kappa C_A(\phi^+, t) \quad (5.3.6)$$

where κ is the extraction constant. The diffusion coefficient of A in the film, D_S, is also different from that in solution, D. This type of film can

5.3. PERMEATION OF SOLUTION SPECIES THROUGH FILMS

also be investigated electrochemically by potential step and RDE techniques.

Chronoamperometry. Typical concentration profiles for a potential step experiment where the concentration of A at the electrode–film interface, $C_A(x = 0) \approx 0$ are shown in Figure 5.3.3. The expression for the current, normalized to that at the bare electrode is (19)

$$\frac{i(\tau)}{i_d(\tau)} = a\left\{1 + 2\sum_{j=1}^{\infty}\left(\frac{1-a}{1+a}\right)^j \exp\left(\frac{-j^2}{\tau}\right)\right\} \quad (5.3.7)$$

where $\tau = D_S t/\phi^2$ and $a = \kappa(D/D_S)^{1/2}$. Plots of the normalized current versus log τ, for different values of a, are shown in Figure 5.3.2b. Note the similarity to the pinhole model curves in Figure 5.3.2a. At short times, when the diffusion layer thickness is small compared to the film thickness [i.e., $(D_S t)^{1/2} \ll \phi$], the electrolysis occurs completely within the film and is characterized by a diffusion coefficient, D_S, and an initial concentration, κC_A. Under these conditions the current ratio approaches $\kappa(D_S/D)^{1/2}$. At long times, the diffusion layer extends well into the solution phase, and the current ratio approaches unity.

Experimental chronoamperometric investigations of polymer films (~micrometers thick) have been carried out and values of κ and D_S reported [e.g., for benzoquinone and methyl viologen in poly(vinyl-

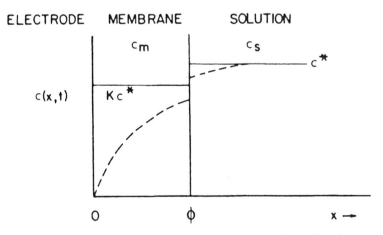

FIGURE 5.3.3. Concentration profiles for membrane model of thin film (thickness, ϕ). Solid lines, initial concentrations; dashed lines, profile after a potential step. The situation considered is for an extraction constant, κ, of less than one (<1).

ferrocene) films] (21). It may be difficult to use these kinds of data to choose between the pinhole and membrane models, because the appropriate time range needed to span a sufficiently large range of τ values may not be available.

Rotating-Disk Electrode. The application of the membrane model to the RDE (22, 26, 27) results in the following equation for the steady-state limiting current:

$$\frac{1}{i_{\lim}} = \frac{1}{i_A} + \frac{1}{i_S} \qquad (5.3.8)$$

where

$$i_S = \frac{nFA\kappa C_A D_S}{\phi} \qquad (5.3.9)$$

and i_A is defined in Eq. (5.3.3). The form of Eq. (5.3.8) is the same as that for the pinhole model, Eq. (5.3.4), and as that for slow heterogeneous kinetics (28). In all cases the equations have the form with one term dependent on $\omega^{-1/2}$ (solution mass transfer) and the other representing a slow process in series with the mass transfer. Thus a plot of $i_{\lim}^{-1}$ versus $\omega^{-1/2}$ will yield the desired rate-limiting term (e.g., i_S^{-1} in the case here) as the intercept.

It is instructive to sketch the derivation of this equation, since it illustrates the general approach to be used in following sections to elucidate processes occurring in films on electrodes by employing RDE steady-state measurements. Consider the concentration profiles in Figure 5.3.4 after steady state is attained at the RDE. The concentration of A at the electrode–film interface is essentially zero because the potential is held at a sufficiently negative value that all A reaching the electrode is reduced. In the bulk solution, A remains at the value C_A. At steady state, the flux of A across the film, $D_S C_A(\phi^-)/\phi$, must equal that in solution, $D[C_A - C_A(\phi^+)]/\delta$. Thus

$$\frac{D_S C_A(\phi^-)}{\phi} = \frac{D[C_A - C_A(\phi^+)]}{\delta} = \frac{i}{nFA} \qquad (5.3.10)$$

Combining this equation with Eq. (5.3.6) yields Eqs. (5.3.8) and (5.3.9), or

$$i^{-1} = i_S^{-1} + (0.62nFAD^{2/3}\nu^{-1/6}C_A)^{-1}\omega^{-1/2} \qquad (5.3.11)$$

5.3. PERMEATION OF SOLUTION SPECIES THROUGH FILMS 203

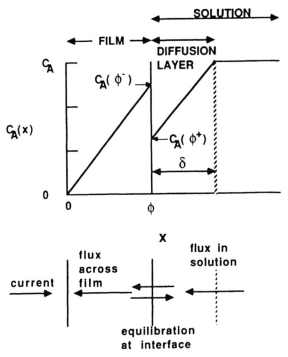

FIGURE 5.3.4. Steady-state concentration profiles for a RDE experiment with a film of thickness ϕ on the electrode surface. The thickness of the diffusion layer in solution is δ. The various fluxes of species A are indicated for the case (case S) where the current is determined by the mass transfer of A in the film and in solution.

A typical analysis of experimental data according to Eq. (5.3.11), based on an RDE study of the reduction of benzoquinone on an electrode coated with a film of poly(vinylferrocene), is shown in Figure 5.3.5 (22). Note that the slopes of the curves of $i_{\lim}^{-1}$ versus $\omega^{-1/2}$ are the same in the presence and absence of film, since they are determined by mass transport only in solution (i.e., i_A) and that the intercepts depend on ϕ and C_A, as predicted by Eq. (5.3.9). While these data could also be considered to fit the pinhole model, Eq. (5.3.4), in such a plot, the resulting intercepts yield pinholes with radii of < 10 nm; such a film can be considered as the equivalent of a homogeneous membrane (21). The case of a filmed RDE in which the steady-state current is completely limited by the extraction of a solution species and its diffusion across the film is designated "case S."

The case where an extraction equilibrium is not attained at the film-solution interface has also been considered (29). In this case, the transport of A across the interface becomes another limiting flux, given by

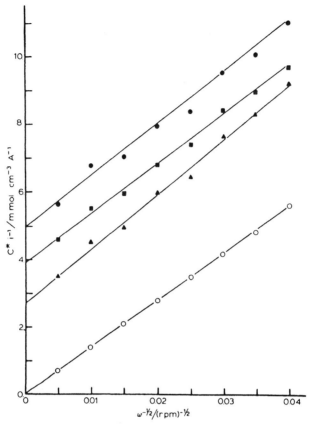

FIGURE 5.3.5. Experimental rotating-disk results for the reduction of benzoquinone (BQ) through a film of poly(vinylferrocene) on a Pt substrate. The reciprocal current (normalized for the concentration of BQ in solution, 5.82, 3.84, and 1.96 mM from top to bottom curves) versus $\omega^{1/2}$ (where ω is the angular velocity of rotation of the RDE). The open circles show the results for a bare Pt electrode immersed in a solution containing 5.82 mM BQ. [Reprinted with permission from J. Leddy and A. J. Bard, *J. Electroanal. Chem.*, **153**, 223 (1983). Copyright 1983 Elsevier.]

the expression $\chi_f C_A(\phi^+) - \chi_b C_A(\phi^-)$, where χ_f and χ_b are the rate constants for transfer of A from solution into film, and from film into solution, respectively. This can be equated to the other fluxes in Eq. (5.3.10) to yield

$$i_{lim}^{-1} = i_A^{-1} + i_S^{-1} + i_P^{-1} \quad (5.3.12)$$

where the permeation current, i_P, is given by

$$i_P = nFAC_A\chi_f \quad (5.3.13)$$

5.4. CHARGE TRANSPORT IN FILMS

Note that when the flux of A across the interface, measured by i_P, is large, the limiting behavior of Eq. (5.3.8) is obtained.

5.4. CHARGE TRANSPORT IN FILMS

The electrical properties of films are controlled by the movement of charged species (electrons, ions) through them. Such charge-transport processes in various types of films (e.g., semiconductors and polymers) have been the subject of extensive studies (see, e.g., Refs. 3, 4, and 30). Only a brief overview, from the point of view of the electrochemical characterization of films, will be attempted here. Charge can be transported through the film in the form of ionic species that can freely diffuse within the film (e.g., Cp_2FeTMA^+ in Nafion, discussed in Section 5.2). In films with fixed redox sites, such as poly(vinylferrocene (PVF), where the sites are part of the polymer chain, the sites cannot carry out large translational movements (although they can undergo short-range motions and reorientations). For such films, charge can be transported by site-to-site electron hopping, that is, by electron transfer between neighboring oxidized and reduced centers (process 2 in Fig. 5.1.1) (31). For example, a film of PVF on a conductive substrate would be an electronic insulator in the dry state. Any attempt to pass charge through the film by contacting the PVF surface with a second conductor would only lead to some charging of the film–conductor interfaces (capacitive behavior). The film would be highly resistive because there are no free ions to move through the film. For the same film immersed in a solution containing dissolved electrolyte (e.g., Na^+X^-), oxidation of a ferrocene (Fc) site at the electrode to form a Fc^+ center can be compensated by the movement of an anion, X^-, into the film. The charge at this center can move within the film by the reaction

$$\sim Fc_A^+ X^- + \sim Fc_B \rightarrow \sim Fc_A + \sim Fc_B^+ X^- \qquad (5.4.1)$$

where subscripts identify two different Fc sites, A and B.

5.4.1. The Electron-Transfer Diffusion Coefficient

The apparent rate at which a species appears to move through the film in this case depends on the rate of the electron-transfer (et) reaction. Considerations of analogous reactions in homogeneous solution showed that such an et process is equivalent to diffusion (32, 33). In this case, the apparent diffusion coefficient observed for a species, D_{app}, would be composed of contributions from the actual movement of the species (governed by its translational diffusion coefficient, D) and the et process. When bimolecular kinetics can be applied to the et process and the species can be considered as points, an equation of the form

$$D_{app} = D + k\delta^2 C_A b = D + D_E \qquad (5.4.2)$$

applies, where δ is the distance between sites for et, b is a numerical constant (frequently taken as $\pi/4$ or as $\frac{1}{6}$ for three-dimensional diffusion), and C_A is the total concentration of sites, oxidized and reduced. Similar and equivalent representations were given for polymer films on electrodes, where charge hopping was again treated as a diffusional process (17, 34, 35). Thus the et movement of charge through the polymer can be treated in terms of a diffusion coefficient, D_E (sometimes also written in the literature as D_{ET} or D_{ct}), which is related to the et kinetics, and should be distinguished from actual mass-transfer diffusion coefficients, such as D and D_S. Even for a polymer film that contains redox centers that are mobile (e.g., electroactive cations in Nafion or another cation-exchange polymer, or in clay), electron hopping can contribute to the charge transfer through the film. This relative et contribution, as expressed in Eq. (5.4.2), will be more important in polymer or clay films than in homogeneous solutions, since the mass-transfer diffusion coefficients will be much smaller in such films.

The value for D_E can be obtained for fixed-site materials by electrochemical measurements (e.g., chronoamperometry or cyclic voltammetry) in a manner similar to that for determinations of D_S, discussed above (i.e., at times sufficiently small that mass-transport limiting currents are obtained). Determination of the diffusion coefficient does require knowledge of the concentration of electroactive sites in the film. The total amount of electroactive substance can usually be obtained from the amount of charge contained under the CV curve in a slow sweep experiment, Eq. (5.2.3). Determination of the concentration, however, requires measurement of the film thickness, ϕ, as well. For mobile species within a film, additional information must be used to separate the measured diffusion coefficient, written as D_{app} or D_{expl}, into the component attributable to mass transfer, D_S, and that to et, D_E. One approach is to estimate D_S by measuring the actual rate of permeation of the electroactive species into the film (8). In this case, a film initially devoid of any electroactive species is immersed into a solution containing the permeating species, and the time required for the permeating species to reach the electrode–film interface is noted. This permeation

5.4. CHARGE TRANSPORT IN FILMS

time is conveniently measured by noting the onset of current when the electroactive species arrives at the electrode, which is held at a potential where an et reaction occurs. This number can then be compared with D_{app} and an estimate of the contribution of charge hopping and D_E obtained. This approach was used with micrometer-thick Nafion films with Cp_2FeTMA^+ and $Ru(bpy)_3^{2+}$ as electroactive components. For Cp_2FeTMA^+, $D_S = D_{app}$, which is consistent with a very small contribution from charge hopping because of the rather slow et exchange rates observed with ferrocene species. For $Ru(bpy)_3^{2+}$, $D_{app} \approx 4 \times 10^{-10}$ cm^2/s, while $D_S \approx \sim 0.2 \times 10^{-10}$ cm^2/s, suggesting that the charge hopping route contributes to the charge transfer through the film.

Another approach for estimating the charge hopping contribution in polyelectrolyte films is based on the comparative behavior of different species in the same film (36). For example, $Co(bpy)_3^{2+}$ and $Ru(bpy)_3^{2+}$ are roughly the same size and show equal diffusion coefficients in aqueous solution (6 × 10^{-6} cm^2/s). The measured D_{app} values for these two species in a Nafion film are very different, however, with the $Ru(bpy)_3^{2+}$ value being about eight times that of the $Co(bpy)_3^{2+}$. This difference was ascribed to the very different values of the et self-exchange rate constants for these species [20 M^{-1} s^{-1} for $Co(bpy)_3^{2+}$ vs. 10^9 M^{-1} s^{-1} for $Ru(bpy)_3^{2+}$ in aqueous solution], with charge hopping again playing a significant role for the $Ru(bpy)_3^{2+}$ in Nafion. A similar case can be made by comparing the CV waves for reduction and oxidation of $Co(bpy)_3^{2+}$ in Nafion (Fig. 5.4.1) (37). Note that the oxidation wave for the +3/+2 couple at +0.1 V is much smaller than that for the +2/+1 couple at −1.19 V, although the concentration of electroactive species must be the same for both waves. This again is consistent with the much higher et self-exchange rate for the +2/+1 couple (>10^{13} M^{-1} s^{-1}) compared to that of the +3/+2 couple.

The actual situation is more complicated than the preceding discussion might suggest (4). Mass transfer through polymers like Nafion may occur by movement via different domains or phases within the material, and et reactions may occur between species residing in these different domains (37). Rates of et reactions within a polymer film are influenced by the film environment and appear to be slower than that of the analogous reaction in a homogeneous, low-viscosity, solution environment. This appears to be true for both heterogeneous and homogeneous et reactions (see, e.g., the discussion of the effects of et on the ESR spectra of methyl viologen radical cation in Section 3.3.3 and Section 5.5 below), although this field is still a developing one (38).

A number of studies have addressed the determination of D_E and the

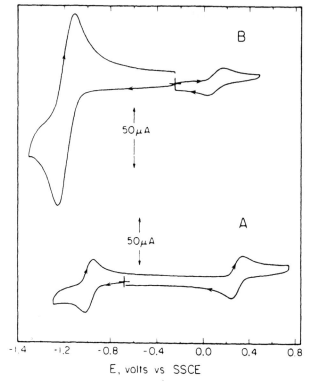

FIGURE 5.4.1. Cyclic voltammograms of $Co(bpy)_3^{2+}$. (A) 0.5 mM $Co(bpy)_3^{2+}$ in acetonitrile containing 0.1 M TEAP and showing equal-size waves for reduction to +1 species and oxidation to the +3 species. (B) $Co(bpy)_3^{2+}$ in a 0.6-μm-thick Nafion coating at a concentration of ~1 M, in 0.5 M Na_2SO_4 showing unequal-size waves because the effective diffusion coefficients for the charge hopping processes are different. Scan rate, 100 mV/s. [Reprinted with permission from D. A. Buffy and F. C. Anson, J. Am. Chem. Soc., **105**, 685 (1983). Copyright 1983 American Chemical Society.]

effect of polymer structure and concentration of electroactive species, C_A, on its magnitude (4, 39). Understanding the different factors that contribute to the magnitude of D_E, however, is difficult, since processes other than the et kinetics, implied in Eq. (5.4.2), can be rate-limiting. For example, the movement of counterions that compensate the charge may be the slow step (40–42). Segmental motions of the polymer chains may play a role. Thus changes in C_A, for example, in a fixed-site polymer film by copolymerization of a monomer with the electroactive site with a diluent monomer, or in an ion-exchange polymer by variation of the amount of electroactive ion in the layer, can produce different kinds of variation of D_E; it is almost never that predicted from Eq. (5.4.2). Indeed, in many cases, D_E appears to be almost independent of C_A (8, 37, 43). There are even cases where D_E decreases with an increase of C_A (44). The problem is that changes in C_A can also affect the structure of the polymer film. For example, increases in the concentration of Ru(bpy)$_3^{2+}$ in Nafion appear to increase the degree of "crosslinking" of the chains, making the overall structure less flexible. This is demonstrated, for example, by the decrease in the permeation of hydroquinone through a Nafion film when it contains exchanged Ru(bpy)$_3^{2+}$ (45). With fixed-site polymers, the behavior is also not that predicted by Eq. (5.4.2), even when great care is taken to have the diluent monomer of the copolymer film be structurally very similar to the electroactive one (46). For example, a copolymer of M(bpy)$_2$(p-cinn)$_2^{2+}$ [where (p-cinn is 4-pyNHCOCH=CHPh}, with M = Os(II) and Ru(II)], can be prepared by electroreduction of the monomers to produce films with different Os/Ru ratios. Since the potentials of the Os(+2/+3) and Ru(+2/+3) couples are very different, the electrochemical oxidation of the Os(2+) form can be studied with the structurally similar Ru form behaving as a diluent. The resulting behavior of D_E as a function of C_{Os} is shown in Figure 5.4.2. This behavior has been attributed qualitatively to the importance of site-to-site et in region A, intrinsic mobility of redox sites to bring Os centers into proximity for et in region B, and separation of Os centers to distances where et becomes improbable in region C (40, 46). More complete quantitative models for et in polymer films that take these kinds of factors into account are still lacking.

5.4.2. Rotating-Disk Electrode Studies of Film Charge Transfer

The RDE technique can also be applied to study charge transfer through the polymer film. Consider a polymer film containing a species, P, at a total concentration, C_P^*, that can be reduced at the electrode–film interface to Q:

$$P + e \rightarrow Q \qquad (5.4.3)$$

A steady-state current corresponding to the flux of charge through the film can be established, if a solution species, A, is present which is reduced rapidly by Q. The concentration profiles under these conditions are shown in Figure 5.4.3. Note that it is assumed here that the et reactions at both interfaces are very rapid, so that $C_Q(x = 0) = C_P^*$ and

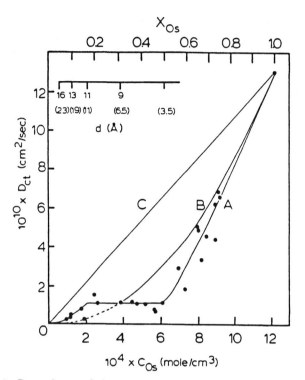

FIGURE 5.4.2. Dependence of charge-transport coefficient, D_{ct} (equivalent to D_E) on mole fraction of Os, X_{Os}, or concentration of Os, C_{Os} in films of copolymers of $M(bpy)_2(p\text{-cinn})_2^{2+}$ [where (p-cinn) is 4-pyNHCOCH=CHPh], with M = Os(II) and Ru(II)]; d is the estimated edge-to-edge separation of the Os sites; D_{ct} measured electrochemically for the reaction Os(II) → Os(III). Curve A is the experimental data. Curves B and C are, respectively, the second-order and first-order theoretical results of a random walk simulation of electron jump mechanisms. [Reprinted with permission from J. S. Facci, Ph.D. dissertation, University of North Carolina, 1982.]

$C_Q(x = \phi^-) = 0$. The general case, where the rate of the et reaction between A and Q is finite, is considered in Section 5.6. By the same reasoning as that leading to the derivation of Eq. (5.3.8), the flux of charge across the film is $D_E[C_Q(x = 0) - C_Q(x = \phi^-)]/\phi$. Hence the limiting current under these conditions is

$$i_E = \frac{nFAD_E C_P^*}{\phi} = \frac{nFAD_E \Gamma_P}{\phi^2} \qquad (5.4.4)$$

This situation is a limiting case of the more general one involving a catalyst system (P, Q) incorporated into a film for the reduction of a solution species, A (47), and is designated "case E."

5.5. CHARGE TRANSFER AT THE ELECTRODE–FILM INTERFACE

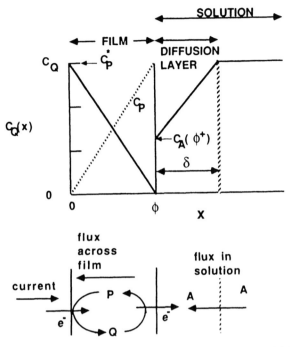

FIGURE 5.4.3. Concentration profiles for an RDE experiment in which a species Q, generated by reduction of P in a film, reacts with a solution component, A. Profiles for P and Q in the film and A in solution are shown.

5.5. CHARGE TRANSFER AT THE ELECTRODE–FILM INTERFACE

The rate of the heterogeneous et reaction of a species in the film on or near the electrode surface (process 1 in Fig. 5.1.1) can be probed by conventional electrochemical techniques, such as cyclic or normal pulse voltammetry or ac methods (1). In making these measurements, special care must be taken to avoid mass-transfer limitations caused by the small diffusion coefficients found in many films and to correct for uncompensated resistance in the films. For example, in a cyclic voltammetric study, the scan rate, v, must be chosen so that the dimensionless parameter, $\Lambda = [(RT/F)/Dv]^{1/2} k°$, is <15. Under these conditions the voltammetric response deviates from that of a nernstian et reaction and kinetic information about the heterogeneous et can be obtained. Application of convolution techniques is useful in correcting for uncompensated resistance and double-layer capacitance effects (1, 38, 48).

Such studies have typically shown that the rate constants for heterogeneous et for reactants in polymer films on electrode surfaces are two to three orders of magnitude smaller than those for the same reactants dissolved in solution (38, 45, 49). The $k°$ values appear to track, at least approximately, the diffusion coefficient, suggesting that the more "viscous" polymer medium affects the rate. Indeed, studies of $k°$ in homogeneous solutions whose viscosity has been varied by the addition of sugar (to minimize changes in other solvent properties, such as the dielectric constant), show similar effects (50). This dependence points to the importance of solvent dynamics, such as the dielectric relaxation and thermal orientational times, in the preexponential factor in the et rate expressions, as considered in theoretical treatments (51). As discussed in Section 5.4.1, although there have not been quantitative studies of homogeneous et reactions in polymer films compared to the same reactions in liquid solutions, there are indications that those rates are similarly decreased.

5.6. CHARGE TRANSFER AT THE FILM–SOLUTION INTERFACE—CATALYSIS

The transfer of charge across the film–solution interface, as represented by process 3 in Figure 5.1.1, must be considered when the polymer film species (P, Q) acts as an electron-transfer catalyst for reduction of a solution species, A. In this scheme, sometimes called *electron-transfer-mediated catalysis*, the reaction P + e → Q occurs at the electrode–film interface; Q moves to the film–solution interface by diffusion and electron hopping; A penetrates the film and partitions, with a partition constant, κ, given by

$$\kappa = \frac{C_A(\phi^-)}{C_A(\phi^+)} \quad (5.6.1)$$

A reacts with Q via the reaction

$$Q + A \xrightarrow{k_1} P + B \quad (5.6.2)$$

and, in the general case, can also diffuse to the electrode surface and react there, in the same way as discussed in Section 5.3.2. This sequence is of interest when species A does not react at the electrode material very rapidly and the P, Q couple is introduced to catalyze its reduction. Numerous studies of this rather complex reaction sequence have appeared, since there is a great deal of interest in the design of electro-

5.6. CHARGE TRANSFER AT THE FILM-SOLUTION INTERFACE—CATALYSIS

catalysts by immobilizing homogeneous catalysts on electrode surfaces (52).

5.6.1. Cyclic Voltammetry

There are a number of examples of CV studies of this scheme. Qualitatively it is easy to recognize this catalytic effect, since the CV behavior will change from that characteristic of the immobilized P, as described in Section 5.2.2, in the absence of A in solution, to one in which the reduction of A also contributes to the current, when A is added to the solution. A typical example is shown in Figure 5.6.1 (53). The oxidation of nicotinamide adenosine dinucleotide (NADH) at carbon electrodes is complicated by a large activation energy and adsorption of the product, NAD^+. A polymer containing hydroquinone (HQ) functionalities (shown below), prepared by reacting dopamine and poly(methacryloyl chloride), coated on a glassy carbon electrode shows the CV surface waves characteristic of an immobilized HQ-quinone(Q) couple. On addition of NADH to the solution, the height of the oxidation wave increases, because electrogenerated Q in the film is reduced by the NADH in solution, regenerating HQ. The increase in the anodic peak current is a measure of the rate of the reaction of NADH with Q. In this example, the HQ-Q couple behaves as the mediator that catalyzes the oxidation of NADH.

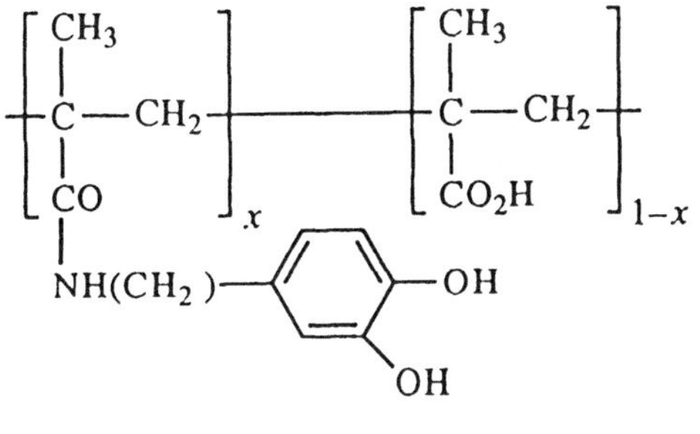

$$\left[\begin{array}{c} CH_3 \\ | \\ -C-CH_2- \\ | \\ CO \\ | \\ NH(CH_2)- \end{array} \right]_x \left[\begin{array}{c} CH_3 \\ | \\ -C-CH_2- \\ | \\ CO_2H \end{array} \right]_{1-x}$$

(with phenyl-OH, OH attached to NH(CH₂)-)

Another example of this scheme involves the oxidation of $Fe(CN)_6^{4-}$ at an ITO electrode, in the absence and presence of a film of Nafion, with and without $Ru(bpy)_3^{2+}$ in the film (45). Typical CVs are shown in Figure 5.6.2. At a bare GC electrode, $Fe(CN)_6^{4-}$ shows a well-defined wave (curve a). When a 0.4-μm Nafion film is cast on the electrode, the wave disappears (curve b), because $Fe(CN)_6^{4-}$ does not penetrate the cation-exchange polymer layer. Indeed, $Fe(CN)_6^{4-}$ can be used to test the quality of a Nafion film and its freedom from pinholes, as discussed in

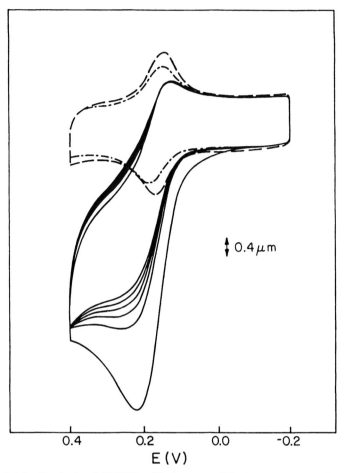

FIGURE 5.6.1. Catalysis of NADH oxidation on a GC electrode with a film of dopamine polymer (0.12 nmol/cm^2) in pH 6.35 buffer. Cyclic voltammograms at 46 mV/s for polymer alone (---); polymer with 0.5 mM NADH in solution (six cycles shown) (solid lines); polymer alone after previous six sweeps (-----). [Reprinted with permission from C. DeGrand and L. L. Miller, *J. Am. Chem. Soc.*, **102**, 5728 (1980). Copyright 1980 American Chemical Society.]

FIGURE 5.6.2. "Catalysis" of oxidation of ferrocyanide by Ru(III) contained in polymer film. Cyclic voltammograms at 10 mV/s in 0.1 M KCl of (a) ITO electrode in 1 mM Fe(CN)$_6^{4-}$; (b) as in (a) with ITO covered by 0.4-μm film of Nafion, showing blocking of oxidation by film; (c) ITO covered with 0.3 μm film of Nafion containing Ru(bpy)$_3^{2+}$; (d) as in (c) with 1 mM Fe(CN)$_6^{4-}$; (e) as (c) with thicker Nafion film (3 μm); (f) as in (d) with thicker Nafion film. [Reprinted with permission from M. Krishnan, X. Zhang, and A. J. Bard, *J. Am. Chem.*, **106**, 7371 (1984). Copyright 1984 American Chemical Society.]

5.6. CHARGE TRANSFER AT THE FILM–SOLUTION INTERFACE—CATALYSIS

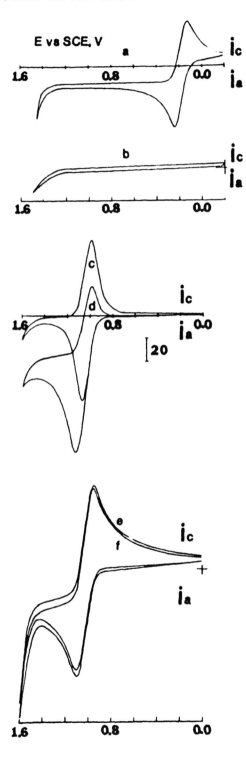

Section 5.3. A Nafion film containing Ru(bpy)$_3^{2+}$ shows the typical CV thin-layer behavior, discussed in Section 5.2 (curve c). In the presence of Fe(CN)$_6^{4-}$ in solution, the anodic current increases [because reaction of Ru(bpy)$_3^{3+}$ with Fe(CN)$_6^{4-}$ regenerates Ru(bpy)$_3^{2+}$] and the cathodic wave decreases. The shape and magnitude of the waves depend on the kinetics of the different processes that occur at the interfaces and within the film. In a thicker film of Nafion, curves e and f, the Ru(bpy)$_3^{2+}$ oxidation wave shows a more diffusion-limited shape and the wave shape is essentially unperturbed by the presence of Fe(CN)$_6^{4-}$ in solution. Note that the behavior in Figure 5.6.2 is described as an example of a "catalytic" mechanism, even though the oxidation of Fe(CN)$_6^{4-}$ occurs *more easily* (i.e., at less positive potentials) at an unmodified ITO electrode than it does at the Ru–Nafion-modified electrode.

5.6.2. Rotating-Disk Electrode Studies

While CV studies can clearly show when a catalytic reaction occurs, quantitative studies by CV are complicated by the non-steady-state mass transfer of the species in both the polymer film and solution as well as possible problems with heterogeneous et kinetics at the electrode–film interface and the presence of charging current, especially at high scan rates. Steady-state measurements at an RDE held at a potential where the reduction of P to Q occurs at the diffusion-controlled rate are easier to interpret. The treatment basically follows the concepts outlined in Sections 5.3.2 and 5.4.2, and the general case (28, 29, 35, 47) applied here actually includes these earlier ones (cases S and E) as special cases. We introduce the general treatment with one more special case. Consider the situation where reactant A exchanges rapidly into the film with a partition coefficient, κ, but diffuses only very slowly through the film, so that the direct reduction of A at the electrode makes a negligible contribution to the steady-state current at the RDE. We also assume that the transport of electrons through the film to the interface with the solution, either by diffusion of Q or electron hopping, is very rapid. In this case, the rate of reduction of A at the interface is governed either by the mass transfer of A in solution or the rate of reaction with Q. The steady-state current under these conditions (called "case R") can be derived by equating the flux of A in solution to the reaction rate, using the same procedures as those for obtaining Eq. (5.3.8) via Eqs. (5.3.10) and (5.3.11). The result is

$$\frac{1}{i_{\lim}} = \frac{1}{i_A} + \frac{1}{i_k} \qquad (5.6.3)$$

5.6. CHARGE TRANSFER AT THE FILM–SOLUTION INTERFACE—CATALYSIS

$$i_k = nFAk_1\Gamma_P\kappa C_A \qquad (5.6.4)$$

where k_1 is the rate constant of the reaction of Q with A, Eq. (5.6.2), and Γ_P is the total amount of P and Q in the film. As before, a plot of i_{lim}^{-1} versus $\omega^{-1/2}$ is linear and yields the limiting kinetic current, i_k, as the intercept.

In the general case, all the processes previously considered and illustrated in Figure 5.1.1 can contribute to the rate of the reaction. The overall general treatment is more complex than those for the various limiting cases and requires a fuller discussion than can be given here (52). Briefly, the different processes are represented by the currents:

i_A—mass-transport rate of A in solution to a bare electrode, Eq. (5.3.3)

i_S—mass-transport rate of A in the film, Eq. (5.3.9)

i_E—effective charge-transfer rate via the mediator Q in the film, Eq. (5.4.4)

i_P—mass-transfer rate of A across the film–solution interface, Eq. (5.3.13)

i_k—rate of et reaction between A and Q, Eq. (5.6.4)

Schematic concentration profiles are shown in Figure 5.6.3. The limiting current in the general case, with all of the processes contributing, can only be obtained by numerical solution of the differential equations governing the system. However, in most experimental systems only some of the processes will be important, and various limiting cases will apply. The relevant expressions for the limiting cases are given in Table 5.6.1. Which limiting case applies, that is, what factors are rate-determining, is a function of the relative magnitudes of the preceding currents, or more explicitly, by the ratios i_S^*/i_k^* and i_E/i_k^*, where

$$i_S^* = i_S[1 - i(i_A^{-1} + i_P^{-1})] \qquad (5.6.5)$$

$$i_k^* = i_k[1 - i(i_A^{-1} + i_P^{-1})] \qquad (5.6.6)$$

With knowledge of these ratios, the proper limiting case can be determined from the zone diagram of Figure 5.6.4. In practice, the problem is the reverse, to determine which case applies from the experimental results, specifically, i_{lim} as a function of ω, Γ_P, C_A, and ϕ. This is discussed and diagnostic criteria for analyzing the experimental results are given in Ref. 52.

When the mediated reaction occurs in a monolayer at the electrode-film interface, that is, when Q is confined as a monolayer on the elec-

TABLE 5.6.1. Expressions of the Plateau Currents

(R + S)

† $\dfrac{1}{i_1} = \dfrac{1}{i_A} + \left\{ \dfrac{1}{i_p} + \dfrac{1}{(i_k i_S)^{1/2} \tanh\left(\dfrac{i_k}{i_S}\right)^{1/2}} \right\}$

$(m) = 1$

† $\dfrac{1}{i_1 + i_2} = \dfrac{1}{i_A} + \left\{ \dfrac{1}{i_p} + \dfrac{\tanh\left(\dfrac{i_k}{i_S}\right)^{1/2}}{(i_k i_S)^{1/2}} \right\}$

$(m) = 1$

(R)

† $\dfrac{1}{i_1} = \dfrac{1}{i_A} + \left\{ \dfrac{1}{i_p} + \dfrac{1}{i_k} \right\}$

$(m) = 1 \qquad (b) = (Fc_p^o c_A^o k_1 \kappa \phi)^{-1} + i_p^{-1}$

† $\dfrac{1}{i_1 + i_2} = \dfrac{1}{i_A} + \left\{ \dfrac{1}{i_p} + \dfrac{1}{i_S} \right\}$

$(m) = 1 \qquad (b) = \phi / Fc_A^o \kappa D_S + i_p^{-1}$

(E + R)

† $\dfrac{1}{i_1} = \dfrac{1}{i_A} + \dfrac{1}{i_p} + \dfrac{i_1}{i_k i_E \tanh^2\left(\dfrac{i_k}{i_E}\left[1 - i_1\left(\dfrac{1}{i_A} + \dfrac{1}{i_p}\right)\right]\right)^{1/2}}$

† $\dfrac{1}{i_1 + i_2} = \dfrac{1}{i_A} + \left\{ \dfrac{1}{i_p} + \dfrac{1}{i_S} \right\}$

$(m) = 1 \qquad (b) = \phi / Fc_A^o \kappa D_S + i_p^{-1}$

(SR)

† $\dfrac{1}{i_1} = \dfrac{1}{i_A} + \left\{ \dfrac{1}{i_p} + \dfrac{1}{(i_k i_S)^{1/2}} \right\}$

$(m) = 1 \qquad (b) = 1/Fc_A^o \kappa (c_p^o D_S k_1)^{1/2} + i_p^{-1}$

$i_2 = 0$

General case

$i = i_S^* \left(\dfrac{da^*}{dy}\right)_1$

Numerical resolution of eqn. (30) with $a_1^* = 1$ and $(da^*/dy)_0 = 0$ at the first wave and $a_0 = 0$ at the second wave

(ER)

¶ $\left[\dfrac{1}{i_1} - \dfrac{1}{i_A}\right] = \left(\dfrac{1}{i_p}\right) + \dfrac{i_1}{\{i_k i_E\}}$

$(m) = (i_k i_E)^{-1} \qquad (b) = i_p^{-1}$

† $\dfrac{1}{i_1 + i_2} = \dfrac{1}{i_A} + \left\{ \dfrac{1}{i_p} + \dfrac{1}{i_S} \right\}$

$(m) = 1 \qquad (b) = \phi / Fc_A^o \kappa D_S + i_p^{-1}$

$$\frac{1}{i_1} = \frac{1}{i_A} + \frac{1}{i_P} + \left[\frac{1}{i_K i_S\left(1 - \frac{i_1}{i_E}\right)}\right]^{1/2}$$

$i_2 = 0$

(SR + E)

† $\frac{1}{i_1} = \left\{\frac{1}{i_E}\right\}$ $(b) = i_E^{-1}$

$(m) = 0$

$i_2 = 0$

(E)

† Exhibits linear Koutecký–Levich behavior with (m) = slope and (b) = intercept

¶ Non-linear Koutecký–Levich behavior, but a linear form is given having (m) = slope and (b) = intercept
$i_P^{-1} = 1/Fc_A^o X_f$

† $\frac{1}{i_1} = \left\{\frac{i_S}{i_S + i_E}\right\} \frac{1}{i_A} + \left\{\frac{1}{i_S + i_E}\left(1 + \frac{i_S}{i_P}\right)\right\}$

$(b)/(m) = \phi/Fc_A^o \kappa D_S + i_P^{-1}$

$i_2 = 0$

(S + E)

¶ $\left[\frac{1}{i_1} - \frac{1}{i_A}\right] = \left(\frac{1}{i_P} + \frac{1}{i_S}\right) + \frac{i_1}{(i_K i_E)}$

$(m) = (i_K i_E)^{-1}$ $(b) = \phi/Fc_A^o \kappa D_S + i_P^{-1}$

† $\frac{1}{i_1 + i_2} = \frac{1}{i_A} + \left(\frac{1}{i_P} + \frac{1}{i_S}\right)$

$(m) = 1$ $(b) = \phi/Fc_A^o \kappa D_S + i_P^{-1}$

(ER + S)

† $\frac{1}{i_1} = \frac{1}{i_A} + \left(\frac{1}{i_P} + \frac{1}{i_S}\right)$

$(m) = 1$ $(b) = \phi/Fc_A^o \kappa D_S + i_P^{-1}$

$i_2 = 0$

(S)

Source: Reprinted with permission from J. Leddy, A. J. Bard, J. T. Malloy, and J. M. Savéant, *J. Electroanal. Chem.*, **187**, 205 (1985). Copyright 1985 Elsevier.

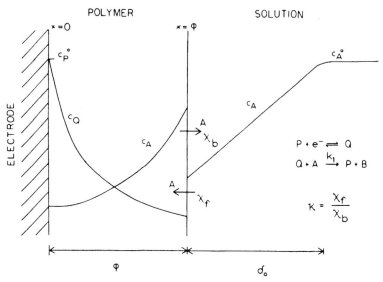

FIGURE 5.6.3. Schematic concentration profiles for the general case of mediated (catalytic) reduction of solution species, A, by electrogenerated film species, Q. The electrode is held at a potential where all P at the electrode surface is reduced to Q so that the concentration of Q at the electrode surface is $C_P^°$ ($= \Gamma_P/\Lambda$). Within the solution ($x > \phi$) the concentration profile for A is approximately linear. χ_f and χ_b are the rate constants for the transport of A into and out of the film, respectively. [Reprinted with permission from J. Leddy, A. J. Bard, J. T. Maloy, and J. M. Saveant, *J. Electroanal. Chem.*, **187**, 205 (1985). Copyright 1985 Elsevier.]

trode surface with no mass transfer of Q or electron hopping through the film (denoted the Er + S case), the limiting current expression is

$$\frac{1}{i_{\text{lim}}} = \frac{1}{i_k^*} + \frac{1}{i_S^*} \quad (5.6.7)$$

An analogous situation arises when the reaction occurs in a monolayer at the film–solution interface. When the solution species, A, cannot penetrate the film (denoted the Sr + E case), the limiting current is given by

$$1/i_{\text{lim}} = 1/i_k^* + 1/i_E = 1/i_A + \left[i_k\left(1 - \frac{i_{\text{lim}}}{i_E}\right)\right]^{-1} \quad (5.6.8)$$

A number of these cases have been observed in RDE studies of actual experimental systems (52). For example, the Nafion/Ru(bpy)$_3^{2+}$/Fe(CN)$_6^{4-}$ system discussed above has also been studied with the RDE;

5.6. CHARGE TRANSFER AT THE FILM–SOLUTION INTERFACE—CATALYSIS

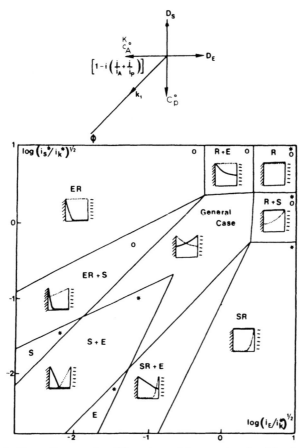

FIGURE 5.6.4. Zone diagram for different special cases involved in the general case of mediated reduction of a solution species at a film-covered electrode. Some of these cases (e.g., cases S, R, E) are discussed in the text. Cases marked * are those which produce linear plots of $(1/i)$ versus $\omega^{-1/2}$ plots. The experimental behavior would be affected by the magnitude of various experimental parameters shown at the top of the figure. Schematic concentration profiles of the species A ($\cdots$) and Q (——) are shown for the different cases. (See also Ref. 28.) [Reprinted with permission from J. Leddy, A. J. Bard, J. T. Maloy, and J. M. Saveant, *J. Electroanal. Chem.*, **187**, 205 (1985). Copyright 1985 Elsevier.]

typical current–potential curves obtained under different conditions (corresponding to the CV curves in Fig. 5.6.2) are shown in Figure 5.6.5 (45). Curve D shows the mediated oxidation at the RDE, with a limiting current representing oxidation of $Fe(CN)_6^{4-}$ (species A) in solution by $Ru(bpy)_3^{3+}$ (species Q) incorporated in the film. Plots of $1/i_{lim}$ versus $\omega^{-1/2}$ obtained at different concentrations of $Fe(CN)_6^{4-}$ are given in Figure 5.6.6. Since $Fe(CN)_6^{4-}$ cannot penetrate the film, this is an ex-

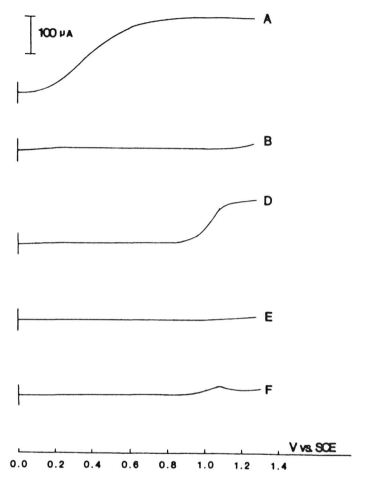

FIGURE 5.6.5. Voltammetry at a rotating GC disk electrode in a 1 mM Fe(CN)$_6^{4-}$ solution at a rotation rate of 1000 rpm; (*A*) bare electrode; (*B*) electrode with 0.3-μm film of Nafion showing blocking of reaction; (*D*) as (*B*) with film containing Ru(bpy)$_3^{2+}$ showing catalytic reaction of electrogenerated Ru(III) in film and Fe(CN)$_6^{4-}$ in solution; (*E*) as (*B*) with 3-μm film; (*F*) as (*E*) with film containing Ru(bpy)$_3^{2+}$. Compare to the CV results in Figure 5.6.2. [Reprinted with permission from M. Krishnan, X. Zhang, and A. J. Bard, *J. Am. Chem. Soc.*, **106.** 7371 (1984). Copyright 1984 American Chemical Society.]

ample of the Sr + E case, governed by Eq. (5.6.8). This can be rearranged to yield

$$\left(\frac{1}{i_{\lim}}\right)^2 - \frac{1}{i_{\lim}}\left(\frac{1}{i_A} + \frac{1}{i_E} + \frac{1}{i_k}\right) + \frac{1}{i_A i_E} = 0 \qquad (5.6.9)$$

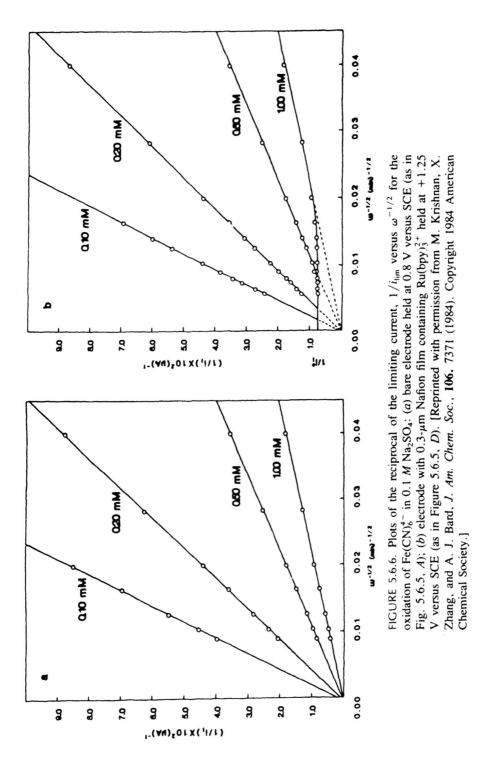

FIGURE 5.6.6. Plots of the reciprocal of the limiting current, $1/i_{lim}$ versus $\omega^{-1/2}$ for the oxidation of $Fe(CN)_6^{4-}$ in 0.1 M Na_2SO_4: (a) bare electrode held at 0.8 V versus SCE (as in Fig. 5.6.5, A); (b) electrode with 0.3-μm Nafion film containing $Ru(bpy)_3^{2+}$ held at +1.25 V versus SCE (as in Figure 5.6.5, D). [Reprinted with permission from M. Krishnan, X. Zhang, and A. J. Bard, J. Am. Chem. Soc., 106. 7371 (1984). Copyright 1984 American Chemical Society.]

Under conditions when the interfacial et reaction is rapid, $1/i_k \to 0$, and Eq. (5.6.9) yields curves of the type shown in Figure 5.6.6. These can be analyzed according to two limiting cases. At low rotation rates, mass transfer in solution is rate-determining and $1/i_{lim} = 1/i_A$. In this region, a straight line with a zero intercept and the same slope as that observed for the uncoated electrode is obtained. At high rotation rates, the limiting current becomes independent of ω and $1/i_{lim} = 1/i_E$; this is equivalent to case E discussed in Section 5.4.2.

5.7. CONCLUSIONS

The models and results discussed in this chapter illustrate the utility of electrochemical methods in characterizing films and structures of integrated chemical systems formed on conductive substrates. Information about et thermodynamics and kinetics, mass transport, electron hopping, film structure (e.g., pinholes), extraction of species into a film, and reactions within a film can be obtained by such techniques. Moreover, as discussed in Chapters 2 and 5, electrochemical methods can be used to construct ICSs, by deposition of thin films or the fabrication of patterns. The electrochemical principles discussed in this chapter should be useful in the characterization of ICSs as well as in the interpretation of their behavior, such as those used as sensors or electronic devices.

REFERENCES

1. A. J. Bard and L. R. Faulkner, *Electrochemical Methods*, Wiley, New York, 1980.
2. P. T. Kissinger and W. R. Heineman, eds., *Laboratory Techniques in Electroanalytical Chemistry*, Marcel Dekker, New York, 1984.
3. R. W. Murray, ed., *Molecular Design of Electrode Surfaces*, Wiley-Interscience, New York, 1992.
4. G. Inzelt, in *Electroanalytical Chemistry*, A. J. Bard, ed., Marcel Dekker, New York, 1994, Vol. 18, pp. 89-241.
5. E. Laviron, in *Electroanalytical Chemistry*, A. J. Bard, ed., Marcel Dekker, New York, 1982, Vol. 12, p. 53.
6. R. W. Murray, in *Electroanalytical Chemistry*, A. J. Bard, ed., Marcel Dekker, New York, 1984, Vol. 13, p. 191.
7. A. J. Bard and L. R. Faulkner, *Electrochemical Methods*, Wiley, New York, 1980, p. 142.
8. H. S. White, J. Leddy, and A. J. Bard, *J. Am. Chem. Soc.*, **104**, 4811 (1982).

REFERENCES

9. A. J. Bard and L. R. Faulkner, *Electrochemical Methods*, Wiley, New York, 1980, p. 213.
10. A. J. Bard and L. R. Faulkner, *Electrochemical Methods*, Wiley, New York, 1980, pp. 409, 521.
11. A. J. Bard and L. R. Faulkner, *Electrochemical Methods*, Wiley, New York, 1980, p. 523.
12. A. N. Frumkin and B. B. Damaskin, in *Modern Aspects of Electrochemistry*, J. O'M. Bockris and B. E. Conway, eds., Butterworths, London, 1964, Vol. 3.
13. P. Delahay, *Double Layer and Electrode Kinetics*, Interscience, New York, 1965.
14. E. Laviron, *J. Electroanal. Chem.*, **100**, 263 (1979).
15. H. Matsuda, K. Aoki, and K. Tokuda, *J. Electroanal. Chem.*, **217**, 1 (1987); Ibid., p. 15.
16. C. P. Smith and H. S. White, *Anal. Chem.*, **64**, 2398 (1992).
17. See, for example: P. J. Peerce and A. J. Bard, *J. Electroanal. Chem.*, **114**, 89 (1980).
18. T. Gueshi, K. Tokuda, and H. Matsuda, *J. Electroanal. Chem.*, **89**, 247 (1978).
19. P. J. Peerce and A. J. Bard, *J. Electroanal. Chem.*, **112**, 97 (1980).
20. A. J. Bard and L. R. Faulkner, *Electrochemical Methods*, Wiley, New York, 1980, p. 288.
21. (a) F. Scheller, S. Muller, R. Landsberg, and H. J. Spitzer, *J. Electroanal. Chem.*, **19**, 187 (1968); (b) F. Scheller, R. Landsberg, and S. Muller, *J. Electroanal. Chem.*, **20**, 375 (1969); (c) R. Landsberg and R. Thiele, *Electrochim. Acta*, **11**, 1243 (1966).
22. J. Leddy and A. J. Bard, *J. Electroanal. Chem.*, **153**, 223 (1983).
23. C. Amatore, J. M. Savéant, and D. Tessier, *J. Electroanal. Chem.*, **147**, 39 (1983).
24. T. Gueshi, K. Tokuda, and H. Matsuda, *J. Electroanal. Chem.*, **101**, 29 (1979).
25. (a) K. Tokuda, T. Gueshi, and H. Matsuda, *J. Electroanal. Chem.*, **102**, 41 (1979); (b) K. J. Vetter, *Z. Physik. Chem.*, **199**, 300 (1952); (c) J. Lindemann and R. Landsberg, *J. Electroanal. Chem.*, **25**, app 20 (1970); (d) R. Landsberg and R. Thiele, *Electrochim. Acta*, **121**, 1243 (1966).
26. (a) D. A. Gough and J. K. Leypoldt, *Anal. Chem.*, **51**, 439 (1979); (b) *AIChE J.*, **26**, 1013 (1980); (c) *Anal. Chem.*, **52**, 1126 (1980); (d) *J. Electrochem. Soc.*, **127**, 1278 (1980).
27. T. Ikeda, R. Schmehl, P. Denisevich, K. Willman, and R. W. Murray, *J. Am. Chem. Soc.*, **104**, 2684 (1982).
28. A. J. Bard and L. R. Faulkner, *Electrochemical Methods*, Wiley, New York, 1980, p. 290.
29. J. Leddy, A. J. Bard, J. T. Maloy, and J. M. Savéant, *J. Electroanal. Chem.*, **187**, 205 (1985).
30. H. Böttger and V. V. Bryksin, *Hopping Conduction in Solids*, VCH Verlagsgesellschaft, Weinheim, 1985.
31. F. B. Kaufman and E. M. Engler, *J. Am. Chem. Soc.*, **101**, 547 (1979).
32. H. Dahms, *J. Phys. Chem.*, **72**, 362 (1968).

33. I. Ruff and V. J. Friedrich, *J. Phys. Chem.*, **75**, 3297 (1971); I. Ruff, V. J. Friedrich, K. Demeter, and K. Csillag, ibid., **75**, 3303 (1971); I. Ruff and I. Korösi-Odor, *Inorg. Chem.*, **9**, 186 (1970); I. Ruff and L. Botár, *J. Chem. Phys.*, **83**, 1292 (1985).
34. E. Laviron, *J. Electroanal. Chem.*, **112**, 1 (1980).
35. C. P. Andrieux and J. M. Savéant, *J. Electroanal. Chem.*, **111**, 377 (1980).
36. D. A. Buttry and F. C. Anson, *J. Electroanal. Chem.*, **130**, 333 (1981).
37. D. A. Buttry and F. C. Anson, *J. Am. Chem. Soc.*, **105**, 685 (1983).
38. J. Leddy and A. J. Bard, *J. Electroanal. Chem.*, **189**, 203 (1985).
39. R. W. Murray, ed., *Molecular Design of Electrode Surfaces*, Wiley-Interscience, New York, 1992, pp. 334FF.
40. M. Majda and L. R. Faulkner, *J. Electroanal. Chem.*, **137**, 149 (1982).
41. J. M. Sevéant, *J. Phys. Chem.*, **92**, 4526 (1988).
42. R. P. Buck, *J. Phys. Chem.*, **92**, 4196 (1988).
43. C. R. Martin, I. Rubinstein, and A. J. Bard, *J. Am. Chem. Soc.*, **104**, 4817 (1982).
44. J. S. Facci and R. W. Murray, *J. Phys. Chem.*, **85**, 2870 (1981).
45. M. Krishnan, X. Zhang, and A. J. Bard, *J. Am. Chem. Soc.*, **106**, 7371 (1984).
46. J. S. Facci, R. H. Schmehl, and R. W. Murray, *J. Am. Chem. Soc.*, **104**, 4959 (1982).
47. C. P. Andrieux, J. M. Dumas-Bouchiat, and J. M. Savéant, *J. Electroanal. Chem.*, **131**, 1 (1982); C. P. Andrieux and J. M. Savéant, ibid., **134**, 163 (1982); ibid., 142 (1982); C. P. Andrieux, J. M. Dumas-Bouchiat, and J. M. Savéant, *J. Electroanal. Chem.*, **169**, 9 (1984); C. P. Andrieux and J. M. Savéant, *J. Electroanal. Chem.*, **171**, 65 (1984); F. C. Anson, J. M. Savéant, and K. Shigehara, *J. Phys. Chem.*, **87**, 214 (1983).
48. (a) J. C. Imbeaux and J. M. Savéant, *J. Electroanal. Chem.*, **44**, 169 (1973); (b) K. B. Oldham and J. Spanier, *J. Electroanal. Chem.*, **26**, 331 (1970); (c) A. J. Bard and L. R. Faulkner, *Electrochemical Methods*, Wiley, New York, 1980, p. 236.
49. N. Oyama and T. Ohsaka, *J. Am. Chem. Soc.*, **105**, 6003 (1983).
50. (a) X. Zhang, J. Leddy, and A. J. Bard, *J. Am. Chem. Soc.*, **107**, 3719 (1985); (b) X. Zhang, H. Yang, and A. J. Bard, *J. Am. Chem. Soc.*, **109**, 1916 (1987).
51. See, for example: (a) D. F. Calef and P. G. Wolynes, *J. Phys. Chem.*, **87**, 3387 (1983); (b) L. D. Zusman, *J. Chem. Phys.*, **49**, 295 (1980); (c) G. Van der Zawn and J. T. Hynes, *J. Phys. Chem.*, **89**, 4181 (1985); (d) R. A. Marcus and H. Sumi, *J. Chem. Phys.*, **84**, 4272 (1986).
52. C. P. Andrieux and J. M. Savéant, in R. W. Murray, ed., *Molecular Design of Electrode Surfaces*, Wiley-Interscience, New York, 1992, p. 207.
53. C. DeGrand and L. L. Miller, *J. Am. Chem. Soc.*, **102**, 5728 (1980).

Chapter 6
PHOTOELECTROCHEMISTRY AND SEMICONDUCTOR MATERIALS

6.1. INTRODUCTION

The electrical and electrochemical properties of ICSs were discussed in Chapter 4. Here we consider photochemical and photoelectrochemical ICSs, where the absorption of light by the system leads to chemical reactions and the flow of current. A number of ICSs of this type have been constructed and investigated; two were introduced in Section 1.2.3. In treating these systems we will discuss semiconductor materials and their interfaces with metals and liquids. We will also address here the important topic of how the properties of materials change with the size of the structure and the differences between bulk properties and those of smaller particles or clusters (i.e., so-called nanoparticle, "quantum," or "Q effects").

Let us first consider the general problem of designing an ICS for conversion of sunlight to useful chemical products. A schematic representation of such a system is shown in Figure 6.1.1 (1, 2). The reaction carried out by this system is

$$Ox + Red' + h\nu \rightarrow Red + Ox' \qquad (6.1.1)$$

where the products Red and Ox' represent, for example, H_2 and O_2 from water, or H_2, Cl_2, and OH^- from a NaCl solution. Some typical half-reactions that might be of some practical interest are given in Table 6.1.1. The different steps shown in the overall process are carried out most efficiently with different components. The key first step is the absorption of a photon of incident light and its conversion to an electron-hole pair (e^-h^+). As discussed below, this energy transduction step is often carried out with a semiconductor material. Often an electrical field exists at the semiconductor–solution interface that promotes separation

227

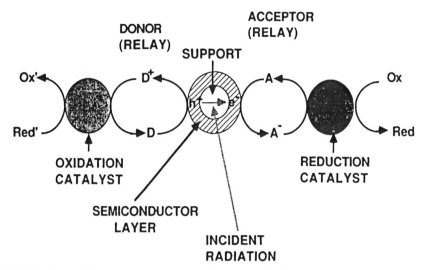

FIGURE 6.1.1. Schematic representation of a generalized semiconductor-based system for the utilization of solar energy to drive the reaction Ox + Red' + $h\nu \rightarrow$ Red + Ox' (e.g., $2H_2O \rightarrow 2H_2 + O_2$). The light is captured by the semiconductor, which may be on a suitable support, to create an electron (e^-) and a hole (h^+). These are captured by the relays, A and D, respectively. The species A^- and D^+ that are produced are used, in conjunction with suitable catalysts, to drive the final reaction. [Reprinted with permission from A. J. Bard, *Ber. Bunsenges. Phys. Chem.*, **92**, 1187 (1988). Copyright 1988 VCH Verlagsgesellschaft.]

TABLE 6.1.1. Representative Half-Reactions of Interest in Photoelectrochemistry

Reductions			Oxidations		
Ox	Red	Application	Red'	Ox'	Application
H^+	H_2	Fuel generation	Cl^-	Cl_2	Disinfection
CO_2	CH_4	Fuel generation	Br^-	Br_2	Energy storage
Cu^{2+}	Cu	Metal removal	Organic	CO_2	Wastewater treatment
Ag^+	Ag	Metal recovery	CN^-	CNO^-	Wastewater treatment
Pt(IV)	Pt	Catalyst preparation	H_2O	O_2	Inexpensive reductant
O_2	H_2O_2	Synthesis	$CH_3CO_2^-$	CO_2, CH_3·	Synthesis

6.1. INTRODUCTION

of e^-h^+ and the movement of e^- and h^+ to sites where the next reactions occur. The electron and hole can then react with relays or mediators, which serve the role of capturing e^- and h^+, thereby preventing the backreaction (recombination of e^- and h^+ to produce heat) and conducting these charges to catalyst centers where the final reactions occur. The neterogeneous catalysts for oxidation and reduction are frequently metals or metal oxides that are selected to promote the reactions of interest. Note that this photoelectrochemical (PEC) ICS is, in many ways, analogous to the biological photosynthetic system discussed in Section 1.2.1.

While this chapter is mainly concerned with PEC systems based on semiconductors, we might note that other types of photochemical ICSs, such as those based on vesicles or micelles, are also possible (3). For example, vesicles of the surfactant dioctadecyldimethylammonium chloride (DODAC)

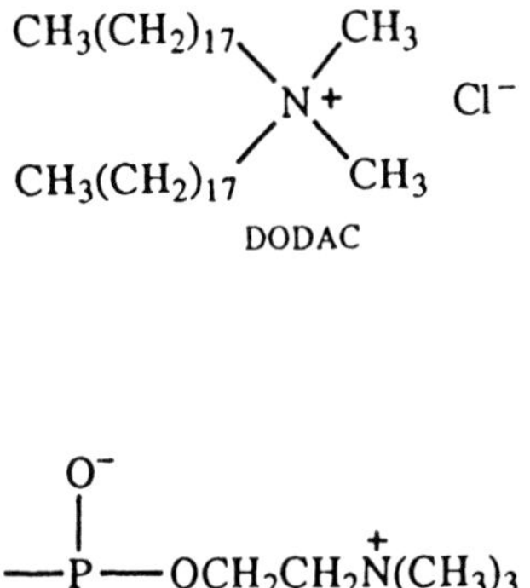

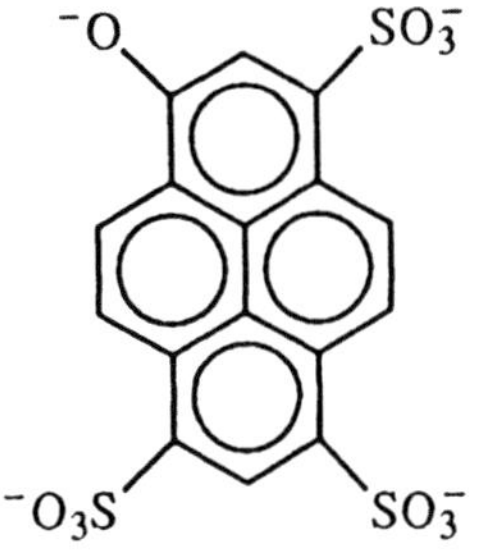

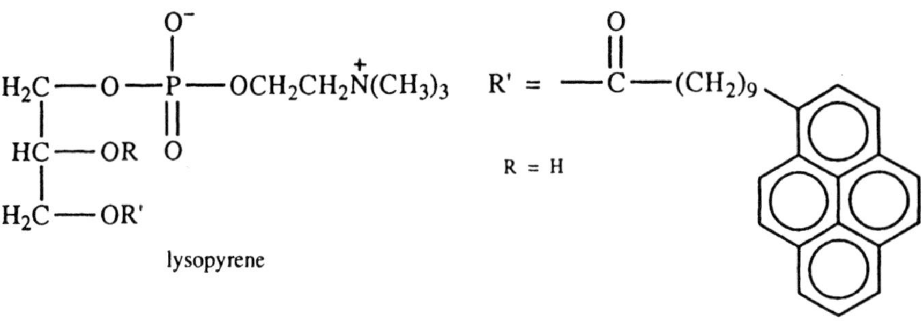

that contain, in the hydrophobic portion of the vesicles, the energy donor lysopyrene can be prepared (4). The soluble energy acceptor, pyranine, that bears four negative charges, is held near the positively charged surface of the vesicle (Fig. 6.1.2a). Excitation of the lysopyrene results in excitation energy transfer (Förster-type) to the pyranine with an efficiency of up to 43%. This can be compared to an energy-transfer

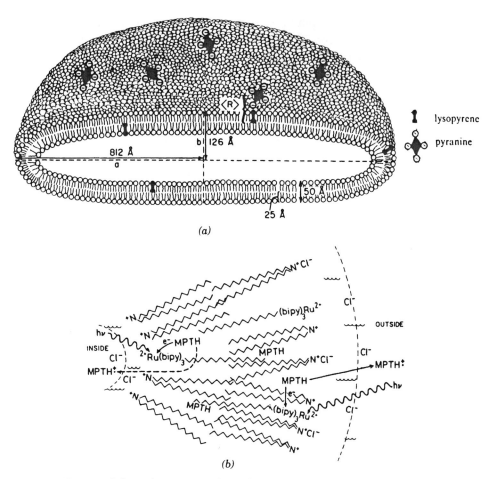

FIGURE 6.1.2. Schematic representations of (a) a DODAC surfactant vesicle containing lysopyrene and pyranine [reprinted with permission from T. Nomura, J. R. Escabi-Perez, J. Sunamoto, and J. H. Fendler, *J. Am. Chem. Soc.*, **102**, 1484 (1980); copyright 1980 American Chemical Society] and (b) DODAC vesicle containing $RuC_{18}(bpy)_3^{2+}$ and MPTH and photoinduced electron transfer and product separation [reprinted with permission from P. P. Infelta, M. Grätzel, and J. H. Fendler, *J. Am. Chem. Soc.*, **102**, 1479 (1980); copyright 1980 American Chemical Society].

efficiency of 3% with similar concentrations of the same reactants in the absence of vesicles. The vesicles thus serve to assemble and localize the energy acceptors and donors and increase the probability of energy transfer.

Vesicles can also improve electron transfer efficiencies. For example, the electron-transfer reaction between an excited long-chain derivative of $Ru(bpy)_3^{2+}$ (where a C_{18} hydrocarbon chain has been attached to one of the bpy groups and is written as $RuC_{18}(bpy)_3^{2+}$) and *N*-methylphenothiazine (MPTH) in DODAC vesicles has also been investigated (5). The following reactions occur on irradiation:

$$RuC_{18}(bpy)_3^{2+} + h\nu \rightarrow RuC_{18}(bpy)_3^{2+*} \qquad (6.1.2)$$

$$RuC_{18}(bpy)_3^{2+*} + MPTH \rightarrow RuC_{18}(bpy)_3^{+} + MPTH^{+} \qquad (6.1.3)$$

The efficiency of the electron-transfer reaction is promoted by the presence of the surfactants that were said to anchor the $RuC_{18}(bpy)_3^{2+}$ molecules to the vesicle surface, with the MPTH distributed in the more hydrophobic regions. The electron-transfer produces a more hydrophilic form, $MPTH^+$, that escapes the vesicle (Fig. 6.1.2*b*). Although the products of reaction (6.1.3) have the same charge and hence are electrostatically repelled from each other, the extent of backreaction with the generation of the ground-state reactants is still quite large, so that the net efficiency of conversion of photons to oxidized and reduced products is modest.

The basic concept involved in these studies is that organized assemblies, bilayers, micelles, and vesicles can serve as supports for the reactants. With proper design, reactants can be preconcentrated in the supports, which might also serve as templates to provide the desired structural arrangement. Further details on systems of this type are given in the references (3).

6.2. SEMICONDUCTORS

In this section we briefly consider the basic concepts of semiconductors, in a somewhat simplified and approximate manner. The theory of solids and semiconductors has been developed extensively and is the subject of numerous texts and references; see, for example, Ref. 6. Here we review band theory and the main concepts involved in consideration of photoeffects at the semiconductor–liquid interface. These same concepts come into play, however, in the application of semiconductors in other devices and sensors and are generally important in the development of ICSs involving these materials.

6.2.1. The Band Model of Solids

Intrinsic Semiconductors. To describe the electronic properties of a solid, we consider the energy of an electron moving in the field of the positive atomic nucleii of the lattice (modified by the other electrons). The crystal lattice is considered to be a periodic array of potential wells set up by the nuclei and the core electrons. The wave function (the *Bloch function*) and the energies of an electron that moves in this field are then calculated. The computation leads to the band model of the solid, where there are allowed energy bands (analogous to the allowed orbitals in an isolated atom) separated by forbidden energy gaps. An alternative (more "chemical") approach to the formation of energy bands can be given by considering the molecular orbitals that form as isolated atoms are assembled into a bulk solid (Fig. 6.2.1). The isolated atoms contain vacant and occupied atomic orbitals. When the atoms combine, molecular orbitals are formed from the atomic orbitals. Consider the formation of a dimer (e.g., of C or Si atoms) and let us be concerned only with the highest occupied and lowest vacant orbitals. The atomic orbitals form molecular orbitals: occupied *bonding* orbitals and unfilled *antibonding* orbitals. This process can be continued with the formation of trimer, tetramer, and so on and the formation of more and more bonding and anitbonding orbitals. If this process is continued with a far greater number of atoms to assemble a lattice, for example, 5×10^{22} Si atoms/cm^3, the numerous bonding and antibonding orbitals form essentially continuous bands. The bonding orbitals form the *valence band* (VB) and the antibonding orbitals form a *conduction band* (CB). The bands are separated by a *forbidden region* or *band gap*; the energy difference between the top of the valence band and the bottom of the conduction band is denoted, E_g, often given in units of electron-volts (eV). The magnitude of E_g depends on the nature of the orbitals that form the bands and the structure and spacing of the atomic lattice. If we consider the case where the VB is completely occupied and the CB is completely vacant, metallic behavior will result when the bands overlap or when $E_g \ll kT$. In these cases, there will be both electrons and vacancies for electrons at the same energy, so that movement of electrons through the lattice occurs readily. Metallic behavior also results from a partially filled valence band (e.g., as in metallic Li). For large values of E_g, with a filled VB and empty CB, the electrical conductivity is very low. The VB electrons cannot contribute to conduction, because this band is filled, and there are essentially no electrons in the conduction band. For example, in silicon, a bonding (valence) electron cannot move into another unbroken, covalent, bond (Fig. 6.2.2). Similarly, there are no electrons in the conduction band, which would represent free (excited or loosely bound) electrons within the Si lattice. At a sufficiently high temperature, however, some of the bonds may be broken, that is, electrons can be excited thermally from the valence to the conduction band. The electron in the

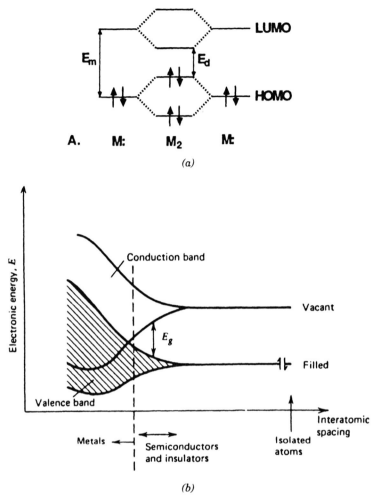

FIGURE 6.2.1. Electronic energy levels: (a) formation of a dimer, M_2, from two monomers, M;, showing formation of four new orbitals and change in energy gap between highest filled and lowest vacant orbitals; (b) formation of bands in solid by assembly of a large number of isolated atoms (characterized by the orbitals on the far right of the interatomic spacing axis) into a lattice with a high density of atoms. [Reprinted with permission from A. J. Bard and L. R. Faulkner, *Electrochemical Methods*, Wiley, New York, 1980. Copyright © 1980 John Wiley & Sons.]

conduction band is free to move to the many unfilled sites in the conduction band. Similarly, the broken Si—Si bond results in a vacancy (or *hole*) that can be filled by a neighboring VB electron. The successive filling of the positive vacancy by bonding electrons is equivalent to the movement of the hole through the lattice, so that the hole is the conducting species in the valence band. Thus promotion of an electron from

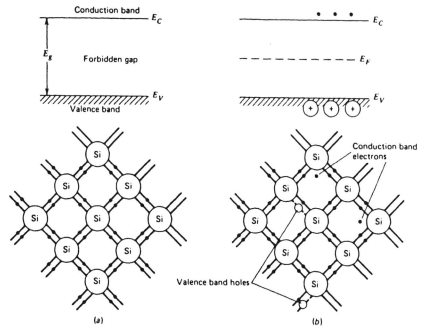

FIGURE 6.2.2. Energy bands (above) and two-dimensional representation of an intrinsic semiconductor (Si) lattice. (a) At absolute zero (or where $E_g \gg kT$). For a perfect lattice (no defects or impurities), no mobile holes or electrons exist. (b) At a temperature where some of the lattice bonds are broken to produce conduction band electrons and VB holes. E_F is the Fermi level in the intrinsic semiconductor. [Reprinted with permission from A. J. Bard and L. R. Faulkner, *Electrochemical Methods*, Wiley, New York, 1980. Copyright © 1980 John Wiley & Sons.]

the valence to the conduction band results in the production of two species of different mobility within the Si lattice: a negatively charged CB electron and a positively charged VB hole.

A pure (undoped) semiconductor material, as described above, is an *intrinsic semiconductor*. The number of electrons (n_i) and holes (p_i) per cubic centimeter that exist at a given temperature, T, is related to the magnitude of E_g by the equation

$$n_i = p_i = (N_C N_V)^{1/2} \exp\left[-\frac{E_g}{2kT}\right] \quad (6.2.1)$$

$$N_C = 2\left(\frac{2\pi m_n kT}{h^2}\right)^{3/2} = (4.83 \times 10^{15})(m_e{}^*T)^{3/2} \quad (6.2.2)$$

$$N_V = 2\left(\frac{2\pi m_p kT}{h^2}\right)^{3/2} = (4.83 \times 10^{15})(m_h{}^*T)^{3/2} \quad (6.2.3)$$

where T is given in kelvins. The terms m_n and m_p are the reduced masses of the electrons (negative) and holes (positive), respectively, and are sometimes written as the relative effective masses of electrons and holes, m_e^* and m_h^*, where $m_e^* = m_n/m_0$, $m_h^* = m_p/m_0$, and m_0 is the rest mass of an electron. In describing electron energies, an electron in free space has an energy related to its momentum by the factor m_0. The same relationship is assumed to hold in a solid, except that the effective masses, which may be larger or smaller than m_0, apply. Values of E_g for some materials of interest in ICSs are given in Table 6.2.1 and some properties of the familiar semiconductors Si and GaAs are given in Table 6.2.2. The electrons and holes (the *mobile carriers*) move in the semiconductor with characteristic mobilities [in square centimeters per volt per second (cm^2 V^{-1} s^{-1})], μ_n and μ_p, respectively; these are related to their diffusion coefficients, D_n and D_p (cm^2 s^{-1}), by the equation

$$D_i = kT\mu_i = (0.0257)\,\mu \quad \text{(at 25°C)} \quad (i = \text{n,p}) \quad (6.2.4)$$

Note that the diffusion coefficients of mobile carriers in solids are usually orders of magnitude larger than those of typical ions in solution.

Extrinsic Semiconductors. Additional electronic energy levels in the solid can be introduced by doping the solid with other species; these are usually at fairly low concentrations (~ 1 ppm) and are called *dopants* or *impurities*. Energy levels of different dopants in Si and GaAs (*impurity levels*) are shown in Figure 6.2.3. Impurity levels for species that are near (within $\sim kT$ of) the conduction band edge (e.g., As in Si) can

TABLE 6.2.1. Energy Gaps (E_g) of Selected Materials

Substance	E_g (eV)	Substance	E_g (eV)
Ge	0.67	Fe$_2$O$_3$	~ 2.3
CuInSe$_2$	0.9	CdS	2.42
Si	1.12	ZnSe	2.58
WSe$_2$	~ 1.1	WO$_3$	2.8
MoSe$_2$	~ 1.1	TiO$_2$ (rutile)	3.0
InP	1.3	TiO$_2$ (anatase)	3.2
GaAs	1.4	ZnO (zincite)	3.2
CdTe	1.50	SrTiO$_3$	3.2
CdSe	1.74	SnO$_2$	3.5
GaP	2.2	ZnS (zinc blende)	3.54
		C (diamond)	5.4

TABLE 6.2.2. Properties of Si and GaAs

Property	Si	GaAs
Atoms/cm^3	5.0×10^{22}	2.21×10^{22}
E_g (eV) at 300 K	1.12	1.43
Crystal structure	Diamond	Zinc blende
Density (g/cm^3)	2.328	5.32
Effective density of states in conduction band, N_C (cm^{-3})	2.8×10^{19}	4.7×10^{17}
Effective density of states in valence band, N_V (cm^{-3})	1.02×10^{19}	7.0×10^{18}
Effective mass (m*/m$_0$)		
Electrons	0.97, 0.19	0.068
Holes	0.16, 0.5	0.12, 0.5
Dielectric constant	11.8	10.9
n_i, p_i (cm^{-3}) at 300 K	6.8×10^9	1.8×10^6
Mobility (cm^2 V^{-1} s^{-1}) at 300 K		
Electrons	1900	8800
Holes	500	400

serve as electron donors. Such *shallow-level impurities* are ionized to a large extent at room temperature with the introduction of electrons into the conduction band, leaving behind a positively charged site (Fig. 6.2.4). The amount of added dopant is usually of the order of a few parts per million in a lightly doped semiconductor; typical donor densities, N_D, are 10^{15} to 10^{17} cm^{-3}. The positive sites (e.g., As$^+$) are spaced far apart and do not contribute to the conduction process. Thus, in diagrams such as that in Figure 6.2.4, the ionized donor levels are often shown as separated segments. Materials doped with donor impurities are called *n-type extrinsic* semiconductors. Thermal activation of VB atoms also occurs in extrinsic materials, so that some holes are also present, at a density, p. The total density of electrons in the conduction band, n, is thus

$$n = p + N_D \quad (6.2.5)$$

or, since in most cases for moderate doping $N_D \gg p$, $n \approx N_D$. For any material, intrinsic or extrinsic,

$$np = N_C N_V \exp \frac{-E_g}{kT} = n_i^2 = p_i^2 \quad (6.2.6)$$

6.2. SEMICONDUCTORS

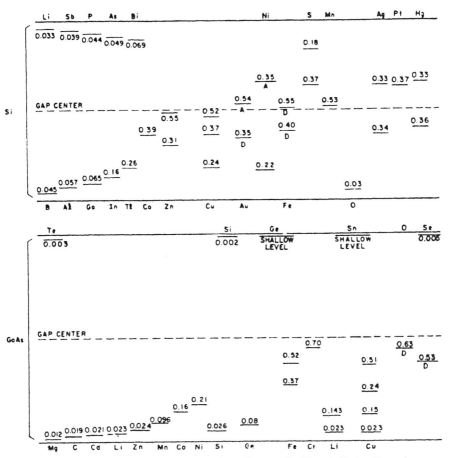

FIGURE 6.2.3. Ionization energies for various dopants in Si and GaAs. Energies are indicated on the different levels. Levels below the gap center are measured from the top of the VB and are acceptor levels (unless indicated by a D for donor). The levels above the gap center are measured from the bottom of the conduction band and are donor levels (unless indicated by an A for acceptor). The band gap energies for Si and GaAs are 1.12 and 1.43 eV, respectively, at 300 K. [Reprinted with permission from S. M. Sze and J. C. Irwin, *Solid State Electron.*, **11**, 599 (1968). Copyright © 1988 Pergamon Press Ltd.]

Thus, for an n-type semiconductor

$$p = \frac{N_C N_V}{N_D} \exp \frac{-E_g}{kT} = \frac{n_i^2}{N_D} \quad (6.2.7)$$

For example, for Si doped with As at a level of 10^{17} cm^{-3}, the electron density is about 10^{17} cm^{-3} and the hole density is about 460. Clearly

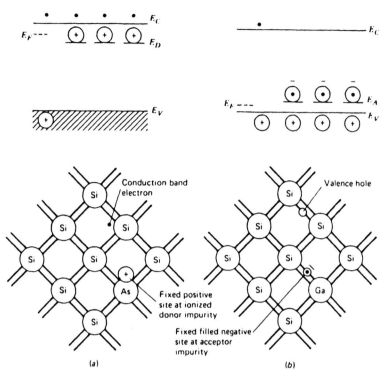

FIGURE 6.2.4. Energy bands and a two-dimensional representation of extrinsic semiconductor (doped-Si) lattices: (a) n-type semiconductor (As dopant); (b) p-type semiconductor (Ga dopant). [Reprinted with permission from A. J. Bard and L. R. Faulkner, *Electrochemical Methods*, Wiley, New York, 1980. Copyright © 1980 John Wiley & Sons.]

the addition of donor impurities greatly decreases the hole density compared to the intrinsic material. In the conduction process, most of the current is carried by the electrons, which are the *majority carriers* in the n-type material; the holes are the *minority carriers*.

If shallow-level acceptor atoms are added to the lattice, for example, by the addition of Ga to Si, an acceptor level at energy E_A near the VB edge is introduced. In this case, electrons are thermally promoted into the acceptor levels from the valence band to form negatively charged acceptor sites and leave behind mobile holes in the valence band (Fig. 6.2.4b). If the dopant (acceptor) density is N_A, then the total density of holes is

$$p = n + N_A \tag{6.2.8}$$

6.2. SEMICONDUCTORS

or, when $N_A \gg n$, $p = N_A$. The number of electrons in the conduction band is given by

$$n = \frac{N_C N_V}{N_A} \exp \frac{-E_g}{kT} = \frac{n_i^2}{N_A} \qquad (6.2.9)$$

Thus the addition of an acceptor dopant decreases n, compared to n_i, and increases p, compared to p_i. In this *p-type* semiconductor, holes are the majority carriers and electrons are the minority carriers. Again, the widely separated, negatively charged, acceptor sites do not contribute to the conduction.

Donors and acceptors with energy levels near the middle of the gap (*deep-level impurities*) do not contribute as effectively as the shallow-level ones, since thermal population from the conduction and valence bands is less probable. It is possible to have both shallow-level donors and acceptors present in the semiconductor lattice. Under these conditions the electroneutrality condition is

$$p + N_D = n + N_A \qquad (6.2.10)$$

The nature of the resulting semiconductor (n- or p-type) will then depend on the relative amounts of the acceptor and donor impurities. An n-type semiconductor can be made intrinsic by adding an acceptor type dopant as a *compensating* impurity.

In a compound semiconductor (e.g., GaAs or TiO_2), as opposed to an elemental semiconductor (e.g., Si), the chemical nature of the impurity atom and the site that it occupies determines whether the impurity is a donor or acceptor. If the impurity atom replaces one of the constituent lattice atoms and provides an additional electron compared to the atom it replaced, it is a donor impurity. If it provides less electrons than the replaced atom, it is an acceptor impurity. Thus, Te on an As site or Si on a Ga site in GaAs would behave as a donor. On the other hand, Zn on a Ga site or Si on an As site behaves as an acceptor. Doping may also occur when an impurity atom lodges in an interstitial position; in this case the dopant electrons are usually available for conduction and the impurity behaves as a donor. A lattice vacancy or broken bond corresponds to a hole, so that these are effectively acceptors. Similarly, a deviation from stoichiometry in a compound semiconductor results in an n-type or p-type material, depending on which atom is in excess. Doping is sometimes carried out by some specific chemical treatment to the pure semiconductor material, rather than the purposeful addition of a specific impurity. For example, pure single-crystal TiO_2 is an insulator. If it is heated in a vacuum at 600–700°C for several hours or heated in a CO or hydrogen atmosphere, n-type TiO_2 is produced.

6.2.2. The Fermi Level

The probability that an electronic level at energy, E, is occupied by an electron at thermal equilibrium, $f(E)$, is given by the Fermi–Dirac distribution function:

$$f(E) = \frac{1}{1 + \exp[(E - E_F)/kT]} \qquad (6.2.11)$$

where E_F is called the *Fermi-level* energy and represents that value of E for which $f(E) = 1/2$ (i.e., where it is equally probable that a level is occupied or vacant). At $T = 0$ K, all levels below E_F ($E < E_F$) are occupied [$\exp[(E - E_F)/kT] \to 0$, $f(E) \to 1$]; all levels above E_F ($E > E_F$) are vacant [$\exp[(E - E_F)/kT] \to \infty$, $f(E) \to 0$]. At 300 K, essentially all levels more than 0.1 eV below E_F are filled and those more than 0.1 eV above E_F are vacant. For an intrinsic semiconductor, E_F lies in the band gap, essentially midway between the edges of the conduction and valence bands. Note that, in contrast to a metal where both occupied and vacant states are present at energies near E_F, for an intrinsic semiconductor, neither electrons nor unfilled sites are present near E_F. For an extrinsic semiconductor, the location of the Fermi level depends on the extent of doping; acceptors move E_F down toward the valence band:

$$E_F = E_V + kT \ln\left(\frac{N_V}{N_A}\right) \quad \text{(p-type semiconductor)} \qquad (6.2.12)$$

and donors move E_F up toward the conduction band:

$$E_F = E_C - kT \ln\left(\frac{N_C}{N_D}\right) \quad \text{(n-type semiconductor)} \qquad (6.2.13)$$

These equations hold as long as $N_A < N_V$ and $N_D < N_C$, that is, under nondegenerate conditions. At higher doping levels, the semiconductor is said to become degenerate, and the Fermi level moves into the valence or conduction band. Under these conditions, the material will start to show metallic conductivity. For example, SnO_2 is a rather wide bandgap semiconductor (E_g = 3.5 eV) and hence is transparent in the visible region of the spectrum. When it is heavily doped with Sb(III) ($N_D > 10^{19}$ cm^{-3}), the material becomes degenerate and conductive.

An alternative definition of E_F for a phase, α, is

$$E_F^\alpha = \bar{\mu}_e^\alpha = \mu_e^\alpha - ze\phi^\alpha \tag{6.2.14}$$

where $\bar{\mu}_e^\alpha$ is the *electrochemical potential* of electrons in phase α, μ_e^α is the chemical potential of electrons in this phase, and ϕ^α is the inner potential of α (related to the electrical potential applied to the phase). Electrochemical potentials can be defined for different species in phases such as metals and solutions and are useful in thermodynamic considerations of reactions and interfaces. When a system is at equilibrium electrically, the electrochemical potential of electrons in all phases must be the same. Thus systems at electronic equilibrium will have Fermi levels at the same energy. This equilibrium usually occurs by the transfer of charge from one phase to another. For example, electrons will tend to transfer from a phase with a higher E_F to one with a lower E_F. The Fermi level in the phase that loses electrons will drop while that in the phase that gains electrons rises until, at equilibrium, they are equal. This concept is useful in considering the nature of junctions between semiconductors, metals, and solutions. As is clear from Eq. (6.2.14), the difference in Fermi levels between two phases is also a function of the applied potential.

The Fermi level is often given with reference to a free electron in vacuum, which is assigned an energy of zero. Under these conditions, E_F for an uncharged phase is closely related to the work function of the material, Φ, which can be taken as the work required to bring an electron from the Fermi level of the material to a point in a vacuum just outside the surface of the material, with the substance containing no excess charge or surface dipoles. Under these conditions

$$\Phi = -E_F \tag{6.2.15}$$

6.2.3. Junctions

In the construction of an ICS, one often considers contacts between semiconductors, metals, and solutions. The electrical nature of such junctions or contacts (i.e., the potential distribution, the charge distribution, and the resistance) is of interest and is considered briefly here. The basic principles governing the formation of a junction are the same for phases of any type. In general, the values of E_F (or Φ or $\bar{\mu}_e$) of the two phases brought into contact are different and charge can flow between the phases when they are brought together.[1] For essentially all phases, E_F is represented by a negative number when given with respect to vacuum, since the electron is usually energetically stabilized by interaction with molecules. For example, the Fermi level of gold is -5.1 eV; the negative number implies that work must be performed to remove an electron from gold to a vacuum (Fig. 6.2.5A). The smaller the magnitude of E_F, that is, the "higher the energy level," the easier it is to remove an electron. We can give some general qualitative rules which are helpful in picturing the formation of junctions (Fig. 6.2.5B): (a) electrons will tend to flow spontaneously from higher to lower energy levels (i.e., "downhill"), while holes tend to flow uphill; (b) adding electrons to (or removing holes from) an energy level will move the energy level up (decrease $|E_F|$); (c) removing electrons from (or adding holes to) an energy level will move the level down (increase $|E_F|$); (d) making the potential of a phase negative (e.g., by use of an external applied voltage source) moves the energy level up; making the potential more positive moves the level down.

Consider what occurs when two phases, α and β, are brought into contact. For the purpose of this discussion it will be convenient to think of α as a metal and β as a semiconductor, as shown in Figure 6.2.6, although the same principles apply for junctions between any phases. When the phases are well separated, equilibrium exists within each phase, but not between them. We consider here the case where E_F^β lies above E_F^α, so when an intimate contact is made between the phases, electrons can flow from the semiconductor to the metal. In the process, the energy levels of both phases will shift until a common Fermi level, and hence electronic equilibrium ($\bar{\mu}_e^\alpha = \bar{\mu}_e^\beta$), is attained. In the example, the metal becomes negatively charged and the semiconductor positively charged, with electrostatic forces holding the charges near the interfacial region. The electrostatic field, ϵ, built up by the charge distribution is just sufficient to prevent further charge transfer across the interface. The nature of the field and charge distribution can be obtained by solving the Poisson equation

$$\frac{d^2\phi}{dx^2} = -\left(\frac{1}{\epsilon}\varepsilon_0\right)\rho(x) = -\frac{d\epsilon}{dx} \qquad (6.2.16)$$

---

[1] In practice, the method of producing the junction may be important in determining its characteristics. Some junctions are "blocking," implying that charge cannot flow across them. We consider here mainly "nonblocking" junctions.

6.2. SEMICONDUCTORS

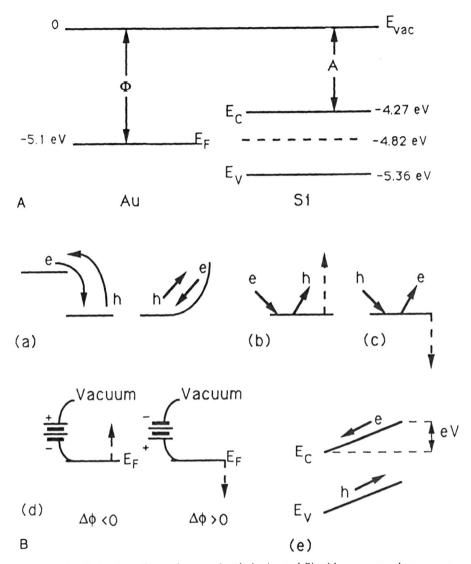

FIGURE 6.2.5. (A) Locations of energy levels in Au and Si with respect to the vacuum level based on the work function (Φ) and electron affinity (A). (B) Schematic representation of general principles involving changes in energy levels and transfer of charge (e, electron; h, hole): (a) spontaneous direction of movement of charge; (b, c) effect of addition or removal of charge on change in energy level; (d) effect of applied potential on energy level; (e) direction of charge flow under an applied potential.

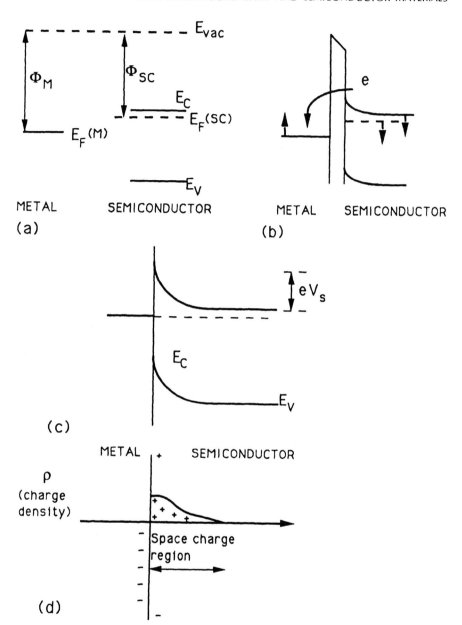

FIGURE 6.2.6. Electron energy-level diagrams showing the formation of a Schottky barrier between a metal and an n-type semiconductor in the absence of surface states: (a) isolated solids (before contact); (b) flow of electron and movement of energy levels when contact is made; (c) at electronic equilibrium on contact, V_s = potential drop across the space charge region; (d) the charge distribution at equilibrium. The negative charge density in the metal is actually in a much thinner layer at the surface than can be represented in the diagram, and the area under the metal (negative) charge density should equal that under the semiconductor (positive) charge density.

6.2. SEMICONDUCTORS

where ϕ is the electrical potential (V); $\rho(x)$ is the charge distribution (C/cm^3) as a function of distance from interface, x; ε_0 is the permittivity of free space; ε is the dielectric constant of the phase; and ϵ is the electric field (V/cm). Equivalent expressions that are useful are

$$\epsilon(x) = \left(\frac{1}{\varepsilon}\varepsilon_0\right) \int \rho(x)\, dx \qquad (6.2.17)$$

$$\phi(x) = -\int \epsilon(x)\, dx \qquad (6.2.18)$$

Solution of (6.2.16) with the proper boundary conditions shows that the charge is distributed in each phase in a *space charge region* whose thickness is related to L, which depends on ε and the total carrier density, n, of the phase, and is given by

$$L \approx \left(\frac{\varepsilon\varepsilon_0 kT}{2e^2 n}\right)^{1/2} \approx 84\,[\varepsilon/n]^{1/2} \text{ cm} \qquad (6.2.19)$$

where the numerical expression, in terms of dielectric constant and total carrier density (cm^{-3}), is valid at 298 K. Actually the thickness of the space charge layer also depends on the potential, but Eq. (6.2.19) is a useful first approximation. For a metal, n is large ($>10^{22}$ cm^{-3}), so that the space region is negligibly thin (<0.5 Å), and the excess charge can be considered to be at the interface. For semiconductors or solutions, however, the carrier densities (i.e., the dopant levels in the semiconductors and the concentration of ionic species in solution) are much smaller, and the charge is distributed over a larger thickness near the interface. For example, for $\varepsilon = 10$ and $n = 10^{13}$ cm^{-3} in a semiconductor, $L = 8.4 \times 10^{-5}$ cm, so that the space charge region is of the order of a micrometer wide.

The situation for a metal–solution interface is shown in Figure 6.2.7, both for an initially uncharged metal (a metal at the "point of zero charge" or pzc) with a layer of oriented solvent dipoles on the surface and for a metal that has been charged negative with respect to the solution (7). The upper portions of Figure 6.2.7 show the charge distributions; the nature of the field and potential distributions can be obtained by integration of the charge distribution, Eq. (6.2.17), and then of the field, Eq. (6.2.18). The negative charge in the metal is compensated by positive ions moving toward the electrode surface and negative ions moving away from it, until the net positive charge in solution in the

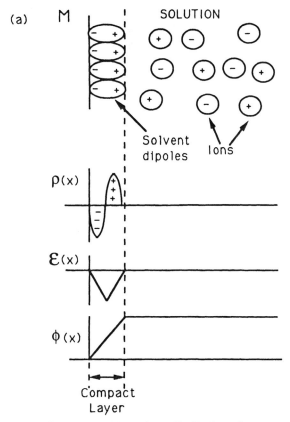

FIGURE 6.2.7. Schematic representation of the distribution of charge and the electric field and potential at the metal-electrolyte interface. (*a*) At point of zero charge (no excess charge in metal or electrolyte). (*b*) With excess negative charge in metal $\rho(x)$ = charge density (C/cm^3); $\epsilon(x)$ = electric field (V/cm); $\phi(x)$ = potential (V). Adsorbed solvent molecules (dipoles) are shown, but bulk solvent and ionic solvation is not indicated.

interfacial region equals that in a metal. Just as in a semiconductor, because the number of ions in solution is relatively low, the excess charge in the solution is distributed over some distance near the metal surface. The space charge layer in solution is called the *diffuse double layer*. Its thickness for a 1:1 electrolyte is also related to L, Eq. (6.2.19), where n here represents the total ionic concentration in ions/cm^{-3}. For the concentration given in the more usual units of mol/L or M, the expression for a 1:1 electrolyte at concentration C at 298 K is

$$L \text{ (cm)} \approx 3.43 \times 10^{-9} \left(\frac{\varepsilon}{C}\right)^{1/2} \tag{6.2.20}$$

6.2. SEMICONDUCTORS

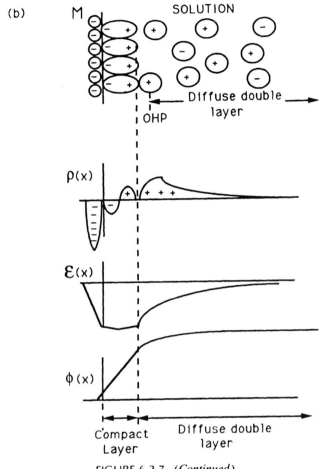

FIGURE 6.2.7. (*Continued*)

Thus for an aqueous solution ($\varepsilon = 78.5$) containing 0.1 M NaCl, $L = 9.6 \times 10^{-8}$ cm. The layer of solvent dipoles on the metal surface forms the *compact* or *Helmholtz layer*. In the presence of ions that are directly (specifically) adsorbed on the metal surface, their charges would also be present within the compact layer.

It is also useful to consider the interfacial capacitances. The differential capacitance, C_d, in $\mu F/cm^2$ (microfarads per square centimeter), can be represented approximately by a simple two-layer capacitor of thickness L (cm) by the expression

$$C_d \, (\mu F/cm^2) \approx \varepsilon \frac{\varepsilon_0}{L} \approx \varepsilon \frac{8.85 \times 10^{-8}}{L} \qquad (6.2.21)$$

Thus the capacitance of the compact layer, C_H, ($L \approx 2$ Å) is ~ 20 $\mu F/cm^2$. The capacitance of the diffuse double layer, C_{dl}, depends on the concentration of ions, just as the capacitance of the space charge region, C_{sc}, depends on the dopant density. The total capacitance of the solution side of the double layer, C_S, is given by the parallel combination of C_H and C_{dl}

$$\frac{1}{C_S} = \frac{1}{C_H} + \frac{1}{C_{dl}} \qquad (6.2.22)$$

and thus is controlled by the *smaller* of the two capacitances. Typically, C_H is the controlling capacitance and C_s is of the order of 10–20 $\mu F/cm^2$. Since the potential drop across a capacitor is $\Delta V = q/C_d$, where q is the charge on the capacitor, the potential drop is also larger across the smaller capacitor. This implies that in most circumstances at a metal–solution interface the potential mainly drops across the thin compact layer.

The same situation applies to the metal–semiconductor interface in Figure 6.2.6. The space charge capacitance in the semiconductor is much smaller than that in the metal, so the potential drop occurs mainly between the interface and the bulk of the semiconductor within the space charge layer. In the absence of surface states, which will be considered later, the difference in potential that arises, V_S, is close to the difference in work functions of metal and semiconductor

$$-eV_S = \Phi_{sc} - \Phi_M \qquad (6.2.23)$$

Since the electronic energy levels are affected by the potential, the band-edge energies show a similar *band bending*, with the formation of an energy barrier at the interface, called the *Schottky barrier*. Because of this barrier, the current–voltage (i–V) behavior of the junction depends on the direction of the applied voltage and is different when the metal is positive with respect to the semiconductor (*forward bias*) than it is when it is negative (*reverse bias*) (Fig. 6.2.8). This i–V behavior is typical of a rectifier, and such junctions are called *rectifying junctions*. For n-type semiconductors, a rectifying junction will be produced when $E_{F,sc}$ lies above $E_{F,m}$, that is, when $\Phi_M > \Phi_{sc}$. For p-type materials, this condition for a barrier to the flow of holes (the majority carrier) is that $E_{F,m}$ lies above $E_{F,sc}$, or $\Phi_M < \Phi_{sc}$. If, for the metal/n-type semiconductor junction, $E_{F,m}$ is very near or above $E_{F,sc}$, then no barrier, or only a very small one, will be formed on contact. The i–V behavior in this case is essentially linear and the contact is said to be *ohmic*. For

6.2. SEMICONDUCTORS

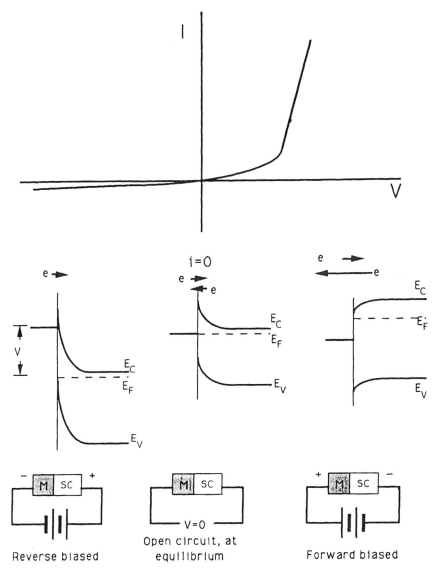

FIGURE 6.2.8. Rectifying metal/n-type semiconductor junction (Schottky barrier), showing bands under different bias conditions and a typical current–voltage (i–V) curve.

metal/p-type semiconductors, ohmic contacts arise when $E_{F,m}$ is at or below $E_{F,sc}$. This can be summarized by saying that ohmic contacts arise in n-type semiconductors when the metal Fermi level lies at energies within the conduction band of the semiconductor and for p-type semiconductors when the metal Fermi level is within the valence band.

While many junctions of semiconductors with metals form Schottky barriers with barrier heights that depend on the work function of the metal, in some cases, for example, for GaAs with a number of metals, the barrier height is almost independent of Φ_M. In this latter case, *surface* or *interface states* have been invoked to explain the behavior. The energies of states at the surface of a semiconductor are different from those in the bulk, because the periodic structure of the lattice is interrupted at the surface. The surface atoms do not have the same environment as those in the bulk, and they may have "dangling bonds." Thus localized surface states with energies within the band-gap region may arise. These may absorb a large part of the charge that has been transferred on making contact with the metal and completely control the barrier height.

6.2.4. Semiconductor–Solution Junctions

The formation of a junction between a semiconductor and a solution will generally follow the same principles as those given above. In this case, however, in addition to electron transfer across the interface, chemical reactions are possible. For example, decomposition of the semiconductor or formation of an oxide film may occur when the semiconductor is immersed in the solution; these chemical effects can complicate the behavior. If a clean Si electrode is immersed in an aqueous solution, especially if the solution contains oxygen or another oxidant, a film of SiO_2 will form at the interface; this insulating film can hinder electron transfer between Si and the solution.

The distribution of charge (electrons and holes in the semiconductor and ions in the solution) and potential in the two phases will depend on their relative Fermi levels as governed by the inherent properties of each phase and the potential difference between them. The Fermi level in the solution, defined as the electrochemical potential of electrons in the solution phase, $\bar{\mu}_e^s$, is usually governed by the nature and concentration of the redox species present in the solution and is directly related to the solution redox potential as calculated by the Nernst equation. At the point of zero charge (and in the absence of surface states and specifically adsorbed ions), no excess charge exists in either phase, the distribution of carriers (e^-, h^+, anions, cations) is uniform from surface to bulk, and the energy bands are flat. The potential at which this condition occurs is called the *flat-band potential*, E_{fb}. Under these conditions, there is no space charge layer in the semiconductor and no diffuse layer in the solution.

When a potential difference exists between semiconductor and solution, either because of an externally applied voltage or because the Fermi levels of the two phases were initially different and the two phases came

6.2. SEMICONDUCTORS

into electrical equilibrium by passage of charge, the interface becomes charged. As before, the excess charge in the semiconductor is contained in the space charge layer. The thickness of this layer, W, depends on the potential difference, ΔV, and the dopant density, N_D, and can be written in terms of the parameter L [Eq. (6.2.19)]:

$$W \approx 2L \left[\frac{\Delta V}{(kT/e)}\right]^{1/2} \approx \left[\left(\frac{2\epsilon\epsilon_0}{e}\right)\frac{\Delta V}{N_D}\right]^{1/2} \quad (6.2.24)$$

$$W \text{ (cm)} \approx 1.05 \times 10^3 \left[\frac{\epsilon \Delta V(\text{V})}{N_D(\text{cm}^{-3})}\right]^{1/2} \quad (6.2.25)$$

A nonuniform carrier density in the semiconductor causes the energy levels or bands to be bent, just as at the semiconductor–metal junction, upward (with respect to the bulk semiconductor) for a positively charged semiconductor and downward for a negatively charged one. Recall that these bent bands represent the electric field that exists in the space charge region; this field governs the direction of motion of any electrons or holes that are created there, such as by absorption of a photon, as discussed in Section 6.3.

Consider an n-type semiconductor placed in contact with a solution (Fig. 6.2.9). The nature of the space charge layer and the distribution of the carriers at the semiconductor–liquid interface depend on the potential applied between the two phases (i.e., via an ohmic contact to the semiconductor and an inert electrode immersed in the solution phase). At the flat-band condition, the holes and electrons are uniformly distributed. The example considered is for $N_D = 10^{18}$ cm^{-3} and $n_i = 10^{13}$ cm^{-3}. When the semiconductor is made positive relative to the solution, the majority carriers (electrons) are repelled (or depleted) from the interface and the minority carriers (holes) move toward the interface. The bands (referenced to the bulk semiconductor) bend upward. The layer at the surface is called a *depletion layer*. If the potential of the semiconductor is made even more positive, a situation may eventually be reached where the concentration of holes at the surface is greater than that of electrons. Under these conditions, the surface becomes p-type and an *inversion layer* is said to have formed. On the other hand, when the semiconductor is made negative with respect to the solution, even more electrons move to the surface, the electric field in the semiconductor points toward the interface, and the bands bend downward. An *accumulation layer* has formed. When the density of majority carriers at the surface approaches the density of states in the material, the surface

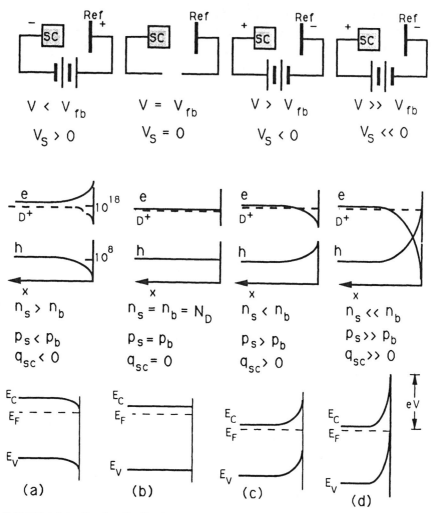

FIGURE 6.2.9. Carrier distributions (electrons, e^-, and holes, h^+) and band bending for an n-type semiconductor ($n_i = 10^{13}$, $N_D = 10^{18}$) in contact with a solution: (a) accumulation layer; (b) flat-band condition; (c) depletion layer; (d) inversion layer.

becomes degenerate. This occurs when the potential of a moderately doped n-type material is made only slightly negative of the flat-band potential. From this point, the potential drop across the space charge layer becomes almost independent of the applied potential and the semiconductor surface approaches metallic behavior. Similar behavior occurs at the junction of a solution with a p-type semiconductor. Here depletion and inversion layers form when the semiconductor is made negative with respect to the solution, and an accumulation layer forms when the semiconductor is made positive (Fig. 6.2.10).

6.2. SEMICONDUCTORS

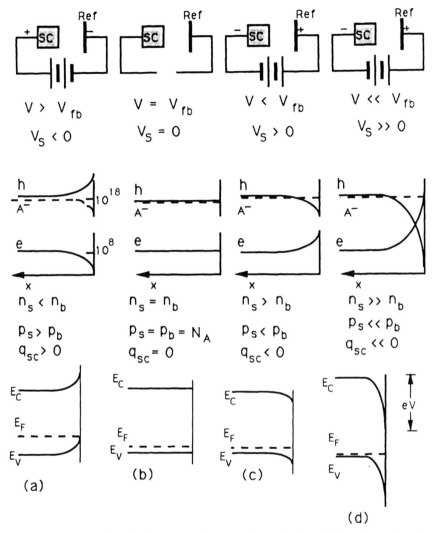

FIGURE 6.2.10. Carrier distributions (electrons, e⁻, and holes, h⁺) and band bending for an p-type semiconductor ($n_i = p_i = 10^{13}$, $N_A = 10^{18}$) in contact with a solution: (a) accumulation layer; (b) flat-band condition; (c) depletion layer; (d) inversion layer.

The capacitance of the space charge layer is given by the expression

$$C_{sc} = (2kTn_i \varepsilon \varepsilon_0)^{1/2}$$

$$\cdot \frac{e}{2kT} \frac{-\lambda e^{-Y} + \lambda^{-1} e^Y + (\lambda - \lambda^{-1})}{[\lambda(e^{-Y} - 1) + \lambda^{-1}(e^Y - 1) + (\lambda - \lambda^{-1})]^{1/2}}$$

(6.2.26)

where $\lambda = n_i/N_D$ and $Y = e\Delta\phi/kT$. This equation can be simplified under the conditions that a depletion layer exists (i.e., $\lambda e^{-Y} \ll \lambda^{-1}$). For an n-type semiconductor, when $\lambda^{-1} \gg \lambda$, this equation can be written with some rearrangement as

$$\frac{1}{C_{sc}^2} = \frac{2}{e\varepsilon\varepsilon_0 N_D}\left(-\Delta\phi - \frac{kT}{e}\right) \qquad (6.2.27)$$

which at 25°C, for C_{sc} in $\mu F/cm^2$, N_D in cm^{-3}, and $\Delta\phi = E_{fb} - E$ in volts is

$$\frac{1}{C_{sc}^2} = \frac{1.41 \times 10^{20}}{\varepsilon N_D}[E - E_{fb} - 0.0257] \qquad (6.2.28)$$

This equation is known as the Mott–Schottky (MS) equation and is useful in characterizing the semiconductor–liquid interface, where a plot of $(1/C_{sc}^2)$ versus E should be linear and yield values for E_{fb} and N_D from the intercept and slope. Note that this applies only to a rather ideal semiconductor–liquid junction and that surface states, nonuniform dopant distributions, and other factors can contribute to deviations from the predicted MS behavior.

6.3. PHOTOELECTROCHEMISTRY

6.3.1. Photoeffects at the Semiconductor–Liquid Interface

The field of semiconductor photoelectrochemistry (PEC) has grown enormously since the first studies in the early 1970s, and a number of monographs and reviews in this area have appeared (8–22). The basic principles of PEC and some applications to ICSs will be discussed here. Consider a photoelectrochemical cell based on an n-type semiconductor immersed in a liquid containing the redox couple D,D^+ (Fig. 6.3.1). When the interface is irradiated with light of energy greater than E_g, electron–hole (e^-h^+) pairs are formed within the semiconductor. These photogenerated carriers perturb the carrier concentrations that existed in the dark semiconductor, increasing both n and p by the amounts Δn^* and Δp^*. Since in the n-type semiconductor $n \gg p$, the extent of perturbation of the hole concentration is much greater than that of the electrons. This nonequilibrium (steady-state) situation is sometimes represented by quasi-Fermi levels for electrons and holes, $E_{F,n}^*$ and $E_{F,p}^*$,

6.3. PHOTOELECTROCHEMISTRY

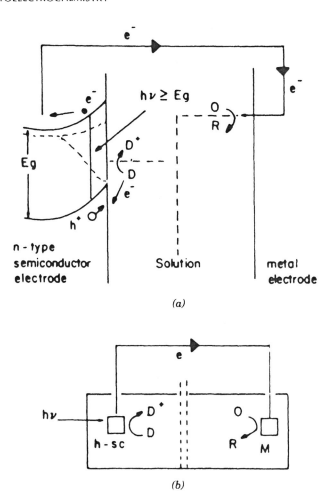

FIGURE 6.3.1. Photoelectrochemical (liquid junction) cell based on an n-type semiconductor. (a) The bands and electronic energy levels under irradiation. The dashed lines represent the quasi-Fermi levels of holes and electrons. (b) Schematic of cell with the n-type semiconductor immersed in a solution of a donor, D, and separated with a porous barrier (indicated by the dashed lines) from a chamber containing an oxidant, O, in contact with a metal electrode.

respectively, where

$$E^*_{F,n} = E_F + kT \ln \frac{n + \Delta n^*}{n} \qquad (6.3.1)$$

$$E^*_{F,p} = E_F - kT \ln \frac{p + \Delta p^*}{p} \qquad (6.3.2)$$

These quasi-Fermi levels represent the potentials (or redox power) of the electrons and holes at the semiconductor–liquid interface under irradiation. For the n-type semiconductor considered here, $\Delta n^* \ll n$, so the potential of the photogenerated electrons is essentially that of the dark Fermi level near the conduction band edge. The potential (oxidizing power) of the holes is greatly changed, however, since $\Delta p^* \gg p$ and under high light intensity will approach that of the VB edge.

The system will tend to relax back to equilibrium conditions by recombination of the electrons and holes (producing heat) or by movement of the carriers in the electric field at the interface in the space charge region. If a depletion region originally existed at the interface, as shown in Figure 6.3.1, the field direction is such that electrons will move away from the interface into the bulk semiconductor and holes will move toward the interface. The holes that arrive at the interface are characterized by an energy, $E^*_{F,p}$, that is near that of the VB edge, E_V. This is equivalent to a redox potential that is more positive than that of the couple D, D^+, so that the reaction $D + h^+ \rightarrow D^+$ can occur at the interface. The electrons can move through a wire contacting the bulk semiconductor to a metal electrode contacting a solution containing species O, where they can cause a reduction reaction, $O + e \rightarrow R$. Electrons and holes that are generated deeper in the semiconductor mainly recombine, except for the carriers that diffuse into the space charge region and then migrate in the electrical field that exists in this region.

An analogous picture can be given for irradiation of the p-type semiconductor–liquid junction (Fig. 6.3.2). Here the field direction in the space charge region promotes the migration of electrons to the interface, with an effective potential near E_C, where reduction of solution species A occurs. The holes move into the bulk semiconductor and through the external circuit to carry out the oxidation of R to O at the metal counter electrode.

The energetics of the semiconductor–liquid interface depends on the location of the semiconductor bands with respect to the redox potentials of the solution species. Diagrams of the type shown in Figure 6.3.3 are often used to represent the relative locations of E_V, E_C, and E_{redox} and allow one to predict what photoreactions are possible for a given semiconductor and solution species. Note that the location of the semiconductor band edges depends on the nature of the interfacial region. Thus a change in pH will affect E_{fb} and the band locations of oxide semiconductors because sites on the semiconductor surface can be protonated. Similarly, adsorption of ions from the solution, such as sulfide onto the surface of CdS, will affect the band positions. In general, specific ad-

6.3. PHOTOELECTROCHEMISTRY

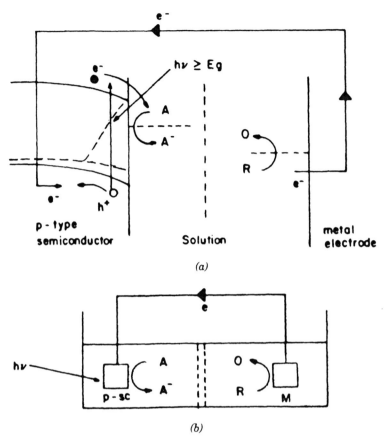

FIGURE 6.3.2. Photoelectrochemical (liquid junction) cell based on a p-type semiconductor. (a) The bands and electronic energy levels under irradiation. The dashed lines represent the quasi-Fermi levels of holes and electrons. (b) Schematic of cell with p-type semiconductor immersed in a solution of an acceptor, A, and separated with a porous barrier (indicated by the dashed lines) from a chamber containing a reductant, R, in contact with a metal electrode.

sorption of an anionic species will move E_{fb} toward more negative potentials and adsorption of cationic species will move E_{fb} toward more positive ones. The solvent will also affect the band locations.

6.3.2. Photoelectrochemical Photovoltaic Cells

When the semiconductor and counter electrodes are immersed in the same solution containing the redox couple D,D^+, no net chemical reaction occurs in the solution during irradiation. The holes oxidize D to

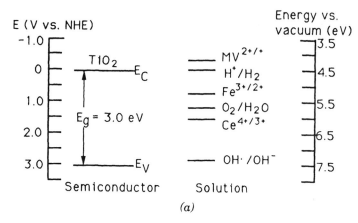

(a)

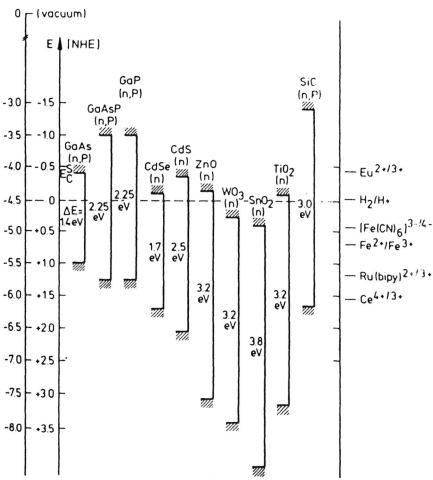

(b)

6.3. PHOTOELECTROCHEMISTRY

D^+ while the electrons reduce D^+ to D. The net effect is a conversion of the radiant energy that generates e^- and h^+ to electrical energy in the external circuit connecting the two electrodes. As with solid-state photovoltaic cells, the conversion efficiency depends on the quantum efficiency of the carrier generation, the rates of interfacial electron transfer, the rates of carrier recombination, and internal cell resistances. Experimentally the cell is frequently characterized by obtaining a power curve, that is, a plot of the cell voltage, V, versus cell current, i (Fig. 6.3.4). When the i is zero, one obtains the open-circuit photovoltage, V_{oc}. Under high light intensities and with rapid electrode reactions, V_{oc} approaches $|E_{fb} - E_{redox}|$, since the maximum potential at the semiconductor under illumination is E_{fb} and the counter electrode is poised at the solution redox potential. The maximum current is the short-circuit photocurrent, i_{sc}, that flows when the two electrodes are shorted together. The power output of the cell, P, is given by $i \cdot V$, so under both open-circuit and short-circuit conditions, $P = 0$. The maximum power output of the cell occurs at a particular point on the curve i_m-V_m and is related to the open- and short-circuit parameters by the fill factor, $f\!f$:

$$f\!f = \frac{i_m V_m}{i_{sc} V_{oc}} \tag{6.3.3}$$

The power efficiency of the cell, η_p, is given by

$$\eta_p = \frac{i_m V_m}{P_r} = \frac{i_{sc} V_{oc} f\!f}{P_r} \tag{6.3.4}$$

where P_r is the radiant power input to the cell. Values for a photovoltaic cell based on WSe_2 in an aqueous I_2, I^- solution are given in Figure 6.3.4 (23).

Within the context of ICSs, miniature photovoltaic cells could be used to generate local electric currents between semiconductive and conduc-

---

FIGURE 6.3.3. Band-edge positions of semiconductors with respect to several redox couples in aqueous solution at pH 1. (a) TiO_2 in rutile form. Reduction by a CB electron can take place when the redox couple lies below E_C; oxidation by a VB hole occurs when the couple lies above E_V. (b) Other semiconductors; here TiO_2 is in the anatase form. Positions are given both as potentials versus NHE and as energies versus the electron in vacuum. [Reprinted with permission from M. Grätzel, in *Photocatalysis—Fundamentals and Applications*, N. Serpone and E. Pelizzetti, eds., Wiley, New York, 1989. Copyright © 1989 John Wiley & Sons.]

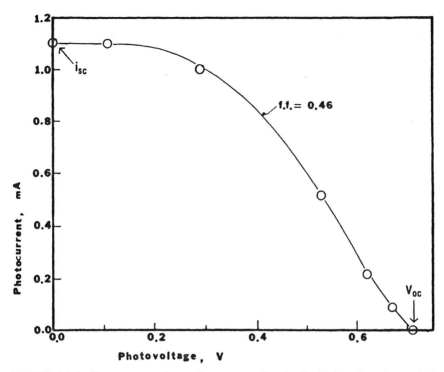

FIGURE 6.3.4. Photocurrent–photovoltage curve for the liquid junction photovoltaic cell n-WSe$_2$/0.50 M Na$_2$SO$_4$, 1.0 NaI, 0.025 M I$_2$/Pt under irradiation from a 450-W Xe lamp. [Reprinted with permission from F-R. F. Fan, H. S. White, B. Wheeler, and A. J. Bard, *J. Electrochem. Soc.*, **127**, 518 (1980). Copyright 1980 The Electrochemical Society.]

tive elements. Purely solid-state junctions, such as p-Si/n-Si, could also be used. A number of PEC photovoltaic cells have been devised; representative examples are given in Table 6.3.1.

6.3.3. Photoelectrosynthetic and Photocatalytic Reactions

We now return to the question of the utilization of radiant energy for the production of chemical products at ICSs based on semiconductors. One can envision two types of reactions (10). In reactions where light is employed to drive a reaction in the nonspontaneous direction (i.e., where the $\Delta G°$ for the dark reaction >0), radiant energy is stored as chemical energy in the products. Such reactions are termed *photoelectrosynthetic* (or simply photosynthetic). The prototypical reaction of this type is the photosplitting of water to hydrogen and oxygen. Light can also be used to drive a slow reaction in the spontaneous direction ($\Delta G°$

TABLE 6.3.1. Representative Liquid Junction Photovoltaic Cells

Semiconductor	E_g (eV)	Redox System	Efficiency (%)	Ref.[a]
n-GaAs (xyl)	1.4	Se_2^{2-}, Se^{2-}	12 (solar)	1, 2
n-GaAs (poly)	1.4	Se_2^{2-}, Se_2^{2-}	7.8 (solar)	1, 3
n-CdTe (xyl)	1.4	Te_2^{2-}, Te_2^{2-}	10 (632.8 nm)	4
n-Si (xyl)	1.1	$Fc^{+/0}$(MeOH)	10 (solar)	5
p-WS_2 (xyl)	1.3	$Fc^{+1/0}$(MeCN)	7 (652.8 nm)	6
p-InP (xyl)	1.4	$V^{3+/2+}$	9.4 (solar)	7

[a]References:
1. A. Heller, H. J. Lewerenz, and B. Miller, *Ber. Bunsenges. Phys. Chem.*, **84**, 592 (1980).
2. R. Noufi and D. Tench, *J. Electrochem. Soc.*, **127**, 188 (1980).
3. A. Heller, B. Miller, S. S. Chu, and Y. T. Lee, *J. Am. Chem. Soc.*, **101**, 7633 (1979).
4. A. B. Ellis, S. W. Kaiser, and M. S. Wrighton, *J. Am. Chem. Soc.*, **98**, 6418 (1976).
5. C. M. Gronet, N. S. Lewis, G. Cogan, J. Gibbons, *Proc. Natl. Acad. Sci. USA*, **80**, 1152 (1983).
6. A. J. Ricco, M. S. Wrighton, G. P. Zosla, *J. Am. Chem. Soc.*, **105**, 2246 (1983).
7. A. Heller, B. Miller, H. J. Lewerenz, K. J. Bachmann, *J. Am. Chem. Soc.*, **102**, 6556 (1980).

Source: Reprinted with permission from A. J. Bard, *Proceedings Robert A. Welch Foundation Conferences on Chemical Research XXVIII*, Houston, Texas, 1984, p. 94.

< 0) in a *photocatalytic* reaction. Here the radiant energy overcomes the energy of activation of the process. The photodecomposition of acetic acid to methane and CO_2 is an example of photocatalysis. Such reactions can be carried out in PEC cells of the type illustrated in Figure 6.3.5. In such cells, the semiconductor and the counter electrodes are shown as being shorted together, since it is the production of the maximum amount of photoproducts, rather than external electrical energy, that is desired. The criteria for selection of the semiconductor for such applications is that the conduction band be located at sufficiently negative potentials to drive the desired reduction half-reaction and that the valence band be located at sufficiently positive potentials to drive the oxidation half-reaction (Fig. 6.3.3). The counter electrode is chosen to be one in which the desired half-reaction proceeds rapidly (i.e., with a low overpotential); this might involve the incorporation of the appropriate electrocatalyst on the electrode surface. Similarly, a catalyst might be employed on the semiconductor electrode surface to promote the reaction that occurs there.

Another consideration with semiconductors immersed in solutions, which applies to PEC photovoltaic cells as well, is the possible instability of the semiconductor under irradiation. For example, with an n-type semiconductor, when a hole is photogenerated at the surface, it may

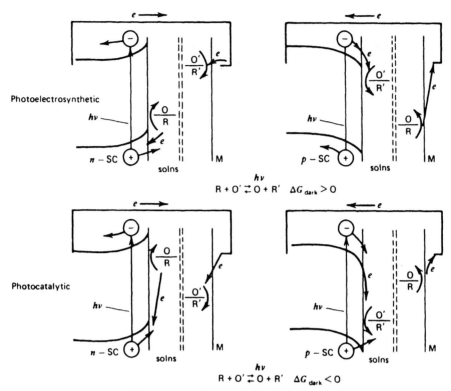

FIGURE 6.3.5. Schematic representations of energy levels and band positions in photoelectrosynthetic (photosynthetic) and photocatalytic cells with n- and p-type semiconductors. Photoelectrosynthetic cells (a) n-type [e.g., n-SrTiO$_3$/H$_2$O/Pt (H$_2$O → H$_2$ + O$_2$)]; (b) p-type [e.g., p-GaP/CO$_2$ (pH 6.8)/C (CO$_2$ reduction)]. Photocatalytic cells (c) n-type [e.g., n-TiO$_2$/CH$_3$COOH/Pt (2CH$_3$COOH → 2CO$_2$ + C$_2$H$_6$)]; (d) p-type [e.g., p-GaP/AlCl$_3$, N$_2$, dimethoxyethane/Al (reduction of N$_2$)]. [Reprinted with permission from A. J. Bard and L. R. Faulkner, *Electrochemical Methods*, Wiley, New York, 1980. Copyright © 1980 John Wiley & Sons, Inc.]

oxidize the semiconductor material itself in addition to oxidizing a solution species. Thus at an irradiated n-Si electrode immersed in an aqueous solution, the following reaction can occur under illumination:

$$Si + 2H_2O + 4h^+ \rightarrow SiO_2 + 4H^+ \qquad (6.3.5)$$

with the surface of the silicon becoming covered with an insulating SiO$_2$ layer. Similarly, irradiation of n-CdS can lead to formation of a layer of S on the electrode surface:

$$CdS + 2h^+ \rightarrow Cd^{2+} + S \qquad (6.3.6)$$

6.3. PHOTOELECTROCHEMISTRY

A number of approaches have been used to decrease the extent of this semiconductor photodecomposition pathway. One is to employ redox couples in the solution that rapidly react with the photogenerated holes and often also interact strongly with the surface to protect it. Some of the redox couples shown in Table 6.3.1 were selected for this purpose. The use of a nonaqueous solvent may also lead to stabilization of the semiconductor. Another approach is to add a catalyst to the semiconductor surface to speed up the desired hole-transfer reaction with solution species relative to the photodecomposition reaction. Finally, thin transparent films of metals (e.g., Pt, Au) or conductive polymers, which allow the photogenerated charge to be transported to the surface but separate the semiconductor surface from the liquid environment, can be employed. Note that reductive decomposition reactions at irradiated p-type semiconductors are also possible, although they appear to be less important in practice.

The design of a photoelectrosynthetic or photocatalytic integrated system thus requires choice of a semiconductor with the appropriate band gap and positions, solvent and solution conditions for the desired reactions, consideration of stability of the semiconductor, and selection of heterogeneous catalysts for the redox reactions. As an example of such design considerations, let us consider the use of an n-type silicon electrode for the photogeneration of chlorine. As shown in Eq. (6.3.5), Si tends to be unstable under irradiation; indeed, if chlorine were generated at the surface of Si immersed in an aqueous solution, corrosion of the Si would certainly occur. Thus the surface of the electrode must be protected from the aqueous environment. One approach is to form a conductive silicide layer on the Si (24). A layer of iridium silicide can be formed by evaporating or sputtering a thin layer of Ir (~ 40 Å) on the Si surface followed by annealing (400°C in vacuum). When such an electrode, represented as n-Si(Ir), is irradiated in contact with a solution containing Cl^-, chlorine is evolved and the electrode is considerably more stable than a bare Si electrode under the same conditions. However, as shown in Figure 6.3.6, the observed photocurrent at this electrode decays over several hours, probably because the photogenerated holes partition between two reactions: chlorine evolution and iridium silicide degradation. The addition of a chlorine evolution catalyst, RuO_2, to the surface speeds up the relative rate of hole capture by Cl^-, and stable operation is observed over many days (Fig. 6.3.6). Another strategy adopted in this system is the use of a very concentrated aqueous solution, 11 M LiCl, which provides a high concentration of the desired reactant and simultaneously decreases the activity of water (25). Finally, note that the photovoltage that allows Cl_2 evolution to start at potentials

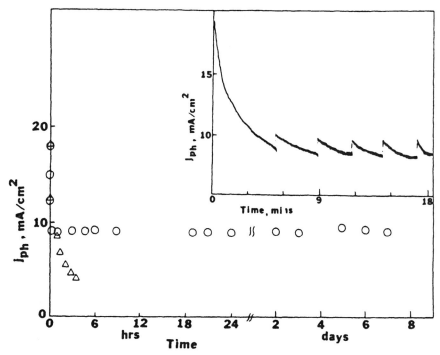

FIGURE 6.3.6. Photocurrent versus time for an n-type Si electrode covered with a film of iridium silicide [denoted n-Si(Ir)] in 11 M LiCl at pH 5 irradiated with a tungsten-halogen lamp at 65 mW/cm² [Δ = untreated n-Si(Ir); O = surface treated with RuO_2 to promote oxidation of Cl^-]. Insert shows early stages and the fluctuations in the photocurrent due to build up of chlorine gas bubbles on the electrode surface. [Reprinted with permission from F-R. Fan, R. G. Keil, and A. J. Bard, *J. Am. Chem. Soc.*, **105**, 220 (1983). Copyright 1983 American Chemical Society.]

about 0.4 V less positive than the standard potential for the reaction is developed at the Si/Si(Ir) interface. This system can be considered an example of the oxidation half (i.e., the left side) of Figure 6.1.1. Many other systems involving semiconductors with stabilized surfaces, catalysts, and suitably chosen solution conditions for photoassisted oxidations have been described.

Similar approaches have been applied with p-type systems. For example, MV^{2+} in solution can be photoreduced at a p-GaAs electrode at potentials about 0.45 V less negative than the reversible MV^{2+}/MV^+ potential (26). Hydrogen evolution will occur in such a solution by the introduction of a heterogeneous catalyst, for example, as a colloidal dispersion of Pt, that speeds up the reaction of MV^+ with H_2O:

$$MV^+ + H_2O \xrightarrow{Pt} MV^{2+} + OH^- + \tfrac{1}{2} H_2 \qquad (6.3.7)$$

This system can also be fabricated with a polymeric form of MV^{2+}, such as MV^{2+} contained in Nafion or as a layer of the polymer PBV (27), so that the reactant resides only on the electrode surface. This decreases the amount of redox relay required and decreases the thickness of the light-absorbing layer of MV^+ produced. The Pt catalyst can be incorporated directly into the PBV layer by soaking it in a $PtCl_6^{2-}$ solution and photoreducing this to metallic Pt particles. Such a modified semiconductor electrode will also generate H_2 on irradiation and represents an example of the reduction half (i.e., the right side) of Figure 6.1.1. Similar systems can be constructed based on p-Si, viologen, and Pt catalyst (28) and a number of other p-type semiconductors.

6.3.4. Surface Modification of Semiconductor Electrodes

The examples discussed above demonstrate the usefulness of chemical modification of the semiconductor electrode surface in the design of photoelectrochemical systems. The surface modifications described were carried out to catalyze electron-transfer reactions, to immobilize a reactant on the electrode surface, and to protect and stabilize the semiconductor. Surface modification can also serve to passivate surface states and decrease the extent of e^-h^+ recombination at the surface. As discussed briefly in Section 6.2.3, surface states often exist on the semiconductor surface at energies within the band-gap region. These states can affect the properties of the semiconductor–liquid interface and the photoelectrochemical reactions that occur there. For example, a high density of surface states (e.g., $\sim 10^{12}$ cm^{-2}, representing about 1% of the surface coverage) can hold an amount of charge that is appreciable compared to that in the space charge region. Under these conditions the surface behaves in a manner analogous to a metal film at the surface and the potential difference between the bulk semiconductor and the surface becomes independent of the nature of the redox couple in solution. This phenomenon, called *Fermi-level pinning*, limits the available photopotential to some fraction of the band gap. In a p-type semiconductor, for example, this is the difference between the valence band edge and the surface state energy (29). In a sense, the surface states play the role of the metal in semiconductor–metal junctions. Even if the surface states are not at a sufficient density to cause pinning, they may serve as recombination centers and decrease the efficiency of photoreactions. Surface treatments are often used to passivate surface states and decrease their effects on junction properties. For example, the oxidation of a Si surface to form a thin layer of SiO_2 to passivate such states is important in the construction of solid-state devices. Similarly, it is found that treatment of the surface of n-type GaAs immersed in a Se_2^{2-}/Se^{2-} solution with $RuCl_3$ or other metal-ion species improves the photoefficiency of PEC photovoltaic cell based on this system (30). This has been ascribed to interaction with surface states, although the surface metal species can also play a catalytic role, promoting hole transfer to solution species. Another example of surface state passivation involves layer-type semiconductors like MoS_2 and WSe_2. These materials, which can be prepared with a very smooth, fresh surface by peeling off the surface layer with adhesive tape, can be used to construct efficient PEC cells (31). However, these electrodes sometimes show large energy losses via recombination processes that occur at crystal imperfections (e.g., exposed edge planes perpendicular to the crystal C axis). Surface treatments to block these recombination centers include treatment with bulky molecules that can intercalate at these edges (31e) and the electrodeposition of insulating polymer at such sites (31f).

Another form of surface modification involves photoactive dyes adsorbed or covalently attached to the electrode surface. These can be used to sensitize the electrode to visible-wavelength (vis) light and produce photocurrents under irradiation with light of smaller energy than that corresponding to the semiconductor band gap (32). The principles of operation of such a sensitized system are shown in Figure 6.3.7A. Absorption of light in the dye layer, presumably of energy smaller than the semiconductor band gap, leads to injection of an electron into the CB of the semiconductor. The hole in the dye layer is filled by electron transfer from a solution species, R, leading to its oxidation. Note that the relative locations of the semiconductor, dye, and solution electronic energy levels are important for successful photoinduced hole transfer to R, and that, although the energy of the radiation needed is smaller ($\Delta E_{dye} < E_g$), the potential of the photogenerated hole in the dye is less positive than that of a hole in the CB of the semiconductor.

An example of a liquid junction photovoltaic cell based on a dye-sensitized semiconductor, TiO_2 with a layer of $RuL_2[\mu\text{-}(CN)Ru(CN)(bpy)_2]_2$, where L = 2,2'-bipyridine-4,4'-dicarboxylic acid, is shown in Figure 6.3.7B (33). The TiO_2 layer (10 μm thick) consisted of nanometer-size particles coated with a monolayer of dye, and the electrolyte was 0.5 M TPAI and 0.04 M iodine in ethylene carbonate-acetonitrile. Iodide was oxidized at the irradiated dye layer and iodine was reduced at the tin oxide counter electrode. The overall light-to-electricity conversion efficiency was about 7-8% with simulated solar radiation.

6.3. PHOTOELECTROCHEMISTRY

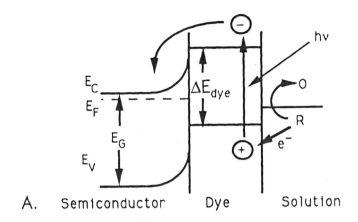

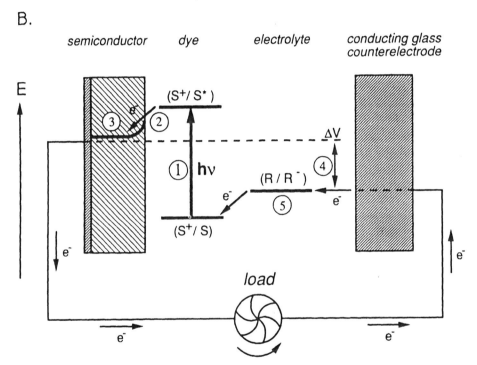

FIGURE 6.3.7. Dye sensitization of semiconductor electrode. (*A*) Principles. (*B*) Schematic representation of $TiO_2/dye/I^-$, I_2/ITO cell (see text) (ΔV corresponds to the open-circuit photovoltage developed in the cell; S, dye sensitizer; S*, electronically excited sensitizer; S^+, oxidized sensitizer; R, iodine; R^-, iodide). [Reprinted, with permission from *Nature*, from B. O'Regan and M. Grätzel, *Nature*, **353**, 737 (1991). Copyright 1991 Macmillam Magazines Limited.]

6.4. PARTICULATE (MICROHETEROGENEOUS) SYSTEMS

6.4.1. Principles

The principles that govern the operation of semiconductor electrode photoelectrochemical cells can be employed in a general way to develop systems based on small semiconductor particles. Such particles can range from those of micrometer dimensions down to colloidal dispersions at the angstrom level (34). Photoreactions have long been investigated on such particles, for example, ZnO and the silver halides, in connection with their use in photographic processes or as pigments. Recent interest has focused on possible applications to the use of solar energy to produce useful materials, such as fuels, or to remove pollutants from water (11). The basic idea is that the particle behaves as a microelectrode that can carry out both the anodic and cathodic half-reactions at different sites on the particle surface (10). In many cases, where both half-reactions can occur readily on the semiconductor surface (i.e., take place with low overpotentials), unmodified semiconductor particles can be used. An example of this type is the use of TiO_2 particles to photocatalyze the oxidation of CN^- by O_2 (34). The photons incident on the particle create e^-h^+ pairs. These carriers react with species on the TiO_2 surface; the ultimate reactions are oxidation of CN^- and reduction of O_2. The general principles of such particle processes are shown in Figure 6.4.1. The dimensions of the particles typically used in these applications are

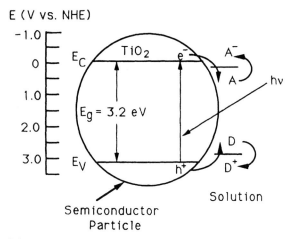

FIGURE 6.4.1. Schematic representation of photogeneration of electron–hole pair on semiconductor (TiO_2–anatase) particle and charge transfer to solution species A and D.

6.4. PARTICULATE (MICROHETEROGENEOUS) SYSTEMS

small compared with the widths of the space charge layers that might exist at the low doping levels of the materials, so separation of the photogenerated charges by electric fields near the particle surface probably does not play a major role in promoting e^--h^+ separation. Fields may exist right at the surface because of specific adsorption of ions. Trapping of charge at surface states may also be important in charge separation. For example, a chemical way of looking at the absorption of light in TiO_2 is to consider the absorption of a photon in a surface $Ti^{IV}-OH^-$ to lead to Ti^{III} $OH^{\cdot}$ in a ligand-to-metal transition. The Ti^{III} is the center for reduction and the hydroxyl radical ($OH^{\cdot}$) acts as an electron acceptor to form OH^- or reacts directly with a solution species (i.e., causes hydroxylation).

The reduction of O_2 can occur at a reasonable rate on an unmodified TiO_2 surface, so the oxidation of CN^- and many organic compounds by O_2 can be promoted by irradiation of a TiO_2 dispersion. If the desired half-reaction does not occur rapidly at the semiconductor surface, it is necessary to add a heterogeneous catalyst to speed up the reaction. For example, Pt can be added to the surface of TiO_2 to promote the reduction of H^+. Indeed, in many photoprocesses significant reaction rates are found only with platinized semiconductor particles. Consider the photodecomposition of acetic acid (an example of the Photo-Kolbe reaction) (35). Here the oxidation process is $CH_3COO^- + h^+ \rightarrow CH_3\cdot + CO_2$. The reduction process, $H^+ + e^- \rightarrow H^{\cdot}$, does not occur readily at an unmodified TiO_2 surface (i.e., TiO_2 shows a high hydrogen overpotential). However, if the surface is platinized, the reaction occurs, since Pt is a good material for proton reduction. Thus irradiation of platinized TiO_2 in the presence of acetic acids leads to formation of methane, CO_2, and smaller amounts of ethane and hydrogen. Many different processes have been studied on irradiated catalyzed and uncatalyzed semiconductor particles (21); some representative examples are given in Table 6.4.1.

A number of approaches have been used to characterize semiconductor particles, for example, to estimate their band-gap and band-edge positions and identify processes occurring at their surfaces. A detailed description of such techniques is beyond the scope of this discussion. However, methods that have been used include electrochemical (36–40), electrophoretic (36), photochemical (41–43), ESR (44), and luminescence (45, 46) methods. In addition to these, X-ray diffraction, photothermal and photoacoustic spectroscopy, surface spectroscopy, microscopy, and other techniques for characterizing solids and interfaces (as discussed in Chapter 3) are often employed.

TABLE 6.4.1. Representative Photoreactions at Particulate Semiconductors

Reactants	Products	Semiconductors	Refs.[a]
CN^-, O_2	CNO^-	TiO_2, CdS, ZnO	1
CO_2, H_2O	CH_3OH, HCHO	TiO_2, CdS, GaAs	2
CH_4, NH_3, H_2O	Amino acids	Pt/TiO_2	3
Many organics, H_2O	H_2, CO_2	Pt/TiO_2	4
Many organics, O_2	H_2O, CO_2	Pt/TiO_2	5
CH_3COOH	CH_4, CO_2	Pt/TiO_2	6
N_2, H_2O	NH_3	TiO_2, WO_3, Fe_2O_3	7
$Ph_2C=CH_2$, O_2	$Ph_2C=O$	TiO_2	8
H_2S	H_2, S	Pt/CdS, RuO_2/CdS	9

[a]References:
1. S. N. Frank and A. J. Bard, *J. Phys. Chem.*, **81**, 1484 (1977).
2. (a) T. Inoue, A. Fujishima, S. Konishi, and K. Honda, *Nature*, **277**, 637 (1979); (b) M. Halmann and B. Ourian-Blajeni, ibid., **275**, 115 (1978).
3. (a) W. W. Dunn, Y. Aikawa, and A. J. Bard, *J. Am. Chem. Soc.*, **103**, 6893 (1981); (b) H. Reiche and A. J. Bard, ibid., **101**, 3127 (1979).
4. T. Kawai and T. Sakata, *Chem. Phys. Lett.*, **80**, 341 (1981); ibid. *Nouv. J. Chim.*, **5**, 279 (1981).
5. I. Izumi, W. W. Dunn, K. Wilbourn, F-R. Fan, and A. J. Bard, *J. Phys. Chem.*, **84**, 3207 (1980).
6. B. Kraeutler and A. J. Bard, *J. Am. Chem. Soc.*, **100**, 5985 (1978).
7. (a) G. N. Schrauzer and T. D. Guth, *J. Am. Chem. Soc.*, **99**, 7189 (1977); (b) E. Endoh, J. K. Leland, and A. J. Bard, *J. Phys. Chem.*, **90**, 6223 (1986).
8. (a) M. A. Fox and C. C. Chen, *J. Am. Chem. Soc.*, **103**, 6757 (1981); (b) other applications to organic synthesis are reviewed in M. A. Fox, *Acc. Chem. Res.*, **16**, 314 (1983).
9. (a) A. J. Nozik, *Appl. Phys. Lett.*, **30**, 567 (1977); (b) E. Borgarello, K. Kalyanasundaram, and M. Grätzel, *Helv. Chim. Acta*, **65**, 243 (1982); (c) M. Matsumura, Y. Saho, and H. Tsubomura, *J. Phys. Chem.*, **87**, 3807 (1983).

Source: Reprinted with permission from A. J. Bard, *Ber. Bunsenges. Phys. Chem.*, **92**, 1187 (1988). Copyright 1988 VCH Verlagsgesellschaft.

6.4.2. Supported Semiconductor Particles

There are several reasons why it is advantageous to form the semiconductor particles that are used for photoelectrochemical reactions within or on a suitable matrix or support. It prevents the flocculation and settling out of very small particles. When the support material is transparent to the wavelengths of interest, it can provide a core for high-area thin films or small particles of the semiconductor. In such an assembly, reactant electrons and holes are formed only near the surface of the semiconductor (within the diffusion length of the carriers) where they can react with species in solution. With larger semiconductor particles, light absorbed inside the particles results in formation of carriers that mainly recombine with subsequent loss of efficiency. The support can also provide a fixed bed or panel of small particles, which can serve as an alternative to particle slurries for use in practical systems with flowing streams. The support can serve as a means of controlling particle size, and perhaps morphology, as described in Section 6.4.4. Most importantly, in terms of ICSs, the support can serve as a means of assembling, and sometimes organizing, a number of components along with the semiconductor particles to form a more complex system that shows improved behavior.

A number of different materials have served as supports (see Section 2.2.1). Polymer films or membranes have been used frequently. As discussed briefly in Section 1.2.3, the ion-exchange polymer Nafion can take up metal cations, such as Cd^{2+} and Zn^{2+}, which can then be precipitated by a suitable treatment (47). Thus particles of CdS or CdSe can be formed in the polymer matrix by treatment of Cd^{2+}-containing Nafion with H_2S or H_2Se, respectively. The matrix in this case allows control of the size of the particles and provides a rigid environment for addition of suitable catalysts, such as Pt particles, or redox species, such as MV^{2+}. Similarly TiO_2 can be incorporated into a Nafion membrane by soaking it in a solution of Ti(III) and then oxidation in moist air (47). A similar strategy has been employed to incorporate PbS into a film of ethylene 15% methacrylic acid copolymer (48).

Inorganic matrices (e.g., SiO_2, clays, and zeolites) can similarly be used to exchange cations and then form the desired semiconductor particle. For example, TiO_2 can be incorporated into a film of hectorite employing essentially the same procedure as used with the Nafion membrane (49). TiO_2 in both Nafion and clay films can be used to carry out the photoreduction of MV^{2+}. Zeolites are useful as inorganic matrices for semiconductors because of their well-defined structures and the possibility of controlling particle size and relative arrangement of the components within and on the surface of the particles by taking advantage of the relative sizes of reactants and those of the zeolite channels and holes. Thus very small particles of CdS, CdSe, PbS, and GaP have been formed by precipitation within the sodalite cage of zeolite Y and related materials (50). Another approach to the production of small particles or thin layers on a support involves synthesis from a monolayer. For example, a layer of WO_3 on SiO_2 particles can be prepared by treatment of SiO_2 with WCl_6 in dry methylene chloride to form a W-containing monolayer that was covalently bound to the SiO_2 surface, followed by hydrolysis (51). Semiconductor particles (e.g., PbS, CdS, CdSe) have also been prepared in silicate glasses by mixing aqueous colloidal dispersions with tetramethoxysilane followed by treatment with ammonia or by incorporating metal ions in silicate gels followed by treatment with the precipitating agent (e.g., H_2S or H_2Se) (52). A similar approach can be employed with porous Vycor glass (53).

Related systems involve the use of dye sensitizers coated on a semiconductor or support. An example of a photochemical ICS with a zeolite support where zeolite L is used for spatial organization is one based on a metal porphyrin as sensitizer (54). The preparation and basic mode of operation of the system are shown in Figure 6.4.2. Zeolite L is characterized by channels with a diameter of about 7.1–7.8 Å. These are large enough to accommodate MV^{2+}, which serves as the redox relay in this scheme, but are much too small to allow the sensitizer, zinc tetra(N-methyl-4-pyridyl)porphyrin ($ZnTMPyP^{4+}$), to enter. However, the $ZnTMPyP^{4+}$, because of its positive charge, will adsorb onto the zeolite particles (diameter ~1 μm). Before exchange of $ZnTMPyP^{4+}$ and MV^{2+}, Pt is introduced into the channels of zeolite L by treatment with platinum acetylacetonate followed by reduction with sodium borohyride; this procedure has been shown to deposit elemental Pt exclusively in the inner surface of the zeolite particles (55). Thus the sequence of treatments shown in Figure 6.4.2 yields a system composed of sensitizer–relay–catalyst spatially arranged by the zeolite support. Irradiation of these particles in an EDTA solution at pH 4 caused oxidation of the EDTA and evolution of hydrogen via the electron-transfer reactions shown in the figure. If any component of the ICS (ZnTMPyP, MV^{2+}, Pt, EDTA) was left out, only trace amounts of hydrogen were observed under irradiation. A similar ICS involving either ZnTMPyP or zinc tetra(p-sulfophenyl)porphyrin ($ZnTPPS_4$) as sensitizer coated on platinized particles of semiconductors (ZnO, TiO_2) or insulators (Al_2O_3, SiO_2) as supports with triethanolamine as sacrificial donor also produces hydrogen under irradiation (56). Many other systems involving sensitizers and catalysts have been described (3c, d). A typical example involved $Ru(bpy)_3^{2+}$ as sensitizer, MV^{2+} as relay, and Pt and RuO_2 as catalysts for the decomposition of water (57). Another example utilized the layered semiconductor $K_4Nb_6O_{17} \cdot 3H_2O$ internally loaded with Pt or Ni as catalysts that photodecompose water to hydrogen and oxygen when irradiated with light above 3.7 eV (58). A related system can be made to

---

FIGURE 6.4.2. Photochemical integrated chemical system based on a zeolite support. (a) Perspective view of zeolite L showing one-dimensional channel in center. (b) Zeolite L main channel and the $ZnTMPyP^{4+}$ and MV^{2+} ions showing relative sizes. (c) Outline of preparation of Pt in zeolite L channel, assembly of system, and proposed operation as photochemical system. [Reprinted with permission from L. Persaud, A. J. Bard, A. Campion, M. A. Fox, T. E. Mallouk, S. E. Webber, and J. M. White, *J. Am. Chem. Soc.*, **109**, 7309 (1987). Copyright 1987 American Chemical Society.]

6.4. PARTICULATE (MICROHETEROGENEOUS) SYSTEMS

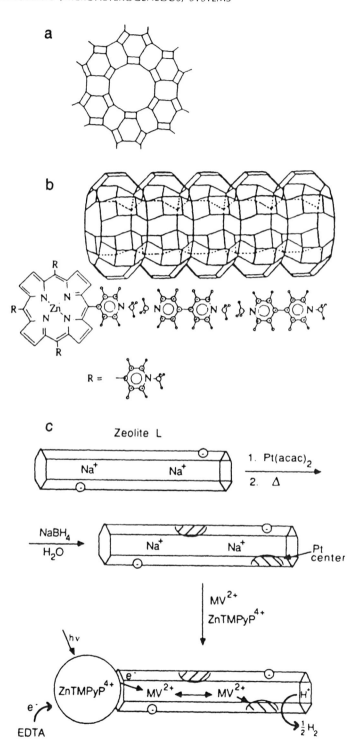

operate in the visible region for the oxidation of a KI solution to hydrogen and triiodide by using the dye sensitization approach discussed in Section 6.3.4 (Fig. 6.4.3) (59). The photosensitizer was RuL_3^{2+}, and the $K_4Nb_6O_{17}$ support was internally platinized to provide sites for hydrogen evolution.

Organic-based supports for semiconductors include membranes, micelles, and vesicles. A number of systems of this type have been described. For example, vesicles formed in aqueous solutions from the surfactant dihexadecyl phosphate (DHP), which are negatively charged,

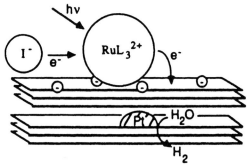

Layered Oxide Semiconductor

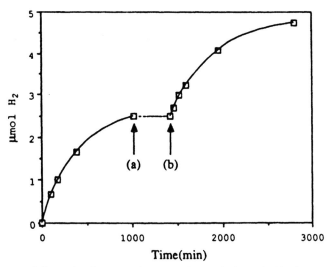

FIGURE 6.4.3. Schematic diagram of system for evolution of triiodide and hydrogen from an aqueous KI solution with particles of dye-sensitized $K_{4-x}H_xNb_6O_{17} \cdot nH_2O$ (L = 4,4'dicarboxy-2,2'-bipyridine). [Reprinted with permission from Y. I. Kim, S. Salim, M. J. Huq, and T. E. Mallouk, *J. Am. Chem. Soc.*, **113**, 9561 (1991). Copyright 1991 American Chemical Society.]

or from dioctadecyldimethylammonium chloride (DODAC), which are positively charged, have been used as supports for the generation of CdS particles (60). An ICS system based on the DHP/CdS structure and incorporating MV^{2+} as relay and Rh as catalyst is shown in Figure 6.4.4 (60a). The vesicles are prepared by sonication in the presence of the desired components (e.g., Cd^{2+} and MV^{2+}) (61). CdS is precipitated with H_2S and Rh is produced by photodeposition in the presence of a sacrificial electron donor, such as benzyl alcohol. The vesicles, with radii in the range of 80–100 nm, produced MV^+ and hydrogen on irradiation. A similar approach was taken with positively charged DODAC vesicles that incorporated a surface-active electron donor, $(n\text{-}C_{18}H_{37})_2N^+(CH_3)(CH_2CH_2SH)\ Br^-$, also illustrated in Figure 6.4.4 (60c). Semiconductor particles can also be prepared in inverse micelles (62) and bilayer lipid membranes (63). Similarly, a monolayer of CdS particles was formed by treating an organized monolayer of cadmium arachidate coated on a glass surface by the Langmuir–Blodgett technique with H_2S (64).

6.4.2. Multijunction Photoelectrochemical Systems

A problem with the practical application of photoelectrochemical systems, for example, for splitting of water to hydrogen and oxygen, is that the maximum driving force of a single junction, for example, semiconductor–solution or p-Si/n-Si, is usually small. The largest open-circuit photovoltages developed are rarely larger than 0.8 V, and the Si p/n junction only produces about 0.55 V. Thus to drive reactions of interest and produce voltages above the thermodynamic requirements of the reaction, and, in addition, to supply needed overpotentials to overcome kinetic barriers, 1.5–2 V may be required. If an external electrical bias is not supplied, then multiple junctions connected in series must be used. Several such systems have been described. The biological analogy here is that of the electric eel. In most biological systems, the potentials established by ionic gradients across membranes are of the order of 0.1 V. However, an electric organ in the electric eel, containing specialized cells (electrocytes) in series, is capable of producing voltages above 500 V.

One system, which probably represented an ICS photoelectrochemical system that was closest to commercial introduction, was the Texas Instruments Solar Energy System (TISES) (65). This system employs panels made up of small (250-μm) silicon spheres of two different kinds (p-Si spheres with an n-Si coating, and n-Si spheres with a p-Si coating) embedded in a glass matrix (Fig. 6.4.5). These are etched and the cores

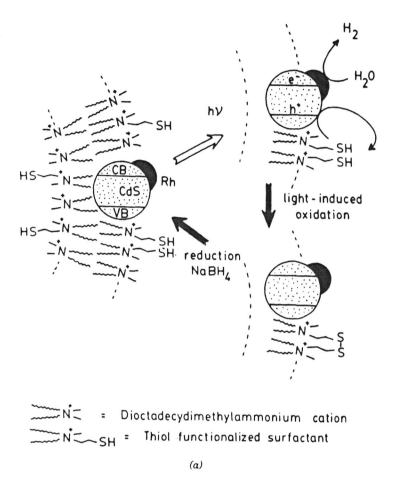

FIGURE 6.4.4. Schematic representations of CdS-based systems in vesicle supports. (a) CdS/Rh in DODAC vesicle containing thiol-functionalized surfactant and mode of hydrogen generation under irradiation with formation of S—S bond (60c). (b) CdS/Rh in DHP vesicle showing MV^{2+} reduction, hydrogen formation, and benzyl alcohol oxidation. [Reprinted with permission from H-C. Youn, Y-M. Tricot, and J. H. Fendler, *J. Phys. Chem.*, **91**, 581 (1987). Copyright 1987 American Chemical Society.]

of the same type are interconnected with a conductive backing. Metals to catalyze the desired reactions and protect the Si layers are coated on the Si that is exposed to the HBr solutions in which the panels are immersed. On irradiation, the potentials of the p/n junctions add in series to produce about 1.1 V, which is sufficient to decompose HBr to H_2 and Br_2. In other parts of the TISES, the H_2 and Br_2 were stored and used, when needed, in a fuel cell to produce electricity. This multi-

6.4. PARTICULATE (MICROHETEROGENEOUS) SYSTEMS

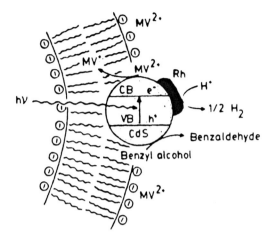

(b)

FIGURE 6.4.4. *(Continued)*

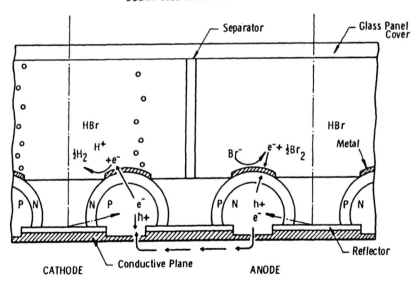

FIGURE 6.4.5. Texas Instrument solar array based on Si spheres embedded in glass for the photodecomposition of HBr to H_2 and Br_2 (65). [Reprinted from A. J. Bard, *J. Electroanal. Chem.*, **168**, 5 (1984). Copyright 1984 Elsevier.]

junction, multiphase system, involving semiconductors, metal catalysts, support, and solution is clearly an example of an ICS.

Another approach is the utilization of multiple semiconductor–solutions junctions in series (66). For example, bipolar TiO_2/Pt panels can be produced by electrochemical anodization of 25-μm-thick Ti foils in 2.5 M H_2SO_4 after a 350-nm-thick layer of Pt has been deposited by sputtering. These panels were then arranged with intervening solution layers (O_2-saturated aqueous NaOH) in a series configuration, so that each TiO_2/NaOH soln/Pt group represents a single photoelectrochemical cell, like that shown in Figure 6.3.1. By placing several panels in a series configuration, a sufficient potential is developed to decompose water. This same approach can be employed with smaller band-gap semiconductors, like CdSe. For example, the system shown in Figure 6.4.6 utilizes CdSe/Ti/CoS panels. The internal CdSe/S, S^{2-},

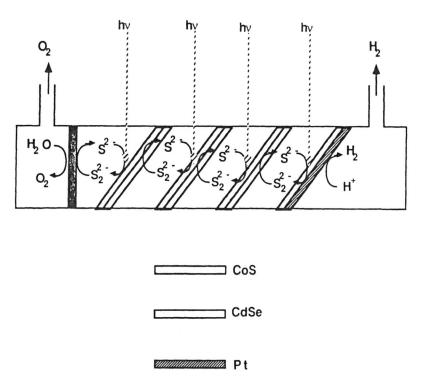

FIGURE 6.4.6. Schematic diagram of a multielectrode array consisting of CdSe/CoS panels in series arranged in a cell for the photosplitting of water to H_2 and O_2 without an external bias. All solutions contain 1 M NaOH. The inner solutions also contain 1 M S and 1 M Na_2S (66). [Reprinted from A. J. Bard, *Ber. Bunsenges. Phys. Chem.* **92**, 1187 (1988). Copyright 1988 VCH Verlagsgesellschaft.]

NaOH/CoS cells behave as photovoltaic cells (S^{2-} oxidized at the CdSe and S reduced at the CoS) that provide a bias for driving the water decomposition reaction at the outer Pt electrodes, as shown. The challenge is to construct true ICSs like these, somehow arranging supported particles with associated catalysts in series on a small scale to produce the desired driving forces (a "photoelectrochemical eel").

6.5. QUANTUM (OR NANO-) PARTICLES

6.5.1. Size Quantization Effects in Small (Q) Particles

As discussed in Section 6.2.1, the band structure of semiconductors arises from the assembly of a large number of atoms into a lattice. Once a sufficiently large particle has been formed, its properties are the same as that of a large piece of bulk material. Thus the electronic properties of a 0.1-μm-diameter particle of CdS, such as the absorption spectrum and the work function, are the same as those of a large single crystal of CdS. However, if the CdS particle sizes are sufficiently small (< 50 Å), their electronic properties will differ from that of the bulk material. Such small particles, which are small in three dimensions, are sometimes called Q (or quantum) particles or quantum dots. Similarly, structures that are small in two dimensions (quantum wires) or thin films that are small in one dimension (quantum wells) show properties that are different from the same material in structures of larger dimensions. There has been particular interest in multiple-quantum-well structures, in which thin (nanometer-scale) layers of one semiconductor material are sandwiched between layers of a second material. Such structures are known as *superlattices*.

6.5.2. Preparation of Q-Particles

There have been a number of studies on the preparation and properties of Q particles, and several reviews in this field have appeared (67). The change in the absorption spectrum of a material as the particle size decreases in the Q region can be understood qualitatively by consideration of the energies of the highest filled and lowest vacant orbitals, as shown in Figure 6.2.1. Thus the electronic transition in Q particles is at higher energies (i.e., is blue-shifted) compared to the bulk material. The effect is analogous to the behavior found when one considers assembling benzene rings to form larger hydrocarbons. The electronic energy transitions move to longer and longer wavelengths as one proceeds down the sequence benzene, naphthalene, anthracene, tetracene, and so on. A similar effect occurs with polyolefins as the chain length increases. The magnitude of the energy of the electronic transition, $E(R)$, as a function of the particle radius, R, can be estimated from an equation based on a "particle in a box" model (68)

$$E(R) = E_g + \frac{\hbar^2\pi^2}{2R^2}\left[\frac{1}{m_e} + \frac{1}{m_h}\right] - \frac{1.8e^2}{\varepsilon R} \qquad (6.5.1)$$

where E_g is the bulk band gap, m_e and m_h are the electron and hole effective masses, and ε is the bulk optical dielectric constant. The particle size at which deviations from the bulk properties start depends on the effective masses and ε; semiconductors with small values for these parameters will show Q effects for larger particles. Although m_e and m_h can depend on the particle size, the values used in calculations with Eq. (6.5.1) are usually those for the bulk material. Alternative models have also been proposed (69).

A number of different methods have been employed to synthesize Q particles. A difficulty with the preparation of very small particles is their tendency to undergo Ostwald ripening in solvents where they show any appreciable solubility. Ostwald ripening is the tendency of smaller particles to dissolve and larger particles to grow, thus minimizing the total surface energy of the system. Moreover, the very small colloidal particles tend to aggregate or flocculate into larger clusters and eventually precipitate from the solution. Thus the preparation of Q particles requires the use of a method that prevents particle growth. One approach is to utilize rapid mixing in an organic solvent, such as MeCN, where the inorganic species, such as CdS, has a very low solubility (70). Another is the use of stabilizers, such as sodium hexametaphosphate or poly(vinylalcohol), that adsorb on the particle surface and prevent dissolution and aggregation (71). With this approach, powders of the Q particles can be isolated by evaporation. For example, powders of semiconductors with small band gaps, such as Cd_3P_2 (0.5 eV) and Cd_3As_2 (0.1 eV), which are black as bulk materials, can be prepared as Q particles of 2–10 nm with different colors, including white (e.g., with $E_g > 3.0$ eV) (71c, d). Thus it is possible to "tune" the band gap of a given material by preparing structures in the size region where Q effects become important. These particles can be redispersed in water to produce transparent sols of the Q particles.

An alternative approach is to prepare the Q structures in a matrix that holds them in place and prevents aggregation. Indeed, many of the

6.5. QUANTUM (OR NANO-) PARTICLES

structures described in Section 6.4.2 involved Q particles prepared and entrapped in micelles, vesicles, polymers, and inorganic matrices, such as silica and zeolites. An interesting approach of this type involves the use of inverse micelles as a medium to prepare "capped" particles (72). The approach is shown schematically in Figure 6.5.1. A microemulsion of water and heptane, stabilized by the surfactant Aerosol-OT (AOT) is prepared. This is composed of inverse micelles: small water "pools" surrounded by the AOT surfactant dispersed in heptane. Added $Cd(ClO_4)_2$ dissolves in the water pools. Treatment with $Se(TMS)_2$, where TMS is trimethylsilyl, causes formation of CdSe, which is protected from aggregation by the AOT surfactant. The particle surface is controlled by the reagent in excess. For example, in the presence of excess Cd^{2+} the CdSe surface is Cd^{2+}-rich. Addition of additional $Se(TMS)_2$ causes further growth of the CdSe particle, rather than nucleation of new particles. Particle size can thus be increased by alternate addition

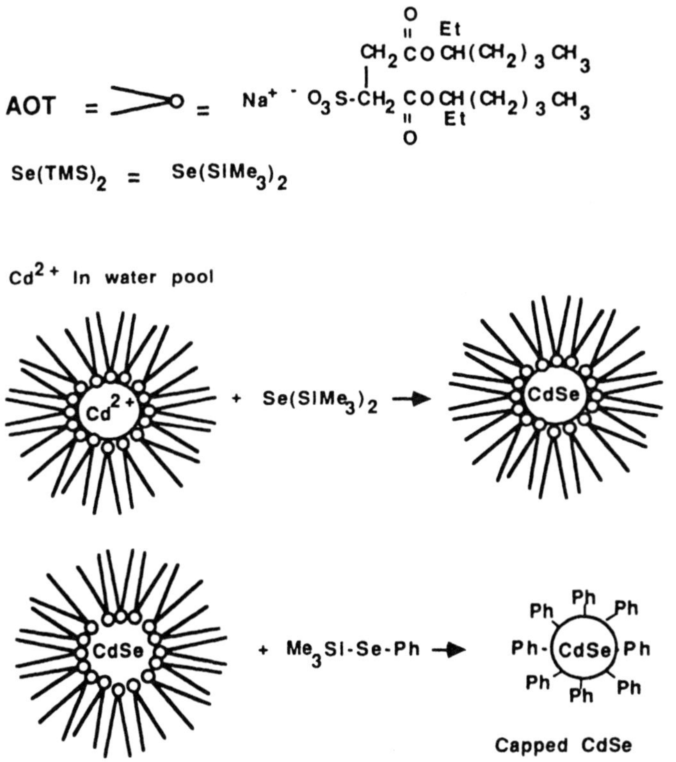

FIGURE 6.5.1. Schematic representation of the preparation of phenyl-capped CdS Q particles in a microemulsion (see Ref. (72)).

of Cd^{2+} and $Se(TMS)_2$. The particles can also be capped in the Cd^{2+}-rich state by addition of PhSeTMS, which produces a Ph-capped CdSe particle. Such particles can be isolated as powders and redispersed in organic solvents. The system demonstrates important principles in the construction of ICSs, namely, the ability to control growth and surface chemistry of a material on the nanometer to micrometer scale. Moreover, as the authors point out, the AOT-encapsulated CdSe can be considered analogous to a "living polymer," so that growth of more complex structures, such as multilayers of different semiconductor materials, is also possible (73).

6.5.3. Particle Charging and Band Energetics in Q Particles

In addition to the increase in the band gap, the energies (i.e., the redox potentials) of electrons and holes in Q particles are also different than those in larger particles or the bulk material. This arises from at least two different effects. The band locations are shifted as implied by Eq. (6.5.1) and the "particle in a box" model. In general, the conduction band (or the lowest quantized vacant orbital) is moved upward to more negative potentials, and the valence band (or the highest quantized filled orbital) is moved downward to more positive potentials, as the particle size decreases. Thus electrons are better reductants and holes are better oxidants in very small particles compared to the bulk material. A second factor relates to the effect of excess charge on a particle on the potential. In small particles, a large excess charge can arise from relatively few additional electrons or holes and can lead to partial filling of the conduction band or emptying of the valence band.

For example, enhanced photoredox chemistry was found at small particles of PbSe and HgSe (74). The bulk E_g of these materials is about 0.3 eV, so large particles of these materials are black and would not be effective for photoelectrochemical reactions. However, small particles ($\sim$20–50 Å) formed by precipitation in water or MeCN in the presence of suitable stabilizers showed large blue shifts in their absorption edges, say, to about 2.8 eV. Photogenerated carriers in these particles showed enhanced redox potentials because of the expansion of the band gap and an increase in the energies of the band edges; H_2 evolution under irradiation was reported with both HgSe and PbSe. Several other studies also reported enhanced photochemical activity for semiconductors as the particle size decreased, as, for example, in photocatalytic hydrogenation at TiO_2 (75).

Excess charging effects in even rather large semiconductor particles can cause electrons to have energies that are higher than those of the bulk semiconductor immersed in the same solution (36). The idea here is that as excess electrons accumulate in a particle, for example, because the charge transfer rate of electrons to an acceptor at an irradiated semiconductor is much smaller than the rate of hole transfer to a donor (Fig. 6.4.1), the potential of the electrons for reduction becomes more negative. These effects will be much more important in very small particles, where even one excess electron or hole produces a high excess charge density (76). For example, one excess electron in a 15-Å-diameter particle is formally equivalent to an excess charge density of 6×10^{20} cm^{-3}. High excess charge densities and dopant densities at levels that cause degeneracy in the semiconductor can affect the observed optical transitions as well. This effect has been studied mostly with small-bandgap bulk semiconductors, such as InSb. For example, intrinsic InSb shows an absorption band edge at 7.2 μm. However, n-InSb with a donor dopant level of 5×10^{18} cm^{-3} shows an absorption edge at 3.2 μm. This shift in absorption edge to higher energies, called the "Burstein shift," is attributed to movement of the Fermi level into the conduction band (77). With this "filling" of the conduction band with electrons, light absorption requires a higher photon energy to promote electrons from the valence band to higher energy levels in the conduction band. While this effect is mainly seen with very small E_g bulk semiconductors, it has been proposed that similar effects would be of importance in Q particles of materials with much larger E_g values (76). Indeed, irradiation of Q particles, such as CdS, especially in the presence of a hole acceptor, leads to a blue shift in the absorption edge (41, 42). This can be explained by photocharging of the particles with excess electrons. However, it is not clear how well a bulk semiconductor model applies to small particles with very large surface areas and large numbers of surface states. Other explanations for the observed blue shifts under irradiation involve polarization of the exciton generated by light absorption by the electric field generated by the excess charge (42) or by a decrease in the oscillator strength of the exciton transition by interaction with the excess charge (78). A blue shift in the absorption edge has also been seen on electrochemical charging of CdS particles in a thin film (79).

6.6. CONCLUSIONS

Semiconductor materials are important components of integrated chemical systems, because of their optical and electrical properties. They serve as useful transducers of radiant, electrical, and chemical energy, and thus can play a role in the development of different types of sensors, detectors, and photocatalytic systems. Photoelectrochemical systems, based on either semiconductor electrodes in cells or particulate systems, are being widely studied as a means of characterizing the semiconductor–liquid interface and for potential practical applications. Semiconductors also illustrate the importance of size effects in the regime of nanometer-size structures in determining the properties of the materials used in the construction of ICS.

REFERENCES

1. A. J. Bard, *J. Phys. Chem.*, **86**, 172 (1982).
2. A. J. Bard, *Ber. Bunsenges. Phys. Chem.*, **92**, 1187 (1988).
3. See, for example: (a) H. Kuhn, *J. Photochem.*, **10**, 111 (1979); (b) J. H. Fendler, *Membrane Mimetic Chemistry*, Wiley, New York, 1982, pp. 492–505; (c) M. Grätzel, *Angew. Chem. Intl. Ed. Engl.*, **84**, 981 (1980); (d) K. Kalyanasundaram, *Photochemistry in Microheterogeneous Systems*, Academic Press, Orlando, 1987 and references cited therein.
4. T. Nomura, J. R. Escabi-Perez, J. Sunamoto, and J. H. Fendler, *J. Am. Chem. Soc.*, **102**, 1484 (1980).
5. P. P. Infelta, M. Grätzel, and J. H. Fendler, *J. Am. Chem. Soc.*, **102**, 1479 (1980).
6. (a) A. K. Honscher, *Principles of Semiconductor Device Operation*, Wiley, New York, 1960; (b) A. Many, Y. Goldstein, and N. B. Grover, *Semiconductor Surfaces*, North-Holland, Amsterdam, 1965; (c) D. Madelung, in *Physical Chemistry—an Advanced Treatise*, W. Jost, ed., Academic Press, New York, 1970, Vol. 10, Chapter 6; (d) G. Ertl and H. Gerischer, ibid., Chapter 7; (e) S. M. Sze, *Physics of Semiconductor Devices*, Wiley, New York, 1981.
7. A. J. Bard and L. R. Faulkner, *Electrochemical Methods*, Wiley, New York, 1980, pp. 500–515.
8. F. Cardon, W. P. Gomes, and W. Dekeyser, *Photovoltaic and Photoelectrochemical Solar Energy Conversion*, Plenum, New York, 1981.
9. Yu. V. Pleskov and Yu. Ya. Gurevich, *Semiconductor Photochemistry*, Consultants Bureau, New York, 1986.
10. A. J. Bard, *J. Photochem.*, **10**, 59 (1979).
11. A. J. Bard, *Science*, **207**, 139 (1980).
12. A. J. Nozik, *Annu. Rev. Phys. Chem.*, **29**, 189 (1978).
13. R. Memming, in *Electroanalytical Chemistry*, A. J. Bard, ed., Marcel Dekker, New York, 1979, Vol. 11, p. 1.
14. M. Wrighton, *Acc. Chem. Res.*, **12**, 303 (1979).
15. A. J. Bard, *J. Electroanal. Chem.*, **168**, 5 (1984).
16. R. Memming, *Topics in Current Chemistry*, Springer-Verlag, Berlin, 1988, Vol. 143, p. 79.
17. A. Heller, *Acc. Chem. Res.*, **14**, 154 (1981).
18. H. Gerischer, in *Physical Chemistry—an Advanced Treatise*, H. Eyring, D. Henderson, and W. Jost, eds., Academic Press, New York, 1970, Vol. IXA, p. 463.

REFERENCES

19. A. J. Bard, *Proceedings Robert A. Welch Foundation Conferences on Chemical Research XXVIII*, Houston, Texas, 1984, p. 94.
20. S. R. Morrison, *Electrochemistry at Semiconductor and Oxidized Metal Electrodes*, Plenum, New York, 1980.
21. N. Serpone and E. Pelizzetti, eds., *Photocatalysis—Fundamentals and Applications*, Wiley, New York, 1989.
22. K. S. V. Santhanam and M. Sharon, eds., *Photoelectrochemical Solar Cells*, Elsevier, Amsterdam, 1988.
23. F-R. F. Fan, H. S. White, B. Wheeler, and A. J. Bard, *J. Electrochem. Soc.*, **127**, 518 (1980).
24. (a) F-R. Fan, G. Hope, and A. J. Bard, *J. Electrochem. Soc.*, **129**, 1647 (1982); F-R. Fan, R. G. Keil, and A. J. Bard, *J. Am. Chem. Soc.*, **105**, 220 (1983).
25. C. P. Kubiak, L. F. Schneemeyer, and M. S. Wrighton, *J. Am. Chem. Soc.*, **102**, 6898 (1980).
26. F-R. Fan, B. Reichman, and A. J. Bard, *J. Am. Chem. Soc.*, **102**, 1488 (1980).
27. H. D. Abruna and A. J. Bard, *J. Am. Chem. Soc.*, **103**, 6898 (1981).
28. (a) D. C. Bookbinder and M. S. Wrighton, *J. Am. Chem. Soc.*, **102**, 5123 (1980); (b) D. C. Bookbinder, N. S. Lewis, M. G. Bradley, A. B. Bocarsly, and M. S. Wrighton, *J. Am. Chem. Soc.*, **101**, 7721 (1979); (c) D. C. Bookbinder, J. A. Bruce, R. N. Dominey, N. S. Lewis, and M. S. Wrighton, *Proc. Natl. Acad. Sci. USA*, **77**, 6280 (1980).
29. (a) A. J. Bard, A. B. Bocarsly, F-R. Fan, E. G. Walton, and M. S. Wrighton, *J. Am. Chem. Soc.*, **102**, 3671 (1980); (b) A. J. Bard, F-R. Fan, A. S. Gioda, G. Nagasubramanian, and H. S. White, *Disc. Faraday Soc.*, **70**, 19 (1980).
30. (a) A. Heller, B. A. Parkinson, and B. Miller, *Appl. Phys. Lett.*, **33**, 521 (1978); A. Heller, *ACS Symposium Series 146*, American Chemical Society, Washington, DC, 1981, p. 57; (c) I. L. Abrahams, B. J. Tufts, and N. S. Lewis, *J. Am. Chem. Soc.*, **109**, 3472 (1987).
31. (a) H. Tributsch and J. C. Bennett, *J. Electroanal. Chem.*, **81**, 97 (1977); (b) H. Tributsch, *Ber. Bunsenges. Phys. Chem.*, **81**, 361 (1977); **82**, 169 (1978); (c) W. Kautek, H. Gerischer, and H. Tribusch, ibid., **83**, 1000 (1979); (d) F-R. Fan, H. S. White, B. L. Wheeler, and A. J. Bard, *J. Am. Chem. Soc.*, **102**, 5142 (1980); (e) D. Canfield and B. A. Parkinson, ibid., **103**, 1281 (1981); (f) H. S. White, H. D. Abruna, and A. J. Bard, *J. Electrochem. Soc.*, **129**, 265 (1982).
32. (a) H. Gerischer, *Photochem. Photobiol.*, **16**, 243 (1972); (b) R. Memming, ibid., p. 325; (c) M. T. Spitler and M. Calvin, *J. Chem. Phys.*, **66**, 4294 (1977); (d) T. Miyasaka, T. Watanabe, A. Fujishima, and K. Honda, *J. Am. Chem. Soc.*, **100**, 6657 (1978); (e) A. Giraudeau, F-R. Fan, and A. J. Bard, ibid., **102**, 5137 (1980).
33. B. O'Regan and M. Grätzel, *Nature*, **353**, 737 (1991).
34. (a) S. N. Frank and A. J. Bard, *J. Am. Chem. Soc.*, **99**, 303 (1977); (b) *J. Phys. Chem.*, **81**, 1484 (1977).
35. (a) B. Kraeutler and A. J. Bard, *J. Am. Chem. Soc.*, **100**, 2239 (1978); (b) ibid., p. 5985.
36. W. W. Dunn, Y. Aikawa, and A. J. Bard, *J. Am. Chem. Soc.*, **103**, 3456 (1981).
37. M. D. Ward, J. R. White, and A. J. Bard, *J. Am. Chem. Soc.*, **105**, 27 (1983).
38. J. R. White and A. J. Bard, *J. Phys. Chem.*, **89**, 1947 (1985).

39. M. D. Ward and A. J. Bard, *J. Phys. Chem.*, **86**, 3599 (1982).
40. J. K. Leland and A. J. Bard, *J. Phys. Chem.*, **91**, 5076 (1987); ibid., p. 5083.
41. W. J. Albery, G. T. Brown, J. R. Darwent, and E. Saievar-Iranizad, *J. Chem. Soc., Faraday Trans. I*, **81**, 1999 (1985).
42. A. Henglein, A. Kumar, E. Janata, and H. Weller, *Chem. Phys. Lett.*, **132**, 133 (1986).
43. D. Duonghong, E. Borgarello, and M. Grätzel, *J. Am. Chem. Soc.*, **103**, 4685 (1981).
44. B. Kraeutler, C. D. Jaeger, and A. J. Bard, *J. Am. Chem. Soc.*, **100**, 4903 (1978); C. D. Jaeger and A. J. Bard, *J. Phys. Chem.*, **83**, 3146 (1979); J. R. Harbour and M. L. Hair, *J. Phys. Chem.*, **81**, 1791 (1977).
45. P. C. Lee and D. Meisel, *J. Am. Chem. Soc.*, **102**, 5477 (1980).
46. M. F. Finlayson, K. H. Park, N. Kakuta, A. J. Bard, A. Campion, M. A. Fox, S. E. Webber, and J. M. White, *J. Luminescence*, **39**, 205 (1988).
47. (a) M. Krishnan, J. R. White, M. A. Fox, and A. J. Bard, *J. Am. Chem. Soc.*, **105**, 7002 (1983); (b) A. W-H. Mau, C. B. Huang, N. Kakuta, A. J. Bard, A. Campion, M. A. Fox, J. M. White, and S. E. Webber, ibid., **106**, 6537 (1984); (c) N. Kakuta, J. M. White, A. Campion, A. J. Bard, M. A. Fox, and S. E. Webber, *J. Phys. Chem.*, **89**, 48 (1985); (d) J. P. Kuczynski, B. H. Milosavljevic, and J. K. Thomas, ibid., **88**, 980 (1984).
48. (a) Y. Wang, A. Suna, W. Mahler, and R. Kasowski, *J. Chem. Phys.*, **87**, 7315 (1987); (b) W. Mahler, *Inorg. Chem.*, **27**, 436 (1988).
49. F-R. F. Fan, H. Y. Liu, and A. J. Bard, *J. Phys. Chem.*, **89**, 4418 (1985).
50. (a) Y. Wang and N. Herron, *J. Phys. Chem.*, **91**, 257 (1987); (b) K. Moller, M. M. Eddy, G. D. Stucky, N. Herron, and T. Bein, *J. Am. Chem. Soc.*, **111**, 2564 (1989); (c) J. E. MacDougall, H. Eckert, G. D. Stucky, N. Herron, Y. Wang, K. Moller, T. Bein, and D. Cox, ibid., p. 8006; (d) R. D. Stramel, T. Nakamura, and J. K. Thomas, *J. Chem. Soc., Faraday Trans. I*, **84**, 1287 (1988).
51. J. K. Leland and A. J. Bard, *Chem. Phys. Lett.*, **139**, 453 (1987).
52. T. Rajh, M. I. Vucemilovic, N. M. Dimitrijevic, O. I. Micic, and A. J. Nozik, *Chem. Phys. Lett.*, **143**, 305 (1988).
53. J. Kuczynski and J. K. Thomas, *J. Phys. Chem.*, **89**, 2720 (1985).
54. L. Persaud, A. J. Bard, A. Campion, M. A. Fox, T. E. Mallouk, S. E. Webber, and J. M. White, *J. Am. Chem. Soc.*, **109**, 7309 (1987).
55. L. Persaud, A. J. Bard, A. Campion, M. A. Fox, T. E. Mallouk, S. E. Webber, and J. M. White, *Inorg. Chem.*, **26**, 3825 (1987).
56. T. Shimidzu, T. Iyoda, Y. Koide, and N. Kanda, *Nouv. J. Chim.*, **7**, 21 (1983).
57. K. Kalyanasundaram and M. Grätzel, *Angew. Chem. Intl. Ed. Engl.*, **18**, 701 (1979).
58. K. Sayama, A. Tanaka, K. Domen, K. Maruya, and T. Onishi, *J. Phys. Chem.*, **95**, 1345 (1991).
59. Y. I. Kim, S. Salim, M. J. Huq, and T. E. Mallouk, *J. Am. Chem. Soc.*, **113**, 9561 (1991).
60. (a) H-C. Youn, Y-M. Tricot, and J. H. Fendler, *J. Phys. Chem.*, **91**, 581 (1987);

(b) H. J. Watzke and J. H. Fendler, ibid., p. 854; (c) R. Rafaeloff, Y.-M. Tricot, F. Nome, and J. H. Fendler, ibid., **89**, 1236 (1985).
61. Y-M. Tricot and J. H. Fendler, *J. Am. Chem. Soc.*, **106**, 7359 (1984).
62. (a) M. Meyer, C. Wallberg, K. Kurihara, and J. H. Fendler, *J. Chem. Soc. Chem. Commun.*, 1984, 90; (b) P. Lianos and J. K. Thomas, *Chem. Phys. Lett.*, **125**, 299 (1986); (c) T. Daunhauser, M. O'Neil, K. Johansson, D. Whitten, and G. McLendon, *J. Phys. Chem.*, **90**, 6074 (1986).
63. X. K. Zhao, S. Baral, R. Rolandi, and J. H. Fendler, *J. Am. Chem. Soc.*, **110**, 1012 (1988).
64. E. S. Smotkin, C. Lee, A. J. Bard, A. Campion, M. A. Fox, T. E. Mallouk, S. E. Webber, and J. M. White, *Chem. Phys. Lett.*, **152**, 265 (1988).
65. (a) J. S. Kilby, J. W. Lathrop, and W. A. Porter, U.S. Patent 4,021,323 (1977); U.S. Patent 4,100,051 (1978); U.S. Patent 4,136,436 (1979); (b) E. L. Johnson, *Proc. Intersoc. Energy Convers. Eng. Conf.*, **16**, 798 (1981); (c) E. L. Johnson, in *Electrochemistry in Industry*, U. Landau, E. Yeager, and D. Kortan, eds., Plenum, New York, 1982, pp. 299–306.
66. (a) E. S. Smotkin, A. J. Bard, A. Campion, M. A. Fox, T. Mallouk, S. E. Webber, and J. M. White, *J. Phys. Chem.*, **90**, 4604 (1986); (b) E. S. Smotkin, S. Cervera-March, A. J. Bard, A. Campion, M. A. Fox, T. Mallouk, S. E. Webber, and J. M. White, *J. Phys. Chem.*, **91**, 6 (1987); (c) S. Cervera-March, E. S. Smotkin, A. J. Bard, A. Campion, M. A. Fox, T. Mallouk, S. E. Webber, and J. M. White, *J. Electrochem. Soc.*, **135**, 567 (1988).
67. (a) A. Henglein, *Topics in Current Chemistry*, Springer-Verlag, Berlin, 1988, Vol. 143, p. 113; (b) A. Henglein, *Chem. Rev.*, **89**, 1861 (1989); (c) M. L. Steigerwald and L. Brus, *Acc. Chem. Res.*, **23**, 183 (1990).
68. L. E. Brus, *J. Chem. Phys.*, **80**, 4403 (1984).
69. (a) Y. Wang, A. Suna, W. Mahler, and R. Kasowski, *J. Chem. Phys.*, **87**, 7315 (1987); (b) L. Spanhel, M. Haase, H. Weller, and A. Henglein, *J. Am. Chem. Soc.*, **106**, 5649 (1987); (c) H. Weller, H. M. Schmidt, U. Koch, A. Fojtik, S. Baral, A. Henglein, W. Kunath, K. Weiss, and E. Dieman, *Chem. Phys. Lett.*, **124**, 557 (1986).
70. (a) R. Rossetti, J. L. Ellison, J. M. Gibson, and L. E. Brus, *J. Chem. Phys.*, **80**, 4464 (1984); (b) N. Chestnoy, R. Hull, and L. E. Brus, ibid., **85**, 2237 (1986); (c) J. J. Ramsden, S. E. Webber, and M. Grätzel, *J. Phys. Chem.*, **89**, 2740 (1985); (d) J-M. Zen, F-R. Fan, G. Chen, and A. J. Bard, *Langmuir*, **5**, 1355 (1989).
71. (a) A. Fojtik, H. Weller, U. Koch, and A. Henglein, *Ber. Bunsenges. Phys. Chem.*, **88**, 969 (1984); (b) L. Spanhel, A. Henglein, and H. Weller, ibid., **91**, 1359 (1987); (c) H. Weller, A. Fojtik, and A. Henglein, *Chem. Phys. Lett.*, **117**, 485 (1985); (d) A. Fojtik, H. Weller, and A. Henglein, ibid., **120**, 552 (1985).
72. M. L. Steigerwald, A. P. Alivisatos, J. M. Gibson, T. D. Harris, R. Kortan, A. J. Muller, A. M. Thayer, T. M. Duncan, D. C. Douglass, and L. E. Brus, *J. Am. Chem. Soc.*, **110**, 3046 (1988).
73. A. R. Kortan, R. Hull, R. L. Opila, M. G. Bawendi, M. Steigerwald, P. Carroll, and L. E. Brus, *J. Am. Chem. Soc.*, **112**, 1327 (1990).

74. J. M. Nedelijkovic, M. T. Nenadovic, O. I. Micic, and A. J. Nozik, *J. Phys. Chem.*, **90,** 12 (1986).
75. M. Anpo, T. Shima, S. Kodama, and Y. Kubokawa, ibid., **91,** 4305 (1987).
76. C-Y. Liu and A. J. Bard, *J. Phys. Chem.*, **93,** 3232 (1989).
77. (a) E. Burstein, *Phys. Rev.*, **93,** 632 (1954); (b) K. Seeger, *Semiconductor Physics*, Springer-Verlag, New York, 1973, p. 342.
78. E. F. Hilinski, P. A. Lucas, and Y. Wang, *J. Chem. Phys.*, **89,** 3435 (1988).
79. C-Y. Liu and A. J. Bard, *J. Phys. Chem.*, **93,** 7749 (1989).

Chapter 7
FUTURE INTEGRATED CHEMICAL SYSTEMS

> For I dipt into the future, far as human eye could see,
> Saw the vision of the world, and all the wonder that would be;
> —Alfred Tennyson

> We still run chemical reactions in much the same way as in the seventeenth and eighteenth centuries. In our age of sophisticated electronic instrumentation, sensors, microprocessors, and robots starting to see use in chemical synthesis, reaction flasks stand on our benches like relics from a distant past.
> —Pierre Laszio
> *Preparative Chemistry Using*
> *Supported Reagents*
> Academic Press, San Diego, 1987

7.1. INTRODUCTION

This monograph began with a brief overview and some examples of existing integrated chemical systems. It is appropriate to conclude with a look toward the future and some admittedly starry-eyed speculation about possible applications of integrated chemical systems. The development of these chemically based systems generally follows the trends of the last decades toward miniaturization, especially in electronics. For example, a number of articles have appeared, many in the popular press, about "microtechnology" or "nanotechnology," and the construction of machines on the micrometer or submicrometer scale (1, 2). Miniature machines can be designed as simple microversions of larger ones, as, for example, with tiny motors to produce motion (3). However, chemical versions of such machines, based on the integrated systems approach and more analogous to biological mechanical systems, are also possible.

In these, for example, piezoelectric materials or polymers whose dimensions can be controlled electrically, rather than motors, might form the basis for motion, that is, would be the integrated systems equivalents of muscles.

We will consider the application of integrated chemical systems for synthetic purposes. We depend on biological systems to synthesize food and fibers on a scale that dwarfs current industrial synthesis of chemicals. In fact, essentially all of the food, cotton, silk, and rubber we use, as well as many drugs, are synthesized in tiny (micrometer-size) cells. Yet in the laboratory, synthesis is carried out in large-volume flasks, and on the industrial scale, huge reactors are used. We will discuss integrated chemical system reactors that might be employed for chemical synthesis. We will also consider possible uses of ICS as sensors, as electrical, optical, or optoelectric devices, and in the fabrication of intelligent materials.

7.2. INTEGRATED CHEMICAL SYNTHESIZERS

7.2.1. Basic Concepts

An ICS synthesizer would consist of small (millimeter- to micrometer-sized) reactors, designed for a particular application, along with associated ICS devices for moving reactant and product streams (pumps), mixing reactants, analyzing streams, and separating products (Fig. 7.2.1). The design of the reactor itself would depend on whether heating was needed or whether heat dissipation was more important. For electrochemical reactions, the reactor would be a small electrochemical cell, while a photochemical reactor would require a window for irradiation by an external light source. It may be useful to modify the walls of the reactor with bound catalysts, such as enzymes, or to add catalyst particles to the reactor. In considering the application of such systems, one could draw on experience in capillary chromatography (4), capillary-zone electrophoresis (5), and hollow-fiber technology, which require moving gas or liquid streams, addition of samples, mixing of streams, and analysis of small amounts of materials contained in small volumes.

Clearly, the main consideration in using an ICS for synthesis is the total amount of material that can be produced in a given system over a reasonable time period. Consider, for example, a cylindrical (capillary) reactor with an internal diameter of 100 μm, 1 cm long, with a cell volume of about 0.08 μL. At a linear flow velocity of 0.1 cm/s, the transit time through the cell would be 10 s, and the volume flow would

7.2. INTEGRATED CHEMICAL SYNTHESIZERS

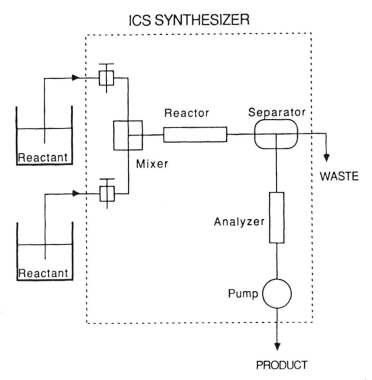

FIGURE 7.2.1. Schematic outline of a simple integrated chemical system synthesizer. The components within the dashed line represent the ICS, specifically, small interconnected elements designed for the reaction system of interest. More complex systems comprising multiple reactors of different types and sizes are also possible.

be 8×10^{-3} μL/s. If conversion of a 1 M solution reactant was complete in this time, then the output of the cell would be 8-nmol product/s. For a product with a molecular weight of 100 g/mol, this would be equivalent to about 3 mg/h or 25 g/year of product. Thus a bench-sized reactor consisting of 1000 ICS units would produce 69 g/day, while a larger reactor with 176,000 units would be needed to produce 11 kg/year.

7.2.2. Advantages and Disadvantages

ICS synthesizers would show a number of advantages compared to conventional systems of larger size. Heat transfer, which depends on ratio of surface area, A, to volume, V, would be much better for the small reactors. This is a major advantage, for example, in capillary-zone electrophoresis compared to large-scale gel electrophoresis. For the 100-μm cylindrical reactor, $A/V \cong 2/r = 400$ cm^{-1} compared to that of a 1-L spherical flask, where $A/V \cong 3/r = 0.5$ cm^{-1}. There would be advantages in both external heating of the reactor and in heat dissipation. Thus the ICS reactor would be characterized by more uniform temperatures throughout the reaction mixture. It would also be much easier to work at high pressures in the small reactors. Recall that typical liquid chromatography systems, which involve larger-diameter tubing, use pumps that can bring the pressure to 5000 psi (lb/in^2) (340 bar). Thus supercritical fluids, even those involving high temperatures and pressures that are difficult to study in larger volumes, could easily be used as working fluids in ICS reactors. For example, studies by our group involving an electrochemical cell for near-critical and supercritical water at temperatures up to 390 °C and pressures of 240 bar could be carried out in a 0.238-cm-i.d. (inner diameter) alumina tube (6). Equivalent studies in a larger volume required the use of a stainless-steel bomb and elaborate safety measures.

The use of small flow reactors should allow precise control of residence time within a reaction zone. This would not only permit better control of reaction conditions but also allow the possibility of quenching reactions at a certain stage to prevent further reaction or the use of serial reactors to allow controlled sequential reactions of intermediates. It should be much easier to scale up reactions based on the ICS approach, since one would simply add additional modules of exactly the same type to increase output. For industrial synthesis, this would eliminate proceeding from a bench-scale reaction through a very different pilot-plant configuration to a full-size reactor. Moreover, the inherent redundancy of multiple parallel ICS reactors implies fewer operation problems with the failure of a few reactors, especially if the system is set up for easy replacement and repair of single ICS lines. Such systems are probably inherently much safer. The rupture of a single ICS, even at a high temperature and pressure, would cause negligible damage, since the total volume and amount of reactants released would be tiny. Overall, the ICS system should result in better yields of products with less waste and disposal problems because of better control of reaction variables. Many of these same considerations apply, of course, to biological syntheses (carried out in small cells), which are characterized by excellent yields under highly controlled conditions. Finally, ICS synthesizers should be easier to control and automate via an external computer.

One must, of course, recognize the problems and limitations of ICSs for synthesis. Since obtaining reasonable yields requires the use of a large number of parallel systems, the unit cost of each must be very small, and the system must be designed in a modular way so that as-

7.2. INTEGRATED CHEMICAL SYNTHESIZERS

sembly is fast and easy. Thus one will have to draw on experience in electronic integrated circuits, circuit-board assembly methods, capillary separation, system component assembly, and related fields to design the proper ICS modules and to work out ways of fabricating these in large numbers at a low unit cost. Biological systems have the advantage of being self-replicating, self-repairing, and self-assembling.

7.2.3. Components and Construction

It is beyond the scope of this work to attempt a detailed description of possible components or methods of construction and assembly. The most straightforward approach would be to utilize silicon substrates and photolithographic techniques to form the desired structures. The Si could be oxidized to form SiO_2 layers that should be reasonably stable as reaction cell walls and flow channels. Approaches of this type have already been taken in the construction of miniature sensors (7). Electrodes, heaters, and other metal or semiconductor elements could be deposited by vacuum evaporation, sputtering, or chemical vapor deposition (CVD). Alternatively, it should be possible to use photolithography and etching techniques directly with glass or plastic wafers, which would be much less expensive than single-crystal Si. Plastic materials might also lend themselves to casting or thermal processing methods.

There are a number of different generic components that would form an ICS synthesizer:

Flow Components. Pumps, flow channels, manifolds, flow restrictors (to raise pressure in system), valves.

Mixers. Flow, ultrasonic.

Reaction Chambers. Thermal, electrochemical, photochemical, pressure.

Separation Chambers. Extraction (counter current flow), chromatographic, electrophoretic, membrane, distillation.

Analyzers. Electrochemical, spectroscopic, fluorescence.

These might be assembled on a single chip or board to form a complete ICS that is then replicated many times to produce the parallel system. Alternatively, especially in the development stage, it might be more useful to have modules of each component that could be assembled on a breadboard. In the latter case, a means for precise alignment of the components and rapid interconnection of the flow channels and electrical parts would have to be developed. Perhaps a system where the component modules plug into a "circuit board" with preexisting flow channels and printed wires in a method analogous to the way integrated-circuit chips and other components are attached to a circuit board would be feasible.

7.3. SENSORS

Several examples of ICS sensors were described in Chapter 1. The design and application of chemical sensors is a very active field, and a number of reviews and monographs have appeared (8). In general, a sensor has two main parts: a transducer that produces the response and output signal and a receptor system that interacts with the species being detected and provides selectivity in the device. Many different transducer principles have been used in sensors. For example, electrochemical sensors generate potentials or currents in response to a chemical species, while optical sensors produce absorbance, reflectance, luminescence, or refractive index signals. Other modes of detection include mass changes, for example, as measured by oscillation frequency changes in piezoelectric crystals, or thermal changes, as measured by resistance changes in a thermistor. The key issue in the design of a sensor is selectivity. Any species adsorbing on a piezoelectric crystal will add mass and cause the frequency of oscillation of the crystal to decrease. How does one distinguish one species from another? The answer seems to lie in the use of the ICS approach to provide a matrix on the sensor that is capable of molecular recognition. The ICS approach should also allow integration of the recognition and detection functions and allow the construction of microfabricated sensors.

7.3.1. Molecular Recognition

Molecular recognition usually implies a specific interaction between two molecules, sometimes called the "host and the guest." Such selective molecular interactions characterize many biological molecules, such as enzymes and antibodies. While molecular recognition has long been the basis of the use of specific ligands in the determination of metal ions (e.g., the classic use of dimethylglyoxime in the determination of nickel), more recent work has been aimed at the purposeful design of host molecules with specific shapes that use chelation and hydrogen bonding to promote selective interaction with a guest metal ion or organic compound. These principles are used, for example, in the synthesis of crown ethers, cryptands, and cavitands, which are macrocyclic ligands with cavity sizes that allow differentiation of metal ions on the basis of size and strength of the metal interaction with the coordination groups of the ligand (9). Many other types of molecules have also been developed that show molecular recognition properties (10). Such recognition is also central to many biological phenomena, such as enzyme–substrate interactions, immune response (antibody–antigen interaction), adhesion, and cell–cell interactions in general (11). The basic concepts in these appear to follow the lock-and-key principle first formulated to explain enzyme–substrate interactions by Fischer in 1897. In general, recognition depends on a stereospecific fit between molecules (a shape factor) reinforced by bonding between the molecules at several sites (an interaction factor). Hydrophobic interactions may also play a role, such as in the intercalation of molecules between the bases of DNA or interactions of metal chelates with surfactants.

Shape-selective sites can be constructed in several ways. One approach involves the use of a template molecule incorporated in a polymer (12). The basic concept is shown schematically in Figure 7.3.1. The template molecule is copolymerized with a backbone monomer to incorporate the template into the crosslinked polymer. The polymer is then treated with a reagent that will cause removal of the template and leave behind the correct shape and binding sites in the polymer. In the example shown, the template is removed by hydrolysis to leave behind $-OH$ or $-NH_2$ binding sites. Such structures have been shown to be selective with respect to molecules of the size and shape of the template. A similar template approach can be used to incorporate a template molecule on a surface (Fig. 7.3.2). This is carried out by a similar approach where a molecule containing the template is bound to the surface and then the template is removed (12a). The remainder of the surface can be blocked [e.g., by treatment with hexamethyldisilane or with a molecule with a longer hydrocarbon chain, such as octadecylsilane (13)]. Shape-selective sites can also be formed on a surface by incorporating template molecules into organized monolayers and then selectively removing the template (14). For example, a glass surface can be treated with a mixture of octadecyltrichlorosilane, which will form a self-assembled monolayer on glass, and the desired template molecule (e.g., a porphyrin, phthalocyanine, cyanine dye) to form a monolayer incorporating the template molecule. The surface is then treated with a reagent that will remove the template but leave the monolayer intact, to produce a "skeletonized monolayer." Such surfaces have been shown to be capable of discriminating between molecules with the shape of the template and closely related structures.

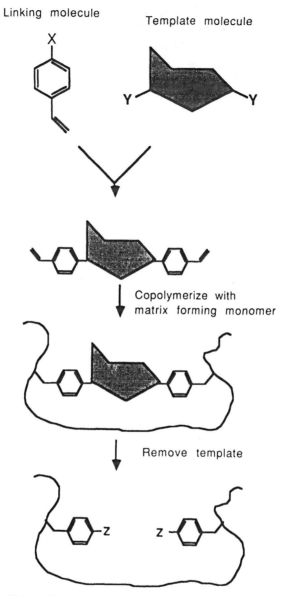

FIGURE 7.3.1. Schematic representation of the formation of a shape-selective site in a polymer by incorporation of a template molecule, polymerization, and removal of template. The polymerization step is assumed to form sufficient crosslinks among the polymer chains to ensure rigidity and maintenance of the structure around the site (see Ref. 12).

7.3. SENSORS

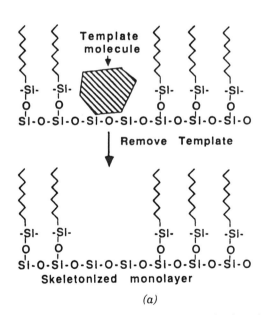

FIGURE 7.3.2. The template approach to produce a selective site on a surface. (a) Outline of chemistry; (b) example of such a site on a monolayer-modified silica surface. [Reprinted with permission from Y-T. Tao and Y-H. Ho, *J. Chem. Soc. Chem. Commun.*, **1988**, 417 (1988). Copyright 1988 The Royal Society of Chemistry.]

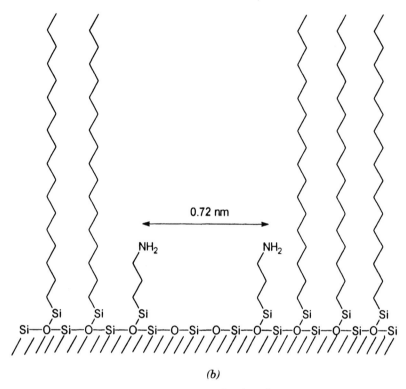

(b)

FIGURE 7.3.2. (*Continued*)

7.3.2. Biosensors

One can utilize the high selectivity of biological molecules, such as enzymes and antibodies, to fabricate sensors. Such biosensors have been investigated for many years and have been of particular interest for the determination of substances like glucose in blood (15). In the design of enzyme-based biosensors, an ICS approach is needed, since several components in a specified arrangement must be used to contact the enzyme, protect the system from components in the environment, and produce the desired signal. Consider an enzyme electrode based on glucose oxidase (GO) immobilized on a conductive electrode surface (16). The response of this system is based on the electron-transfer reaction of the FAD reaction center in GO with glucose, Eq. (7.3.1), to yield the reduced enzyme center ($FADH_2$):

$$\text{Glucose} + \text{FAD (GO)} + 2H^+ \rightarrow FADH_2 \text{ (GO)} + \text{gluconolactone}$$

(7.3.1)

Electrochemical oxidation of the enzyme back to the FAD form regenerates the reactant and provides the signal representing glucose concentration. However, the enzyme, GO, cannot be oxidized directly at a conductive electrode, because the prosthetic group ($FADH_2$) is too far from the surface to allow electrons to tunnel from it to the electrode. However, a mediator or electron relay incorporated between the enzyme and the electrode can behave as a "molecular wire" and penetrate into the enzyme to effect the oxidation. The mediator in this case is a redox polymer (Chapter 5) made by complexing $Os(bpy)_2Cl^+$ with poly(vinylpyridine) (PVP) (16b). Oxidation of this species to the +2 state occurs readily at the electrode surface, with the reaction in Eq. 7.3.2 following:

$$FADH_2 + Os(bpy)_2Cl^{2+} \rightarrow FAD + Os(bpy)_2Cl^+ + 2H^+ \quad (7.3.2)$$

A schematic representation of this system is shown in Figure 7.3.3. In constructing the actual electrode, a protective layer of polyethylene oxide coats the sensor to prevent protein adsorption from the sample. A number of other types of biosensors, based on enzymes, antibodies, whole cells, and tissues have been constructed (17). Catalytic antibodies (18), which can be developed to be selective to a particular antigen, should also prove useful in the fabrication of sensors, particularly if redox-active antibodies can be produced.

7.3.3. Future Sensors

The huge number of potential applications for sensors suggests that this will be an important area for ICS. For example, animals are now used for detection purposes; for instance, dogs sniff out drugs, pigs find truffles, and fish detect pollutants in water. Similarly, human sniffers and tasters are used to evaluate perfumes, wines, and foods. It should be possible to develop chemical sensors for these applications. Medical sensors that could be used for rapid determinations (e.g., of cholesterol and glucose) in drawn samples or even directly and continuously *in vivo*, perhaps with telemetry to a monitoring station, are also on the horizon. Rapid disease diagnosis through direct chemical identification of bacteria and viruses would also be of immense value in medicine. Finally, one can envision sensors that continually monitor air and water for the presence of pollutants, that monitor vehicle emissions, fuel, and air to adjust engine parameters to optimize performance, and that detect drugs and alcohol in blood and urine.

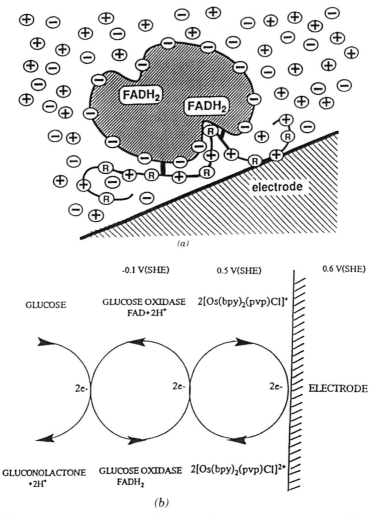

FIGURE 7.3.3. (a) Representation of enzyme electrode based on glucose oxidase (containing $FADH_2$ centers) and a segment of a polycationic polymer incorporating redox groups, R. (b) The electron-transfer steps in the reaction, with the approximate potentials of the electrode and couples indicated above. [Reprinted with permission from A. Heller, *Acc. Chem. Res.*, **23**, 128 (1990). Copyright 1990 American Chemical Society.]

7.4. ELECTRONIC AND OPTICAL DEVICES

Much has been written about the possibility of ultrasmall electronic devices and the development of the field of nanoelectronics. The goals are computers and devices that contain individual elements (switches, diodes) of molecular (nanometer) dimensions, with incredibly powerful

7.4. ELECTRONIC AND OPTICAL DEVICES

machines of small size and reasonable price becoming available. The belief that such a scale of fabrication may be possible is based on an extrapolation of the recent history of electronic devices. For example, the first computers, such as the ENIAC in 1946, were huge machines based on vacuum tubes and relay switches. Although these early computers were very large and expensive and required formidable amounts of power, they had very limited capabilities, below those of the simple hand-held calculators or the Apple II computer. Significant improvements arose with the development of the transistor and more compact memories in the fifties. The invention of the integrated circuit, an ICS, led to the development of complex, yet inexpensive and compact, electronic systems, like computers and video camcorders. Integrated circuits have shown continued improvements in performance and device density. For example, the number of devices on a chip have doubled roughly every 2 years. Current devices have structures with dimensions of the order of 0.8–1 μm. Smaller structures, down to the order of 0.1–0.2 μm should be possible with extensions of current technology, such as by replacing optical lithography with electron-beam and X-ray lithography. If this trend in size reduction continues, and an extrapolation from the past is possible, then nanoelectronic devices are just around the corner.

However, continued extrapolation to smaller and smaller sizes is probably not going to continue. Devices of atomic dimensions, which according to extrapolated ICS size practice might occur as early as 2017, do not seem reasonable. In fact, the small size barrier will be reached at considerably larger structures and there are many reasons why even devices of molecular size are unlikely. The fabrication of devices with structures and geometries below about 0.1 μm (nanoelectronic structures) will probably require approaches that are qualitatively different than those currently used. Moreover, the operation of nanoelectronic devices will involve a number of new factors, such as the onset of quantum effects, the existence of very high current densities that can produce electromigration effects, and heat dissipation requiring cryogenic operation (19).

Nevertheless, electronic devices with structures approaching molecular dimensions (molecular electronics) have been proposed and are being investigated. The belief in the possibility of devices at this size level is probably largely based on our knowledge that the human brain operates on this scale and is capable of storing huge amounts of information. For example, the capacity of the human brain has been estimated as 10^{15} bits (20). This is equivalent to the storage capacity of about 1 million 60-Mbyte magnetic disks. Even in the absence of the ultrafast processing characteristic of state-of-the art Si and GaAs-based electronics, the brain can perform certain operations, such as speech and pattern recognition, that are unattainable by current computers. Systems with this device density and the high speeds of current solid-state devices at these close spacings would indeed be capable of formidable achievements. Although descriptions of potential molecular rectifiers, switches, and counters have been proposed (but never actually reduced to practice) (21), and conferences on molecular electronic devices have been organized (22), it is unlikely that such systems operating on principles similar to current electronic systems will be possible in the foreseeable future.

The problems in the construction of such devices are formidable (23). For example, how can devices of this size be addressed from the external world? Even light with a wave length of 250 nm is hundreds of times larger than the proposed structures and could not be used to address a single unit unless it was spaced so far from the neighboring structure that the advantages of small size would be lost. Molecular "wires," such as single chains of a conductive polymer, have been proposed as connections. However, methods of the controlled preparation and placement of single polymer chains have not been accomplished, and it is not clear that the electrical properties of a single chain would be the same as those extrapolated from larger bundles or particles of the polymeric material. Closely spaced structures would suffer from the possibility of "cross-talk" between them, that is, when a signal appeared on one, a nearby structure would also be affected. This coupling could only be decreased by larger spacings. Finally, single molecules are not very stable, and decomposition of even a few molecules in such a device would be unacceptable. Comparisons with biological systems and proposals for "biochips" that are analogous to IC chips are not really valid, since the means of signal processing and electrical conduction in biological systems is very different. Moreover, biological systems are self-assembling and are capable of self-repair.

With these caveats, however, there may be possibilities for electronic devices based on ICSs. Recall that essentially all of our current electronic devices are based on a few materials, mainly Si and GaAs (and associated insulating, contacts, and packaging). Because we have gained so much knowledge about these substances and have such a wide experience working with them, it may at first seem unlikely that new materials, such as conductive polymers or LB films, will find applications in this area. However, as the size and complexity of electronic devices decreases, there may be special advantages to alternative materials in new arrangements.

Let us consider three recent examples of suggested applications of organic materials as electronic solid-state devices. Many others like these have been described. First consider the possibility of an organic field-effect transistor (FET) (24). We have already touched on unconventional FETs built around polymer-modified electrodes (Sections 1.2.6 and 4.7.4). A solid-state version based on the oligomer or thiophene, α,ωDH6T (sexithiophene substituted with n-hexyl groups in the terminal α positions) is shown in Figure 7.4.1. The terminal hexyl groups in the α,ωDH6T lead to production of an organized film on assembly, as shown, whose conductivity parallel to the plane of the substrate is different from that perpendicular to the plane. The FET constructed with this material showed typical iV characteristics.

An unconventional optoelectronic memory device based on the zinc

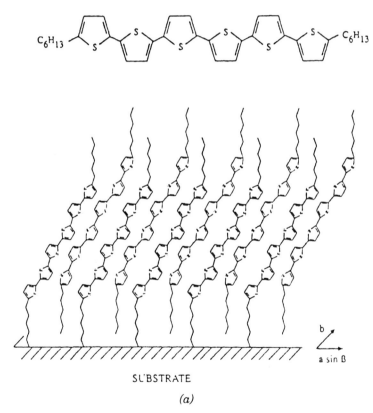

FIGURE 7.4.1. (a) Schematic representation of an α,ωDH6T monolayer film on a substrate. (b) Schematic representation of FET device based on α,ωDH6T. PI represents polyimide insulating layer; source (S), drain (D), and gate (G) electrodes are of gold. (c) Drain current-drain voltage characteristics at various gate voltages for this FET. [Reprinted from F. Garnier, A. Yassar, R. Hajlaoui, G. Horowitz, F. Deloffre, B. Servet, S. Ries, and P. Alnot, *J. Am. Chem. Soc.*, **115**, 8716 (1993). Copyright 1993 American Chemical Society.]

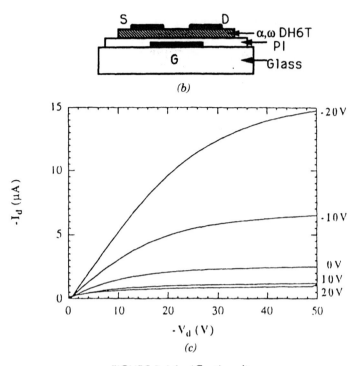

FIGURE 7.4.1. (*Continued*)

octakis(β-decoxyethyl)porphyrin (ZnODEP) has been suggested (25). This material also self-organizes on deposition because of the attached long hydrocarbon chains, and indeed, shows a liquid-crystal phase. A similar solid-state photovoltaic cell based on a liquid-crystal porphyrin has also been described (26). Both the photovoltaic cell and the memory device are based on thin (micrometer) films of ZnODEP deposited on a suitable substrate (e.g., ITO) (Fig. 7.4.2). On irradiation of the film, electrons are injected into the ITO contact and positive charge (holes) move into the bulk of the ZnODEP. Because the material has a very high dark resistance, when the light is turned off, the holes are trapped within the film. This charge-trapping phenomenon can be used to store information with a very high density (> 3 Gbits/cm^2).

Organic polymers have also been proposed for light-emitting diodes (LEDs) and other electroluminescent devices (27). These are typically based on a film of a derivative of poly(p-phenylenevinylene) (PPV) contacted by layers and electrodes that allow injection of electrons and holes, whose recombination within the PPV films leads to light emission. Organic electroluminescent devices based on molecular organic films (e.g., aluminum 8-hydroxyquinolinate) have also been described (28).

7.4. ELECTRONIC AND OPTICAL DEVICES

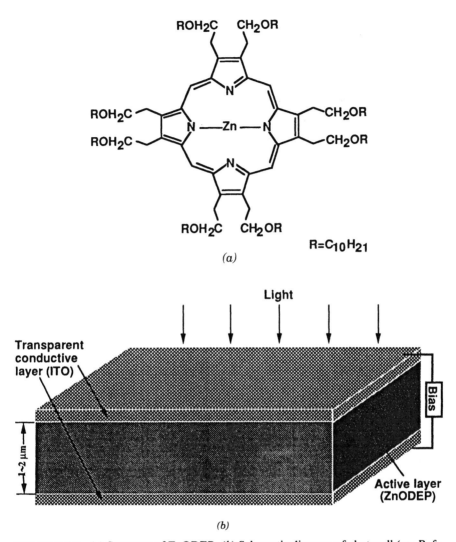

FIGURE 7.4.2. (a) Structure of ZnODEP. (b) Schematic diagram of photocell (see Refs. 25 and 26.) [Reprinted from C. Liu, H. Pan, A. J. Bard, and M. A. Fox, *Science*, **261**, 897 (1993). Copyright 1993 by the AAAS.]

Interest in the organic-based devices revolves around their novel and widely modifiable properties (compared to Si and GaAs). Organic and organometallic solids are molecular materials in which the molecules are held together by weak van der Waals forces, compared to the strongly covalently bonded Si. Thus the band structure and mechanism of charge transport is very different. For example, the onset of quantum effects as the size of a particle decreases, as discussed in Section 6.5, should be less important in molecular materials. Moreover, there is the possibility with these materials of modifying their structures over wide ranges, such as by substitutions on the polymer backbone or by variation of the metal-in-metal complexes and organometallics, to change their properties over wide ranges. Such modifications, based on a wealth of knowledge in organic and organometallic chemistry, could be useful in adjusting the properties (energy gaps, surface structures, organization) to suit desired electronic characteristics. This field is still in its infancy compared to conventional inorganic semiconductors.

Computational devices could be based on neural networks (29). Such networks consist of a number of interconnected "neurons." The state of each neuron depends on the states of the other neurons connected to it as well as the nature ("strength") of the interconnections. Computation in such a system, or the memory content, depends on the collective properties of the system. The collective nature of the system implies that it is fault-tolerant and massively parallel, that is, the contents or properties of the system do not depend strongly on the failure of one or more components in the system. While such systems can be implemented with conventional electronic devices, such as operational amplifiers and resistors, construction by an ICS approach, with conductive polymers and electrochemical interfaces, for example, may be possible. For example, the conductivity of some polymers, such as polypyrrole and polyaniline, can be controlled electrochemically by varying the applied potential and changing the extent of oxidation (or "doping"). Thus components and connections of varying resistivity are possible. A similar approach is possible with other polymers, whose properties can be varied by changing their state of oxidation. Some examples of this approach were described in Chapter 5. It is clear, however, that we are still very much in the learning stage in terms of fabricating electronic devices by these approaches. Although the devices fabricated so far have been rather large and slow, this work has demonstrated that small structures can be fabricated, arranged, and contacted to produce desired electronic effects and suggests that the ICS approach with metals, polymers, and solutions is a possible path to such devices.

The construction of electronic (or ionic) devices from biological materials, such as proteins, to mimic more directly the operation of the brain is a more distant dream (23a). The principles of the construction of such devices are not understood, since we have only a rudimentary knowledge of how the brain and the nervous system operate on the molecular scale. In fact, we do not yet understand how most enzymes function in any detail, and we cannot predict the folding or structure of a protein, even when its complete sequence is known. Moreover, we have not learned how to contact or address single biological molecules or how to manipulate these to carry out "electronic" functions. Perhaps biochips will eventually be possible, but only after we have a considerably better understanding of the fundamental properties and behavior of biological molecules.

7.5. MATERIALS

The concepts important in ICSs may also play a role in the design of structural materials, for example, those used in the automotive, aerospace, and construction industries. Generally, when discussing such materials we imply substances in bulk (metals, ceramics, polymers, semiconductors), such as those made up of fairly homogeneous, but coarse-grained polycrystalline or glassy structures. The properties of such materials, such as toughness, strength, stiffness, weight, hardness, and conductivity, are functions of the molecular structure and the presence of defects and other heterogeneities (30). Recent trends in this area (30, 31) are concerned with the fabrication of advanced materials with more complex structures, such as composites made of several components, often with substructures on the nanometer to submicrometer level. Significant improvement in the properties can be obtained by reducing the size of the microstructures in a material to the nanometer level (e.g., to < 100 nm). Materials of this type include nanocrystalline materials, nanometer-size glasses, and structures that contain nanometer-size fibers or layers (32). One advantage of nanometer-size materials is the possibility of producing compositions that cannot be attained by the usual bulk synthesis methods. For example, to make an alloy of two metals, A and B, the metals must be miscible in a melt. However, a new nanometer-based material can be formed by compression of a mixture of nanometer-size particles of A and B, even if A and B are immiscible. It is mainly in the fabrication of nanocomposites that the principles of ICSs come into play, particularly in the development of specialty materials, such as biomaterials (33) or "intelligent materials" (34).

Composites are composed of two or more substances that maintain their identities when combined (30). For example, a polymer–fiber composite consists of small carbon fibers in a polymer matrix (Fig. 7.5.1) and yields a reinforced plastic with improved strength and stiffness. Such composites can be 10 times stronger than conventional plastics. Similarly, ceramic–ceramic composites are made from ceramic fibers (e.g., SiC) in a ceramic matrix (e.g., Al_2O_3) (Fig. 7.5.2). The fibers in these materials also serve to toughen them by arresting the growth of cracks.

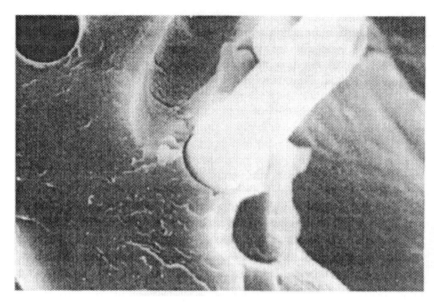

FIGURE 7.5.1. Micrograph of a carbon fiber-reinforced composite fracture surface. [Reprinted with permission from *Materials Science and Engineering for the 1990s: Maintaining Competitiveness in the Age of Materials,* National Academy Press, Washington, DC, 1989. Copyright 1989 E. I. Du Pont de Nemours & Co., Inc.]

Other composites include metal–ceramics (cermets) and some polymer blends. Many of the same considerations that arise in the fabrication of ICSs arise in these composites, such as surface modification of fibers, interfacial processes, adhesion.

Returning to biological systems as analogs, we note that the structural materials found there (bones, shells, ligaments) are often multicomponent ones with improved properties compared to the simple major component [e.g., bones vs. calcium phosphate (apatite)]. For example, the spines of the sea urchin are mainly composed of calcium carbonate (calcite), normally a very brittle material. However, the spines contain proteins, which make up about 0.02% of the crystal by weight. These appear to act in the same way as the fibers in man-made composites to toughen the spines and prevent crack propagation (35). Similarly, in the design of biomaterials, such as implants, the question of biocompatibility and integration with surrounding tissue depends on the surface characteristics of the material and methods for surface modification and adhesion (33). ICSs may also be useful in controlled drug delivery. A film of a polystyrene-based polymer containing isonicotinate and dopamine units cast on an electrode surface will release dopamine when

7.5. MATERIALS

FIGURE 7.5.2. Micrograph of a fracture surface of silicon carbide-reinforced alumina. [Reprinted with permission from *Materials Science and Engineering for the 1990s: Maintaining Competitiveness in the Age of Materials,* National Academy Press, Washington, DC, 1989. Copyright 1989 E. I. Du Pont de Nemours & Co., Inc.]

the electrode potential is switched to a region where reduction of the polymer occurs (36). In an alternative approach, glutamate, incorporated as the counter ion in the oxidized electronically conducting polymer polypyrrole, was released when the polymer film on an electrode was reduced (37).

"Intelligent" materials are those that can sense and respond to their environment, that is, have some degree of self-recovery, self-adjustment, and control (34). These might include materials with memory that can recover their shape after distortion, photochromic materials that darken when exposed to high light levels, and materials that can change their shape or properties in a controlled manner in response to temperature or pressure variations. One example is electrical heating elements (ceramic PTC), which show an abrupt increase in resistance at a given transition temperature. Thus the current through the element will decrease when the transition temperature is attained. In this way, a smart heating device can be fabricated that will maintain a given temperature, without the need for a thermostat (sensor) and switch. Electrochromic materials, that can be made to change color by application of an electric current, might also be considered intelligent materials, especially if they can be incorporated into a system that contains a photosensitive unit to generate the needed electrical signal.

Advanced materials will surely be important in many applications and replace conventional materials like wood and steel. Homes might be constructed of plastic lumber with ceramic roofs and electrochromic windows. Cars can be assembled of composites stronger than steel but 10 times lighter, with ceramic engines operating at high temperatures without a radiator. Artificial muscles will be made of polymers whose dimensions can be varied electrically (e.g., as discussed in connection with polymer modified electrodes). Many of these can be fabricated by using the principles of ICS.

7.6. CONCLUSIONS

For many years chemists have studied reactions in homogeneous phases. We have attained an excellent microscopic understanding of simple chemical reactions in the gas phase and have learned to synthesize complex molecules by solution-phase reactions. We have also learned much about the structure and function of biological molecules. The theme of this book has been the concept of the integrated chemical system, that is, the fabrication and characterization of more complicated assemblies of molecules, frequently on solid surfaces. One might question why one would try to proceed to studies of such complex systems when we don't yet "understand" much simpler ones. The best response, I suppose, is that such integrated systems are of interest, even if a detailed understanding of their principles and interactions (e.g., to the extent that we know about the reaction of two hydrogen atoms in the gas phase) is not possible. For example, we have learned much about electrode reactions at interfaces and have used these in the construction of practical devices. Similarly, integrated circuits have been fabricated and have become the cornerstone for modern electronics. The tools for the construction and characterization of integrated chemical systems are now available. Studies of these systems can lead not only to new applications and devices, but also to a better understanding of the behavior of molecules and intermolecular interactions at interfaces.

REFERENCES

1. S. Brown and T. McCarroll, *Time*, Nov. 20, 1989, p. 108.
2. R. Pool, *Science*, **247**, 26 (1990).
3. R. Pool, *Science*, **242**, 379 (1988).
4. (a) M. J. E. Golay, *Anal. Chem.*, **29**, 928 (1957); (b) M. Novotny, S. R. Springston, P. A. Peaden, J. C. Fjeldsted, and M. L. Lee, ibid., **53**, 407A (1981).
5. (a) J. W. Jorgenson and K. D. Lukacs, *Anal. Chem.*, **53**, 1298 (1981); (b) J. W. Jorgenson and K. D. Lukacs, *Science*, **222**, 266 (1983); (c) A. G. Ewing, R. A. Wallingford, and T. M. Olefirowicz, *Anal. Chem.*, **61**, 292A (1989).
6. W. M. Flarsheim, Y. M. Tsou, I. Trachtenberg, K. P. Johnston, and A. J. Bard, *J. Phys. Chem.*, **90**, 3857 (1986).
7. See, for example: H. Suzuki, E. Tamiya, and I. Karube, *Anal. Chem.*, **60**, 1078 (1988).
8. See, for example: (a) J. Janata, *Anal. Chem.*, **62**, 33R (1990); (b) J. Janata, *Principles of Chemical Sensors*, Plenum, New York, 1989; (c) K. Cammann, U. Lemke, A. Rohen, J. Sander, H. Wilken, and B. Winter, *Angew. Chem. Intl. Ed. Engl.*, **30**, 516 (1991).
9. (a) J-M. Lehn, *Angew. Chem. Intl. Ed. Engl.*, **27**, 89 (1988); (b) J-M. Lehn, ibid., **29**, 1304 (1990); (c) D. Cram, ibid., **27**, 1009 (1988).
10. (a) J. Franke and F. Vögtle, *Top. Curr. Chem.*, **132**, 135 (1986); (b) F. Diederich, *Angew. Chem. Intl. Ed. Engl.*, **27**, 362 (1988); (c) J. Rebek, ibid., **29**, 245 (1990).
11. N. Sharon and H. Lis, *Science*, **246**, 227 (1989).
12. See, for example: (a) G. Wulff, B. Heide, and G. Helfmeier, *J. Am. Chem. Soc.*, **108**, 1089 (1986); (b) K. J. Shea and T. K. Dougherty, ibid., p. 1091; (c) J. Damen and D. C. Neckers, ibid., **102**, 3265 (1980); (d) G. Wulff, in *ACS Symposium Series 186*, W. T. Ford, ed., American Chemical Society, Washington, DC, p. 186.
13. Y-T. Tao, Y-H. Ho, *J. Chem. Soc. Chem. Commun.*, **1988**, 417 (1988).
14. (a) J. Sagiv, *J. Am. Chem. Soc.*, **102**, 92 (1980); (b) J. H. Kim, T. M. Cotton, and R. A. Uphaus, *J. Phys. Chem.*, **92**, 5575 (1988).
15. See, for example: (a) R. P. Buck, W. E. Hatfield, M. Umaña, and E. F. Bowden, eds., *Biosensor Technology—Fundamentals and Applications*, Marcel Dekker, New York, 1990; (b) E. A. H. Hall, *Biosensors*, Prentice-Hall, Englewood Cliffs, NJ, 1991.
16. See, for example: (a) A. Heller, *Acc. Chem. Res.*, **23**, 128 (1990); (b) M. V. Pishko, I. Katakis, S-E. Lindquist, L. Ye, B. A. Gregg, and A. Heller, *Angew. Chem. Intl. Ed. Engl.*, **29**, 82 (1990).
17. (a) A. P. F. Turner, I. Karube, and G. S. Wilson, eds., *Biosensors: Fundamentals and Applications*, Oxford University Press, Oxford, 1988; (b) D. L. Wise, ed., *Applied Biosensors*, Butterworth, Stoneham, 1989.
18. (a) P. G. Schultz, R. A. Lerner, and S. J. Benkovic, *Chem. Eng. News*, May 28, 1990, p. 26; (b) R. J. Massey, *Nature*, **328**, 457 (1987).
19. R. T. Bate, *Nanotechnology*, **1**, 1 (1990).
20. D. Haarer, *Angew. Chem. Intl. Ed. Engl.*, **28**, 1544 (1989).

21. See, for example: (a) A. Aviram and M. A. Ratner, *Chem. Phys. Lett.*, **29**, 277 (1974); (b) J. J. Hopfield, J. N. Onuchic, and D. N. Beratan, *Science*, **241**, 817 (1988).
22. (a) F. L. Carter, ed., *Molecular Electronic Devices*, Marcel Dekker, New York, 1982 (papers presented at the Workshop on "Molecular" Electronic Devices held March 23-24, 1981); (b) F. L. Carter, ed., *Molecular Electronic Devices II*, Marcel Dekker, New York, 1987 (papers from the Second International Workshop on Molecular Electronic Devices held April 13-15, 1983); (c) F. L. Carter, R. E. Siatkowski, and H. Wohltjen, eds., *Molecular Electronic Devices: Proceedings of the 3rd International Symposium on Molecular Electronic Devices*, Arlington, Virginia, October 6-8, 1986, Elsevier, Amsterdam, 1988.
23. (a) R. C. Haddon and A. A. Lamola, " The Sciences," *NY Acad. Sci.*, March, 1983, p. 40; (b) R. Haddon and A. Lamola, *Proc. Natl. Acad. Sci. USA*, **82**, 1874 (1985); (c) R. W. Keyes, *Science*, **230**, 138 (1985).
24. F. Garnier, A. Yassar, R. Hajlaoui, G. Horowitz, F. Deloffre, B. Servet, S. Ries, and P. Alnot, *J. Am. Chem. Soc.*, **115**, 8716 (1993).
25. C. Liu, H. Pan, A. J. Bard, and M. A. Fox, *Science*, **261**, 897 (1993).
26. B. A. Gregg, M. A. Fox, and A. J. Bard, *J. Phys. Chem.*, **94**, 1586 (1990).
27. See, for example: (a) J. H. Burroughes, D. D. C. Bradley, A. R. Brown, R. N. Marks, K. Mackay, R. H. Friend, P. L. Burn, and A. B. Holmes, *Nature*, **347**, 539 (1990); (b) D. Braun and A. J. Heeger, *Appl. Phys. Lett.*, **58**, 1982 (1991).
28. See, for example: (a) C. W. Tang and S. A. VanSlyke, *Appl. Phys. Lett.*, **51**, 913 (1987); (b) M. Era, C. Adachi, T. Tsutui, and S. Saito, *Chem. Phys. Lett.*, **178**, 488 (1991); (c) Y. Hamada, T. Sano, M. Fujita, T. Fujii, Y. Nishio, and K. Shibata, *Chem. Lett.*, **1993**, 905 (1993).
29. (a) J. J. Hopfield, *Proc. Natl. Acad. Sci. USA*, **79**, 2554 (1982); **81**, 3088 (1984); (b) J. J. Hopfield and D. W. Tank, *Science*, **233**, 625 (1986); (c) R. E. Howard, L. D. Jackel, and W. J. Skocpol, *Microelectron. Eng.*, **3**, 3 (1985).
30. *Materials Science and Engineering for the 1990s*, National Academy Press, Washington, DC, 1989.
31. *Research Opportunities for Materials with Ultrafine Microstructures*, NMAB-484, National Academy Press, Washington, DC, 1989.
32. H. Gleiter, in *Encyclopedia of Physical Science and Technology, 1991 Yearbook*, R. A. Meyers, ed., Academic Press, San Diego, 1991, p. 375.
33. J. S. Hanker and B. L. Giammara, *Science*, **242**, 885 (1988).
34. H. Yanagida, *Angew. Chem. Intl. Ed. Engl.*, **27**, 1389 (1988).
35. R. Pool, *Science*, **250**, 629 (1990).
36. A. N. K. Lau and L. L. Miller, *J. Am. Chem. Soc.*, **105**, 5271 (1983).
37. B. Zinger and L. L. Miller, *J. Am. Chem. Soc.*, **106**, 6861 (1984).

INDEX

Acceptor impurity, extrinsic semiconductors, 239
Accumulation layer, semiconductor-solution junctions, 251-253
Adhesion, ICS assembly, 73-74
Adsorption, chemically modified electrodes (CMEs), 135-137
Air-water interface, self-organizing ICS assembly, 55-60
Aluminum oxide, chemically modified electrodes, 147-148
Analyzers, integrated chemical synthesizers, 293
Anodes, heterogeneous catalytic ICS, 16
Anodization, ICS assembly, 82-88
Arginine, electrochemical sensors, 28-29
Array electrode structure, chemically modified electrodes, 163, 170-174
Artificial integrated systems, biological systems as models, 12-15
Assembly of integrated chemical systems:
  self-organizing systems:
    chemically modified electrodes, 53-68
    construction, 54-65
  techniques:
    anodization, 82-84
    chemical vapor deposition (CVD), 75-78
    electrochemical methods, 82-88
    electrodeposition, 84-87
    etching and photoetching, 87
    high-resolution deposition and etching, 88-91
    molecular-beam epitaxy (MBE), 78-81
    plasma processing, 74-75
    spin coating and photolithography, 68-71
    vacuum evaporation and sputtering techniques, 71-73
  template synthesis, 65-68
Atomic force microscopy (AFM):
  ICS assembly, 88-91
  ICS characterization, 124

Attenuated total reflectance (ATR) spectroscopy, 112
Auger electron spectrosopy (AES):
  ICS characterization, 106-109
  spatial information, 98

Band bending, semiconductor junctions, 248
Band energetics, quantum particles, 282-283
Band gap (forbidden region), photoelectrochemistry:
  extrinsic semiconductors, 236-238
  intrinsic semiconductors, 232-235
Band theory of solids, photoelectrochemistry, 231-240
  extrinsic semiconductors, 235-240
  intrinsic semiconductors, 231-235
Bench-scale reaction, integrated chemical synthesizers, 291-293
Benzoquinone (BQ), rotating-disk electrode (RDE) membrane model, 203-204
Bilayer lipid membranes (BLMs), self-organizing ICS assembly, 61-62
Bilayer structure, chemically modified electrodes, 163, 168-170
Binding energy, ICS characterization, 102
Bioconductive films, chemically modified electrodes, 177-180
Biological integrated chemical systems:
  examples of, 7-15
  schematic representation, 1-3
  self-organizing systems, 54-65
Biological materials:
  chemically modified electrodes, 156-157
  polymers, 141
  integrated chemical systems (ICS):
    electronic devices from, 306-307
    structural materials, 308-311
Biosensors:
  chemically modified electrodes, 156-157
  integrated chemical synthesizers, 299-300
Bloch function, photoelectrochemistry, intrinsic semiconductors, 232

313

Block copolymers, ICS assembly, 67-68
Blocking junctions, semiconductor photoelectrochemistry, 242
Blocking polymers, chemically modified electrodes, 141, 143
Bottom-up assembly, ICS, 53
"Breeding solutions," ICS assembly, 67
Burstein shift, quantum particles, 283

Capacitive (charging) current, potential step experiments, 186
Capped quantum particles, 281-282
Carbon, chemically modified electrodes (CMEs), 130-134
Carrier density, photoelectrochemistry, semiconductor junctions, 245
Carrier distributions, semiconductor-solution junctions, 251-254
Cast coating, chemically modified electrodes, 144
Catalysts:
　electron-transfer-mediated catalysis, 212-224
　ICS construction, 39-45
Cavitands, ICS construction, schematic diagrams, 48-49
Cellular automata, self-organizing ICS assembly, 65-66
Ceramic materials, integrated chemical systems (ICS), 307-311
Charge carriers, ICS construction, 45-46
Charge-transfer mechanisms:
　electrode-film interface, 211-212
　film-solution interface, 212-224
　　cyclic voltammetry, 213-216
　　rotating-disk electrodes (RDE), 216-224
　films:
　　electron-transfer diffusion coefficient, 205-209
　　rotating-disk electrodes (RDE), 209-211
Chelating agents, chemically modified electrodes, 158-159
Chemical composition analysis, ICS characterization, 97-98
Chemical sensitivity, chemically modified electrodes, 157-161
Chemical vapor deposition (CVD), ICS assembly, 75-78
Chemically modified electrodes (CMEs):
　biologically related modifiers, 156-157
　complex structures, 161-180
　　array electrodes, 163, 170-174
　　biconductive films, 177-180
　　bilayer structures, 163, 168-170
　　membranes and ion gates, 163, 175-177
　　porous metal films on polymers, 161-165
　　sandwich structures, 163, 165-168
　electrochemical characterization:
　　background, 184-185
　　charge transport mechanisms, 205-211
　　electrode-film charge transfer, 211-212
　　film-solution charge transfer, 212-224
　　monolayers and thin films, 185-196
　　solution species permeation, 196-205
　inorganic film modifiers, 146-156
　　clays and zeolites, 147-154
　　metal oxides, 147-148
　　transition-metal hexacyanides, 154-156
　monolayer modifiers, 134-139
　　adsorption, 135-137
　　covalent attachment, 137-138
　　organized assemblies, 139-141
　overview, 127-129
　polymer modifiers, 139, 141-146
　　preparation, 143-146
　sensitive centers, 157-161
　substrates, 129-134
　　carbon substrates, 130-134
　　conductive polymers and organic metals, 134-135
　　metal substrates, 129-130
　　semiconductors, 134
　surface modification on semiconductors, 265-267
Chemically sensitive centers, ICS construction, 48-51
Chemically-based field-effect transistors (CHEMFETs):
　molecular electronic devices, 31-32
　schematic representation, 26, 28
Chloralkali cells, heterogeneous catalytic ICS, 16
Chlorine evolution reaction, photoelectrochemistry, 263-264
Chlorophyll molecules, role of, in photosynthesis, 9
Chloroplast:
　as integrated chemical system, 8-9
　role of, in photosynthesis, 9-11
Chronoamperometry, solution species electrodes:
　membrane model, 201-202
　pinhole model, 197-199
Clay films:
　chemically modified electrodes, 147-154
　ICS assembly, template synthesis, 66-68
　ICS characterization, electron spin resonance (ESR), 117, 120

INDEX

Compact layers, photoelectrochemistry, semiconductor junctions, 247
Compensating impurity, extrinsic semiconductors, 239-240
Conduction band (CB), photoelectrochemistry, intrinsic semiconductors, 232-235
Controlled drug delivery, integrated chemical systems (ICS) for, 308-309
Convolution techniques, electrode-film charge transfers, 211-212
Copolymers:
 chemically modified electrodes, 141, 157-159
 electron-transfer diffusion coefficient, 208-209
Cottrell equation:
 monolayer and thin film electrochemistry, potential step experiments, 186-188
 solution species electrodes, pinhole model, 197-199
Coupling agents:
 covalent attachment, 47, 52
 ICS construction, 45-48
Covalent attachment:
 chemically modified electrodes (CMEs): monolayers, 137-138
 polymer preparation, 145
 ICS construction, 47, 52
Cross-talking, integrated chemical systems (ICS), 302
Crosslinking agents:
 chemically modified electrodes, 176-177
 electron-transfer diffusion coefficient, 209
 ICS construction, 47-48
Current-time behavior:
 monolayer and thin film electrochemistry:
  Nernstian reactions, 189-191
  potential step experiments, 186-188
 solution species electrodes, pinhole model, 197-199
Current-voltage(i-E) curve:
 ICS assembly, 86-87
 semiconductor photoelectrochemistry: junctions, 248-250
Cyclic voltammetry:
 charge transport in films, 207-209
 chemically modified electrodes (CMEs):
  bilayer structure, 169-170
  bioconductive films, 179-180
  montmorillonite films, 149-151
  transition-metal hexacyanides, 154-156
  zeolite films, 151, 153
 film-solution charge transfer, 213-216
 monolayer and thin film electrochemistry:
  Nernstian reactions, 190-191
  potential sweep experiments, lateral interactions, 193-196
 pinhole electrode models, 199-200
 rotating-disk electrode (RDE): film-solution charge transfer, 220-222
Cyclodextrins:
 chemically modified electrodes, 157-161
 ICS construction, schematic diagrams, 48-51

Deep-level impurities, extrinsic semiconductors, 239-240
Depletion layer, semiconductor-solution junctions, 251-253
Differential capacitance, photoelectrochemistry, semiconductor junctions, 247-248
Diffuse double layer, photoelectrochemistry semiconductor junctions, 246-247
Diffusion coefficient:
 charge transfers:
  electrode-film interface, 212
  electron-transfer diffusion coefficient, 205-209
  rotating-disk electrodes (RDE), 209-211
 membrane electrode model, solution species permeation, 200-205
Dip coating, chemically modified electrodes, 144
DNA replication, ICS assembly, 66-68
DODAC vesicles:
 photoelectrochemical systems, 229-231
 supported semiconductor particles, 274-276
Donor impurities, extrinsic semiconductors, 239
Dopants:
 density, semiconductor-solution junctions, 251
 extrinsic semiconductors, 235-240
Drain electrode, array structure, 170-171
Drop coating, ICS assembly technique, 68
Dye-sensitized semiconductors:
 supported particles, 272
 surface modification, 266-267
Dynamic characterization, integrated chemical systems (ICS), 98

Electroactive cations:
 chemically modified electrodes, 149-151
 polymers, 139, 141
 ICS construction, 48, 51
 electron-transfer diffusion coefficient, 208-209

Electrocatalysis, chemically modified electrodes (CMEs), 128-129
Electrochemical atomic-layer epitaxy (ECALE), 87
Electrochemical techniques:
 electron-transfer diffusion coefficient: charge transport in films, 206-209
 ICS assembly, 82-88
  anodization, 82-84
  electrodeposition, 84-87
  etching and photoetching, 87
  polymerization, 88
 ICS characterization, 100-101
 modified electrode characterization:
  charge transport mechanisms, 205-211
  electrode-film charge transfer, 211-212
  film-solution charge transfer, 212-224
  monolayers and thin films, 185-196
  overview, 184-185
  polymerization, 146
  solution species permeation, 196-205
  switching, 160-162
 potential, semiconductor photoelectrochemistry, 241
 sensors, schematic representation, 26-27
Electrodeposition:
 chemically modified electrodes, polymer preparation, 144-145
 ICS assembly, 84-87
Electrodes, *see also* Chemically modified electrodes (CME)
 charge transfers, film interface, 211-212
 surface modification, 265-267
Electron affinity, photoelectrochemistry, semiconductor junctions, 242-243
Electron-beam-induced current (EBIC), 123
Electron-hole pairs, photogeneration, 268-269
Electron microprobe (EMP), ICS characterization, 123
Electron microscopy, integrated chemical systems (ICS), 98
Electron spin resonance (ESR), ICS characterization, 111, 114-120
Electron transfer (et):
 charge transport in films, 205-211
 diffusion coefficient, 205-209
 ICS characterization, electron spin resonance (ESR), 117
 ICS construction:
  charge carriers and mediators, 45-46
  heterogeneous catalysts, 39, 41
  mitochondrial membrane, 12, 14
 monolayer and thin film electrochemistry:
  potential step experiments, 186-188
  potential sweep experiments, slow heterogeneous kinetics, 192
 photoelectrochemical systems, 230-231
Electronic conductivity:
 ICS support materials, 37-44
 semiconductor photoelectrochemistry, 232-233
Electronic devices, integrated chemical systems (ICS), 300-307
Electronic system, schematic representation, 2, 4
Electronically conductive polymers, CMEs, 141
Electrostatic interactions, ICS construction, 52-53
Elemental analysis, ICS characterization, 97
Ellipsometric methods, ICS characterization, 120-122
Energetic characterization, integrated chemical systems (ICS), 98
Entrapment:
 chemically modified electrodes, 157
 ICS construction, 53
Enzymes:
 electrodes:
  biosensors, 299-300
  schematic representation, 27-29
 integrated chemical systems (ICS):
  as chemical sensors, 42, 44
  electronic devices from, 306-307
Etching techniques, ICS assembly, 87. *See also* Photoetching
 high-resolution deposition and etching, 88-91
 plasma processing, 74-75
Exchange narrowing, ICS characterization, 116-117
Extrinsic semiconductors, photoelectrochemistry, 235-240

Fermi-Dirac distribution function, 240-241
Fermi levels:
 energy measurements, 240
 semiconductor photoelectrochemistry, 240-241
 junctions, 241-250
 semiconductor-liquid interface, 254-257
 semiconductor-solution junctions, 250-254
Fermi-level pinning, electrode surface modification, 265-267

Field-effect transistor (FET):
  chemically modified electrodes, 170-174
  integrated chemical systems (ICS), organic devices, 303-304
  molecular electronic devices, 31-32
  technical principles of, 26, 28
Films:
  balances, self-organizing ICS assembly, 56-58
  charge transfers:
    electrode-film interface, 211-212
    solution interface—catalysis, 212-224
  charge transport, 205-211
    electron-transfer diffusion coefficient, 205-209
    rotating-disk electrodes (RDE), 209-211
  solution species permeation:
    membrane model, 200-205
    pinhole model, 196-200
Flat-band potential, semiconductor-solution junctions, 250-254
Flow components, integrated chemical synthesizers, 293
Fluorescence sensors, schematic representation, 29-30
Forward bias, photoelectrochemistry junctions, 248-249
Fourier transform infrared (FTIR) spectroscopy, ICS characterization, 110, 112-114
Frumkin isotherms, potential sweep experiments, 193-196
Fuel cells, ICS schematic diagram, 51-52

GaAs (gallium arsenide) semiconductor:
  hydrogen photoproduction, 17-18
  ionization energy, 235-237
  photoelectrochemistry, 235-236
Glassy carbon (GC) electrode, film-solution charge transfer, 213-216
Glucose oxidase (GO) electrodes, biosensors, 299-300
Glycolysis, reaction pathway, 12-13
Gold:
  chemically modified electrodes (CMEs), 129-130
  semiconductor photoelectrochemistry junctions, 242-244
Graphite, chemically modified electrodes (CMEs), 130-134
Grazing-angle FTIR, ICS characterization, 112

Half-reactions, photoelectrochemistry, 227-228
Heat transfer, integrated chemical synthesizers, 291-293
Helmholtz layer, photoelectrochemistry, semiconductor junctions, 247
Heteroepitaxy, ICS assembly, 79, 81
Heterogeneous catalysts:
  ICS construction, 39-45
  schematic representation, 15-16
Hierarchical organization, integrated systems, 1-5
High-resolution deposition, ICS assembly, 88-91
Highly ordered pyrolytic graphite (HOPG):
  chemically modified electrodes (CMEs), 131-134
  ICS assembly, anodization, 82-84
Hydrogen:
  ICS construction, heterogeneous catalysts, 41
  photoelectrochemical systems, 16-19
Hydrogen bonding, ICS construction, 53
Hydrophobic interactions:
  ICS construction, 39, 43, 53
  integrated chemical synthesizers, 295-296
Hydroquinone developer, schematic representation, 21-22

Impurities, extrinsic semiconductors, 235-240
Indium tin oxide (ITO), chemically modified electrodes (CMEs), 134
Inorganic films:
  chemically modified electrodes, 146-156
  clays and zeolites, 147-154
  metal oxides, 147-148
  transition-metal hexacyanides, 154-156
  self-organizing ICS assembly, 61-62
  supported semiconductor particles, 271-272
In situ spectroscopy:
  chemically modified electrodes, transition-metal hexacyanides, 154-156
  ICS characterization, 108-121
    electron spin resonance (ESR), 114-120
    ellipsometric techniques, 120-122
    Fourier transform infrared (FTIR) and Raman spectroscopy, 112-114
    photoluminescence spectroscopy, 114
    reflectance spectroscopy, PAS, PTS, 109, 112
    UV-visible spectroscopy, 109
Instant color photographic film, schematic representation, 20-26

Insulators, ICS support materials, 37-39
Integrated chemical systems (ICS):
  assembly principles, 53-68
    self-organizing systems, 54-65
    template synthesis, 65-68
  assembly techniques, 68-92
    chemical vapor deposition (CVD), 75-78
    electrochemical methods, 82-88
    high-resolution deposition and etching, 88-91
    molecular-beam epitaxy (MBE), 78-81
    plasma processing, 74-75
    spin coating and photolithography, 68-71
    vacuum evaporation and sputtering, 71-74
  attachment principles, 52-53
  background and schematic representation, 1-7
  biological systems, 1-3, 7-15
  catalysts, 39-45
  characterization:
    classical techniques, 124-125
    electrochemical techniques, 100-101
    in situ spectroscopic techniques, 108-121
    mass spectrometry techniques, 121
    microscopic techniques, 121-124
    Mössbauer spectroscopy, 121
    nuclear magnetic resonance (NMR), 121
    overview, 96-99
    UHV surface spectroscopy, 100, 102-108
  charge carriers and mediators, 45-46
  chemically sensitive and electroactive centers, 48-51
  components, 51-53
  construction materials, 36-52
  construction overview, 35-36
  defined, 7
  future trends:
    electronic and optical devices, 300-307
    materials, 307-310
    sensors, 294-301
    synthesizers, 290-294
  heterogeneous catalysts, 15-16
  instant color photographic film, 20-26
  linking and coupling agents, 45-48
  microsensors, 26-30
  molecular electronic devices, 30-32
  photoelectrochemical systems, 16-17
  photosensitive centers, 48
  support materials, 36-39
  synthesizers, 290-294

"Intelligent" materials, integrated chemical systems (ICS) construction, 309-310
Interface states, semiconductor photoelectrochemistry, 250
Interfacial capacitances, photoelectrochemistry, 247-248
Intrinsic semiconductors, photoelectrochemistry, 232-235
Inversion layer, semiconductor-solution junctions, 251-253
Ion transport, ICS support materials, 39-41
Ion-beam deposition, ICS assembly, 74, 76
Ion-exchange polymers:
  chemically modified electrodes, 141
  ICS support materials, 39
Ion-gate membrane structure, CMEs, 163, 175-177
Iridium silicide degradation, semiconductor photoelectrochemistry, 263-264

Junctions:
  multijunction photoelectrochemical systems, 275-279
  semiconductor photoelectrochemistry: overview, 241-250
  semiconductor-solution junctions, 250-254

KOH reagent, instant color photographic film, 23-25

Langmuir isotherms, potential sweep experiments, 189-190
Langmuir-Blodgett (LB) films:
  chemically modified electrodes:
    sandwich structure, 167-168
    sensitivity parameters, 158-159
  ICS assembly:
    template synthesis, 67-68
    self-organizing assembly, 55-56, 58-60
  monolayer and thin film electrochemistry, 193-196
  organized assemblies:
    chemically modified electrodes (CMEs), 139-141
  supported semiconductor particles, 275
Lateral interactions, monolayer and thin film electrochemistry, potential sweep experiments, 193-196
Light-driven reactions, schematic representation, 16-17
Light-emitting diodes (LEDs), organic polymers for, 304-306

Limiting current, rotating-disk electrode (RDE):
  pinhole model, 198-199
  reciprocal plots, 223-224
Linking agents, ICS construction, 45-48
Liquid crystalline materials, self-organizing ICS assembly, 60-61
Liver tissue-enzyme electrodes, schematic representation, 27-29
Lock-and-key principle, integrated chemical synthesizers, 295-296
Low-pressure chemical vapor deposition (LPCVD), 78-79

Macrocyclic ligands, integrated chemical systems:
  construction, schematic diagram, 50-51
  synthesizers, molecular recognition, 295
Macrosystem structure, 1-2
Majority carriers, extrinsic semiconductors, 238-239
Membrane electrode model, solution species permeation, 200-205
  chronoamperometry, 201-202
  rotating-disk electrode (RDE), 202-204
Membrane structure:
  chemically modified electrodes, 163, 175-177
  supported semiconductor particles, 274-275
Metal oxide semiconductor field-effect transistor (MOSFET), schematic representation, 27-29
Metal oxides, chemically modified electrodes, 147-148
Metal-organic chemical vapor deposition (MOCVD), 77-79
Metallized dye developers, schematic representation, 21
Metals:
  chemically modified electrodes (CMEs), 129-130
Methyl viologen (MV^+):
  ICS characterization, electron spin resonance (ESR), 117-118
  ICS construction, heterogeneous catalysts, 41
Micelles:
  ICS assembly, template synthesis, 67-68
  photoelectrochemical systems, 229-231
  quantum particles, 281-282
  supported semiconductor particles, 274-275
Microelectrode array structure, chemically modified electrodes, 163, 172-174

Microscopic techniques, ICS characterization, 121-124
Microsensors, schematic representation, 26-30
Miniaturization:
  electronic and optical devices, 300-307
  integrated chemical systems (ICS), 289-290
Minority carriers, extrinsic semiconductors, 238-239
Mitochondria:
  major metabolic reactions, 12-14
  schematic diagram, 11-12
Mixers, integrated chemical synthesizers, 293
Mobile carriers, photoelectrochemistry, intrinsic semiconductors, 235
Molecular-beam epitaxy (MBE), ICS assembly, 78-81
Molecular electronic devices:
  integrated chemical systems (ICS), electronic and optical devices, 301-302
  schematic representation, 30-32
  self-assembly of artificial ICS, 55-56
Molecular engineering, ICS construction, 35
Molecular orientation, ICS characterization, 117, 119
Molecular recognition, ICS construction:
  chemically sensitive and electroactive centers, 48-51
  synthesizers, 294-298
Molecular systems, research methodology for, 5-6
Monolayers:
  chemically modified electrodes (CMEs), 134-139
    adsorption, 135-137
    covalent attachments, 137-138
    organized assemblies, 139-141
    self-assembly, 158, 160
  electrochemistry, 185-196
    lateral film interactions, 193-196
    Nernstian potential sweep experiments, 189-191
    potential step experiment, 186-188
    potential sweep experiments, 189-196
    slow heterogeneous transfer kinetics, 192
  integrated chemical synthesizers, 295-296
Montmorillonite, chemically modified electrodes, 148-149
Mössbauer spectroscopy, ICS characterization, 121
Mott-Schottky (MS) equation, semiconductor-solution junctions, 254

Multielectrode arrays, multijunction photoelectrochemical systems, 278-279
Multijunction photoelectrochemical systems, 275-279

Nafion film:
  chemically modified electrodes, 164-165
  photoelectrochemical systems, 18-19
  supported semiconductor particles, 271
Nanocrystalline materials, integrated chemical systems (ICS), 307-311
Nanostructure, see Integrated chemical system
Near-field scanning optical microscopy (NFSOM), 123
Nernstian reactions, potential sweep experiments, 189-191
Neural networks, integrated chemical systems (ICS), 306
Nonblocking junctions, semiconductor photoelectrochemistry, 242
n-type semiconductor:
  light-driven reaction, 16-17
  photoelectrochemistry:
    extrinsic semiconductors, 236-237
    junctions, 242, 244-245
    photosynthetic and photocatalytic reactions, 261-265
    semiconductor-solution junctions, 251-252
Nuclear magnetic resonance (NMR), ICS characterization, 121

Ohmic contact, semiconductor photoelectrochemistry, 248-249
Optical isomers, chemically modified electrodes, 149-151
Optrodes, schematic representation, 26-30
Orbital formation, photoelectrochemistry, intrinsic semiconductors, 232-233
Organelles:
  as integrated chemical systems, 7-8
  mitochondrion, 11-12
Organic materials, ICS:
  electronic and optical devices, 302-303
  organic metals, 134-135
Orientational parameters, ICS characterization, 98
Ostwald ripening, quantum particles, 280-281
Oxidation-reduction mediators, ICS construction, 45-46

Particle charging, quantum particles, 282-283
"Particle in a box" model, quantum particles, 280, 282-283

Particulate (microheterogeneous) systems, 268-279
  multijunction photoelectrochemical systems, 275-279
  principles, 268-270
  representative photoreactions, 269-270
  supported semiconductor particles, 270-275
Partition constant, film-solution interface, 212-224
Permeation current, rotating-disk electrode (RDE), 204
Permeation time, electron-transfer diffusion coefficient, 206-207
Phase properties, integrated chemical systems (ICS), 99
9,10-Phenanthraquinone (PAQ), CME, monolayer adsorption, 135-136
Photoacoustic spectroscopy (PAS), ICS characterization, 110, 112
Photocatalytic reactions, semiconductors, 260-265
Photochemical polymerization, chemically modified electrodes, 146
Photocurrent-photovoltage curve, semiconductor-liquid interface, 259-260
Photoeffects, semiconductor-liquid interface, 254-257
Photoelectrochemical (PEC) systems, 16-19
  band model of solids, 231-240
  electrode surface modification, 265-267
  Fermi levels, 240-241
  junctions, 241-250
  overview, 227-231
  particulate (microheterogeneous) systems, 268-279
  photoeffects at liquid interface, 254-257
  photosynthetic and photocatalytic reactions, 260-265
  photovoltaic cells, 257-260
  quantum (nano-) particles, 279-283
  solution junctions, 250-254
Photoetching, ICS assembly, 87
Photolithography, ICS assembly technique, 68-71
Photoluminescence spectroscopy, ICS characterization, 110, 114
Photomicrography, self-organizing ICS assembly, 62-63
Photoresists:
  photolithographic ICS assembly, 70-72
  schematic representation, 70
Photosensitive centers, ICS construction, 48
Photosynthesis:
  chloroplast and, 8-9

INDEX 321

model of, 9–11
semiconductor photoelectrochemistry, 260–265
Photosystems (PS I and PS II), function of, 9–11
Photothermal spectroscopy (PTS), ICS characterization, 110, 112
Photovoltaic cells:
  photoelectrochemistry and, 257–260
  representative liquid junctions, 260–261
  surface modification, 266–267
Piezoelectric materials, ICS construction, 51–53
Pinhole electrode model, solution species, 196–200
  chronoamperometry, 197–199
  cyclic voltammetry, 199–200
  rotating-disk electrode (RDE), 198–199
Plasma polymerization, chemically modified electrodes, 146–147
Plasma processing, ICS assembly, 74–75
Plastoquinone molecules (PQ), function of, 9
Plateau currents, film-solution charge transfer—catalysis, 217–219
Platinum:
  chemically modified electrodes, 129–131
  porous metal films, 161–162, 164–165
  ICS construction, heterogeneous catalysts, 41
Point of zero charge (pzc), semiconductor junctions, 245–246
Poisson equation, semiconductor junctions, 242–243
Polaroid SX-70 development system, schematic representation, 20–26
Poly(vinyl alcohol) (PVA), ICS characterization, electron spin resonance (ESR), 117, 120
Polymerase chain reaction, ICS assembly, 66–68
Polymers, *see also specific types of polymers*
  charge transport in films:
    electron-transfer diffusion coefficient, 207–209
    rotating-disk electrodes (RDE), 209–211
  chemically modified electrodes (CMEs), 134–135, 139, 141–146
    porous metal films, 161–162, 164–165
    preparation techniques, 143–146
    types of polymers, 139, 141–143
  heterogeneous catalytic ICS, 15
  integrated chemical systems (ICS):
    assembly, 88
    characterization, 117–118

construction, pressure selectivity, 50–53
organic polymers, 304–306
self-organizing assembly, conjugated components, 56, 58, 60
structural materials, 307–311
support materials, 38–41
synthesizers, molecular recognition, 295–297
matrix, photoelectrochemical systems, 18–19
supported semiconductor particles, 271–275
Polyoxometallates, chemically modified electrodes, 147
Porous metal films, chemically modified electrodes, 161–162, 164–165
Potential step experiments, monolayer and thin films, 186–188
Potential sweep experiments, monolayer and thin films, 189–196
  lateral interactions, 193–196
  Nernstian reactions, 189–191
  slow heterogeneous kinetics, 192
Proteins, integrated chemical systems (ICS), 306–307
Proton transfer:
  ICS construction, heterogeneous catalysts, 39, 41
  mitochondrial membrane, 12, 14
p-type semiconductor:
  photoelectrochemistry, 239
  photosynthetic and photocatalytic reactions, 261–265
  semiconductor-liquid interface, 256–258

Quantitative analysis, integrated chemical systems (ICS), 99
Quantum particles:
  particle charging and band energetics, 282–283
  preparation, 279–282
  size quantization effects in small particles, 279
Quantum wells, molecular-beam epitaxy (MBE), 81

Radiofrequency (RF) field, ICS assembly, 74–75
Raman spectroscopy, ICS characterization, 110, 112–114
Rate constants, charge transfers, 211–212
Reaction chambers, integrated chemical synthesizers, 293
Reactive-ion etching, ICS assembly, 74

Rectifying junctions, semiconductor photoelectrochemistry, 248-249
Redox polymers, chemically modified electrodes, 157, 161
Redox potential:
  ICS construction, electroactive centers, 51
  photosynthetic and photocatalytic reactions, 263-264
  semiconductor-liquid interface, 256, 258-259
Reflectance spectroscopy, ICS characterization, 109, 111-112
Reflection high-energy electron diffraction (RHEED), 80-81
Resonance Raman spectroscopy, ICS characterization, 112-114
Reverse bias, semiconductor photoelectrochemistry, 248-249
Rotating-disk electrode (RDE):
  charge transfers:
    films, 209-211
    film-solution catalysis, 216-224
  solution species electrodes:
    membrane model, 202-204
    pinhole model, 198-199
Roughness factor, integrated chemical systems (ICS) characterization, 99
Ruthenium chelates, integrated chemical systems (ICS):
  characterization, photoluminescence spectroscopy, 114-115
  construction, heterogeneous catalysts, 41

Sacrificial electron donor, photoelectrochemistry, 16-17
Sandwich structures, chemically modified electrodes, 163, 165-168
Scanning acoustic microscopy, ICS characterization, 112
Scanning electrochemical microscopy (SECM):
  ICS assembly, 88-91
  ICS characterization, 123-124
Scanning tunneling microscopy (STM):
  ICS assembly, 88-91
  ICS characterization, 123-124
Schottky barrier, semiconductor photoelectrochemistry junctions, 242, 244-245, 248-250
Secondary-ion mass spectrometry (SIMS), 121
Self-assembly, chemically modified electrodes, 158, 160

Self-organizing ICS assembly, 54-65
  biological vs. artificial systems, 55-56
  evolutionary principles, 63-65
  photolithographic techniques, 71-72
Self-quenching effect, ICS characterization, 114
Semiconductor-liquid interface, photoeffects, 254-257
Semiconductors:
  chemically modified electrodes (CMEs), 134
  ICS assembly, etching and photoetching, 87
  ICS support materials, 38
  photoelectrochemical systems, 16-19
    band model of solids, 231-240
    electrode surface modification, 265-267
    Fermi levels, 240-241
    junctions, 241-250
    overview, 227-231
    particulate (microheterogeneous) systems, 268-279
    photoeffects at liquid interface, 254-257
    photosynthetic and photocatalytic reactions, 260-265
    photovoltaic cells, 257-260
    quantum (nano-) particles, 279-283
    solution junctions, 250-254
Sensors:
  future developments in, 300
  integrated chemical synthesizers, 294-300
  biosensors, 298-301
  molecular recognition, 294-298
Separation chambers, integrated chemical synthesizers, 293
Shadow evaporation, electrode array structure, 172-174
Shallow-level impurities, extrinsic semiconductors, 236-239
Shape-selective sites, molecular recognition, 295-296
Silicate structures:
  heterogeneous catalytic ICS, 15
  ICS support materials, 39, 44
Silicon:
  integrated chemical synthesizer component and construction, 293-294
  ionization energy, 235-237
  photoelectrochemistry, 235-236
  semiconductor junctions, 242-244
Silver halide (AgBr):
  instant color photographic film, 21, 23-24
  self-organizing ICS assembly, 62-66

Size barrier, electronic and optical devices, 301–302
Size scales:
  assembly of ICS, 53–54
  integrated systems, 5–6
Skeletonized monolayers, molecular recognition, 295–296
Slow heterogeneous electron transfer kinetics, 192
Small flow reactors, integrated chemical synthesizers, 291–293
Solar energy systems, semiconductor-based system schematic, 227–228
Solar systems, multijunction photoelectrochemical systems, 275–276
Solid polymer electrolyte (SPE), CMEs, 162, 164–165
Solution species:
  electrochemical characterization, 196–205
  membrane model, 200–205
  pinhole model, 196–200
  rotating-disk electrode (RDE), 217, 220
Source electrode, chemically modified electrode arrays, 170–171
Space charge region:
  semiconductor-liquid interface, 255–257
  semiconductor photoelectrochemistry junctions, 245–247
  semiconductor-solution interface, 253–254
Spatial parameters:
  ICS characterization, 98
  self-organizing ICS assembly, 60–65
Spin coating:
  chemically modified electrodes, 144
  ICS assembly technique, 68–71
Spray pyrolysis or deposition, ICS assembly, 78
Sputtering technique, ICS assembly, 71–73
Stable nonstoichiometric species, 97–98
Stable stoichiometric species, 97–98
Steady-state measurements, rotating-disk electrode (RDE):
  film-solution charge transfer—catalysis, 216–224
  membrane model, 202–204
Stereoselective effects, chemically modified electrodes, 149–151
Structural materials, integrated chemical systems (ICS), 307–311
Substrates, chemically modified electrodes (CMEs), 129–139
  carbon, 130–134
  conductive polymers and organic metals, 134–135
  metals, 129–130
  semiconductors, 134
Superlattices:
  molecular-beam epitaxy (MBE), 81
  quantum particles, 279
Support material:
  ICS construction, 36–39
  semiconductor particles, 270–275
Surface pressure:
  self-organizing ICS assembly, 56, 59
  semiconductor photoelectrochemistry, 250
Surface spectroscopy:
  chemically modified electrodes, 154–156
  ICS characterization, 100, 103–105
Surface-enhanced (SERS) spectroscopy, 112–114
Surfactants:
  ICS construction, 39, 43
  self-organizing ICS assembly, 55–57
Synthesizers, integrated chemical systems (ICS), 290–294
Synthetic reactions, research methodology, 5–6

Tempamine, electron spin resonance (ESR), 117, 119–120
Template synthesis:
  ICS assembly, 65–68
  integrated chemical synthesizers, molecular recognition, 295, 297–298
Texas Instruments Solar Energy System (TISES), 275–278
Thermal polymerization, chemically modified electrodes, 145–146
Thin films, electrochemistry, 185–196
  lateral film interactions, 193–196
  Nernstian experiments, 189–191
  potential step experiment, 186–188
  potential sweep experiments, 189–196
  slow heterogeneous transfer kinetics, 192
Thin layer approximations, potential sweep experiments, 189–191
Three-dimensional structures, electrodeposition techniques, 85–86
Thylakoid membrane, schematic diagram of, 9–10
Tobacco mosaic virus (TMV), self-assembly schematic, 54–55
Top-down assembly, ICS, 53
Transistors, electrochemically-based, 30–31
Transition-metal hexacyanides, chemically modified electrodes, 154–156

Transmission electron microscopy (TEM):
  ICS characterization, 123
  molecular-beam epitaxy (MBE), 81
  self-organizing ICS assembly, 64-66

Ultrahigh vacuum (UHV) surface spectroscopy, ICS characterization, 100, 102-108
  Auger electron spectroscopy, 106-109
  background, 96
  X-ray photoemission spectroscopy (XPS), 102, 106-107
Underpotential deposition (UPD), ICS assembly, 86-87
Unstable intermediates, integrated chemical systems (ICS), 97-98
UV-visible spectroscopy, ICS characterization, 109

Vacuum evaporation, ICS assembly technique, 71-73
Valence band (VB), photoelectrochemistry, intrinsic semiconductors, 232-235
Van der Waals forces, ICS construction, 53
Vectorial charge transfer, photosynthesis, 11
Vesicles:
  photoelectrochemical systems, 229-231
  supported semiconductor particles, 274-275
Viscosity parameters, electrode-film interface, 212

Work function, photoelectrochemistry, 242-243

X-ray photoemission spectroscopy (XPS), ICS characterization, 102, 106-107

Zeolites:
  chemically modified electrodes, 147-154
  ICS assembly, template synthesis, 66-68
  ICS construction, heterogeneous catalysts, 41
  supported semiconductor particles, 272-274
ZnODEP optical device, development of, 303-305

CPSIA information can be obtained at www.ICGtesting.com
Printed in the USA
LVOW100837120812

293906LV00003B/3/A